International Global on Renewable Energies and Energy Efficiency (REEE'2017)

IOP Conference Series: Earth and Environmental Science
Volume 161

Fez, Morocco
8 - 9 November 2017

Editors:

Abdellah Mechaqrane
Mahmoud Bououd

ISBN: 978-1-5108-6479-5
ISSN: 1755-1307

Printed from e-media with permission by:

Curran Associates, Inc.
57 Morehouse Lane
Red Hook, NY 12571

Some format issues inherent in the e-media version may also appear in this print version.

Printed by Curran Associates, Inc. (2018)

Institute of Physics
Dirac House, Temple Back
Bristol BS1 6BE UK

Phone: 44 1 17 929 7481
Fax: 44 1 17 920 0979

techtracking@iop.org

Additional copies of this publication are available from:

Curran Associates, Inc.
57 Morehouse Lane
Red Hook, NY 12571 USA
Phone: 845-758-0400
Fax: 845-758-2633
Email: curran@proceedings.com
Web: www.proceedings.com

Table of contents

Volume 161

International Conference on Renewable Energies and Energy Efficiency (REEE'2017)

8–9 November 2017, FST Fez, Morocco

Accepted papers received: 31 May 2018
Published online: 14 June 2018

Preface

Papers

IOP Conf. Series: Earth and Environmental Science **161** (2018) 011001 doi:10.1088/1755-1315/161/1/011001

Preface

International Conference on Renewable Energies and Energy Efficiency REEE'2017

Overview

The REEE (the International Conference on Renewable Energies and Energy) organized for the first time in Morocco between **8-9 November 2017** on the campus of **Faculty of Sciences and Technology Sidi Mohamed Ben Abdellah University of Fez, the imperial city of Morocco.**

The REEE event was an opened exchange discussion forum, scientific debate and interchange ideas and knowledge between the Maghreb communities and their national and international colleagues. The conference focused on Renewable Energies and energy efficiency, from Resource assessment, prototypes development, to industrial use and exploitation, through sustainable energy systems design. It has been a good opportunity for researchers to establish cooperation relationship, find innovative and adequate solutions for energy saving and production in developing countries, enhance mutual exchanges, and promote industrial research and development, which guarantee widest socio-economic integration of the university and strengthening its role as a nucleus of development and pioneer through research and innovation.

The REEE'2017 program included many topics related to renewable energies and energy efficiency namely: Resource assessment, Modeling and forecasts, Energy Systems Design, Operation and Performance analysis, Smart grids, Advanced Materials for Energy, Energy Storage Systems and materials, Energy Efficiency in building, industry and Transportation, Building Materials and Components.

REEE'2017 has attracted 150 submissions, after the review process, 87 papers have been accepted between oral and posters communications and 29 papers were selected for publication.

The conference was well attended by participants from many different organizations including researchers, PhD students, industrials and academics from different countries.

All our thanks and acknowledgments go to our sponsors who supported REEE'2017 Fez, and all those who contributed to the success of this conference. Especially, we would like to acknowledge ERSI Laboratory members, the organizing and scientific committees for their valuable help, time and effort in the peer-reviewing and the organization.

Editors

IOP Conf. Series: Earth and Environmental Science **161** (2018) 011001 doi:10.1088/1755-1315/161/1/011001

Introduction

Laboratory of Renewable Energies and Intelligent Systems (LERSI) Organizer Lab

Laboratory of Renewable Energies and Intelligent Systems

Electronics and renewable energies are among the strategic development axis, sustained by the high level authorities of the Morocco kingdom. Both of these fields are substantial deposition of direct and indirect jobs.

The ERSI laboratory is coming to support this work. Our university and our institutions have long initiated the training of young graduates in the fields of electronics microelectronics (Master FST), embedded systems (engineers cursus ENSA) and telecommunications systems (engineers cursus FST and ENSA). With their expertise in these areas and given the recent funded research projects in the field of renewable energy as intelligent systems, three teams from both institution (FST) and (ENSA - National School of Applied Sciences) have decided to pool their efforts and establish a research laboratory specializing in renewable energy and intelligent systems.

This laboratory is formed by three teams:

-Systems of renewable energies and energy efficiency (SEREE): this team is formed by researcher coming from the FST, headed by Prof. MECHAQRANE Abdellah, and has a long history and facilities that allow it to qualify solar and wind fields in Fez Boulmane region.

-Telecommunication network and software engineering (MSE), this team is formed by the ENSA researcher, headed by Prof. BENNANI DOSSE Saad. This team has a long research experience in the fields of telecommunications, antennas and related items. Also some of involved researchers work on software engineering to develop some tools for internet of things, data processing and mailing.

-The third team is Microelectronic and embedded systems. This team is a mixed, as it involves researcher from the two establishments (FST and ENSA). It is headed by Prof. A. AHAITOUF and works on microelectronics component, integrated analogue and digital circuits design, VLSI design, embedded systems, video processing etc…

Motivation:

Our skills are varied, diverse and complementary. They include electronics, telecommunications, renewable energy systems design, processing of the signal applied mathematics, numerical methods, and hardware programming languages. We chose to combine renewable energy and intelligent systems already at the name of the laboratory for better visibility and to show the consistency that exists between these two themes and that sometimes goes unnoticed. Indeed, any modern system of renewable energy system incorporates an electronic system for command, sun tracking and/or energy management and optimization. The electronic systems are today incorporating advanced software programs, communications systems ranging from classic wire-connected antennas and networks to the more modern, functioning with all types of waves or wireless systems such Wi-Fi, Zig-Bee, Bluetooth ... etc.

Of course their design responds to a drastic set of specifications both in terms of response time (software design) the space occupation (hardware design). These constraints show how the balance between these various skills is desirable for successful smart and futuristic energy systems. In fact, an intelligent system is not only a system controlled by computer, but it integrates some additional elements including temperature, pressure, humidity and radiation sensors, etc ... However, it can be qualified as smart when it is used to control and manage the multitude of sensors, data and treatments and making the good decision in the convenient time.

Moreover, alongside the renewable energy systems, the development of embedded systems remains a priority for our laboratory, these miniaturized and intelligent systems can make our teams competitive again large international groups in these high technology areas and high added values.

IOP Conf. Series: Earth and Environmental Science **161** (2018) 011001 doi:10.1088/1755-1315/161/1/011001

We have initiated innovative training at national level and hope to move forward in the areas research and develop useful intelligent systems in territorial surveillance, biomedical, video surveillance etc...

It is as natural as any design or invention that will be innovative when it fits perfectly into its environment, this is done in most cases by the communication and exchange of information between a command emitter (master) and a command receiver (slave) and this can only be implemented by a true knowledge of standards and standard telecommunications. The huge potential provided by the RTL team of the ENSA of Fez.

Also in this laboratory we will open our field of investigations on all other issues relating to the progress of our specific research and ranging from optimization algorithms, numerical methods through Web Services, the driven engineering models, and design and implementation of intelligent systems enhancing the concept ABC (Always Best connected) for the next generation of connected vehicles.

Given the complementarily of our skills, given the desire of each member and given the developed research at the university, we decided the common creation and accreditation of this laboratory, to enlarge and diversify the offer of our university and to fix us a new challenging project.

Partnership

National Partners

- Moroccan Foundation for Advanced Science, Innovation and Research (MASCIR)
- National Center for Scientific and Technical Research (CNRST)
- Research Institute for Solar Energy and New Energies (IRESEN)
- International University of Rabat
- Cadi Ayyad University Marrakech
- Faculty of Sciences and Technology Mohammedia
- National School of Applied Sciences Fez, Morocco (ENSAF)
- ENS Casablanca
- ALTEN Maroc
- Zodiac Aerospace
- AIC Metallurgie

International Partners

- Lorraine University LMOPS Laboratory
- Georgiatech Lorraine UMI2958GT-CNRS
- Sherbrooke University Canada
- Lille University 1 IEMN
- INSA Lyon
- INSA EuroMed
- ESEO Angers
- ESIEE Paris
- Bordeaux University
- ENIS de Sfax -Tunisia
- Belfort University
- Limoges University
- Sophia Antipolis Nice LEAT Laboratory
- North China Electric Power, China
- Vilnius Gediminas Technical University, Lithuania
- Iasé Faculty, Romania University

Editors

IOP Conf. Series: Earth and Environmental Science **161** (2018) 011002 doi:10.1088/1755-1315/161/1/011002

International Conference on Renewable Energies and Energy Efficiency REEE'2017

Photographs

8-9 November 2017 on the campus of Faculty of Sciences and Technology
Laboratory of Renewable Energies and Intelligent Systems (LERSI)
Sidi Mohamed Ben Abdellah University, Fez, Morocco

IOP Conf. Series: Earth and Environmental Science **161** (2018) 011002 doi:10.1088/1755-1315/161/1/011002

IOP Conf. Series: Earth and Environmental Science **161** (2018) 011002 doi:10.1088/1755-1315/161/1/011002

IOP Conf. Series: Earth and Environmental Science **161** (2018) 011002 doi:10.1088/1755-1315/161/1/011002

IOP Conf. Series: Earth and Environmental Science **161** (2018) 011002 doi:10.1088/1755-1315/161/1/011002

IOP Conf. Series: Earth and Environmental Science **161** (2018) 011003 doi:10.1088/1755-1315/161/1/011003

List of Committee: REEE'2017

Honour Committee

Prof. Omar ASSOBHEI

President of Sidi Mohamed Ben Abdellah University
Fez, Morocco

Prof. Mustapha IJJAALI

Dean of Faculty of Sciences and Technology
Fez, Morocco

Prof. El Mestafa EL HADRAMI

Vice Dean, in charge of Scientific Research and Cooperation
FST Fez, Morocco

Prof. Taoufiq ACHIBAT

Vice Dean, in charge of pedagogical and student affairs
FST Fez, Morocco

Chairman and Editors

Abdellah MECHAQRANE

Head of Renewable Energies Systems and Energy Efficiency Research Group (RESEE)
Laboratory Renewable Energies and Intelligent Systems (LERSI)
Electrical Engineering Department, Faculty of Sciences and Technology
Sidi Mohamed Ben Abdellah University Fez, Morocco
abdellah.mechaqrane@usmba.ac.ma

Mahmoud BOUOUD

Renewable Energies Systems and Energy Efficiency Research Group
Laboratory of Renewable Energies and Intelligent Systems
Electrical Engineering Department, Faculty of Sciences and Technology
Sidi Mohamed Ben Abdellah University Fez, Morocco
mahmoud.bououd@usmba.ac.ma

Local Organizing Committee

A Mechaqrane
(FST -Fès)

A Ahaitouf
(FST –Fès)

N ES-Sbai
(FST -Fès)

S Bennani Dosse
(ENSA -Fès)

M Jorio
(FST -Fès)

F Errahimi
(FST -Fès)

M Razi
(FST -Fès)

A Jamil
(EST -Fès)

A Zorkani
(FSDM -Fès)

Organizing Committee (PhD. Candidates)

Hicham BENSASSI (FST-Fès)
Ghita BENNIS (FSDM-Fès)
Imane FADLOULLAH (FST-Fès)
Maha KHANFARA (EST-Fès)
Manal MARZOUQ (ENSA-Fès)
Mahmoud BOUOUD (FST-Fès)
Mohamed BOUSSETA (EST-Fès)
Sarah EL HIMER (FST-Fès)
Sara EL YAHYAOUI (FST-Fès)
Zineb BOUNOUA (FST-Fès)
Zahra BEN MOHAMMADI

Website and Online submission

Prof. Abdellah MECHAQRANE (FST -Fès)
Zineb BOUNOUA (FST -Fès)
Hicham BENSASSI (FST -Fès)
Zahra BEN MOHAMMADI

Scientific Committee Members

El Hassan ACHOUYAB, ENSET, Mohammedia, Maroc
Abdelaziz AHAITOUF, FP, Taza, Maroc
Ali AHAITOUF, FST, Fès, Maroc
Thami AITTALEB, FP, Ouarzazate, Maroc
Chakib ALAOUI, UEUROMED, Fès, Maroc
Amine ALLOUHI, EST, Fès, Maroc
Khalid AMECHNOUE, ENSA, Tanger, Maroc
Leonard Kofitse AMEKUDZI, UST, Ghana
Mohamed ASBIK, FS, Meknès, Maroc
El Mahdi ASSAID, UCD, El Jadida, Maroc
El Houssain BAALI, IAV Hassan II, Rabat, Maroc
M'Bark BAKKAS, ENSAM, Meknès, Maroc
Mohamed Najib BARGACH, FS, Rabat, Maroc
Rachid EL BECHTIRI, EST-USMBA, Fès, Maroc
Mustapha EL ALAMI, FS-Ain Chok, Casablanca, Maroc
Fatima ERRAHIMI, FST, Fès, Maroc
Najia ES-SBAI, FST, Fès, Maroc
Abdessamad FAIK, CIC Energigune, Bilbao, Espagne
Gilles FLAMANT, Laboratoire PROMES, France
Sanaa HAYANI MOUNIR, FP-Khouribga, Maroc
Said HOUACHE, FST-USMBA, Fès, Maroc
Ahmed HOURMATALLAH, ENS, Fès, Maroc
Ali IDLIMAM, ENS-UCA, Marrakech, Maroc
Ahmed IHLAL, FS-UIZ, Agadir, Maroc
Lahcen BEJJIT, EST, Meknès, Maroc
Mohamed BENALI, ESCOM, France
Riad BENELMIR, Univ. Loraine-Nancy, France
Saad BENANI DOSSE, ENSA, Fès, Maroc
Brahim BENHAMOU, FSS, Marrakech, Maroc
Saad BENSOUDA, CI, USMBA, Maroc
Badre BOSSOUFI, EST, Oujda, Maroc
Saadi BOUGOUL, Université de Batna1, Algérie
Lahcen BOUKHATTEM, ENSA-UCA, Safi, Maroc
Khalid BOUZIANE, UIR, Rabat, Maroc
Pascal BRUEL, Univ. Pau et PA, France
Hassan CHAIB, FP, Ouarzazate, Maroc
Abdelkader DOUDOU, FP, Nador, Maroc
Najib DROUICHE, CRTSE, Algiers, Algeria
Soumya EL ABED, CI-USMBA, Fès, Maroc
Salma EL AIMANI, FP, Ouarzazate, Maroc
Aumeur EL AMRANI, FST, Errachidia, Maroc
Abdeslam EL BOUARI, FS-BenMsik, Casablanca, Maroc
Abdelaziz EL GHZIZAL, EST, Fès, Maroc
Mustapha MABROUKI, FST, Béni Mellal, Maroc
Anas MANSOURI, ENSA, Fès, Maroc
Vytautas MARTINAITIS, VGTU, Lithuania
Abdellah MECHAQRANE, FST, Fès, Maroc
Abdelaziz MIMET, ENS, Tétouan, Maroc

IOP Conf. Series: Earth and Environmental Science **161** (2018) 011003 doi:10.1088/1755-1315/161/1/011003

Abdelmajid JAMIL, EST, Fès, Maroc
Tarik KOUSKSOU, Univ. Pau et PA, France
Klaus KUHNKE, SAU, Osnabrück, Allemagne
Younes LAKHRISSI, ENSA, Fès, Maroc
Brahim LEKHLIF, EHTP, Casablanca, Maroc
Yang LIUNG, NCEPU, China
Mostafa MAALMI, E. Mines, Rabat, Maroc
Mohamed TAHIRI, UH2, Casablanca, Maroc
Hector VALDES, F. Ingeniería-UCSC, Chile
Hicham ZAITAN, FST, Fès, Maroc
Youssef NAIMI, FS-BM, UH2, Casablanca, Maroc
Mohamed OUAZANI JAMIL, UPF-Fès, Maroc
Fahd OUDRHIRI HASSANI, ENSA-UCA, Safi, Maroc
Abdelhadi RAIHANI, ENSET, Mohammedia, Maroc
Abdellah REZZOUK, FSDM, Fès, Maroc
Touré SIAKA, Univ. Cocody, Côte d'Ivoire
Rachid TADILI, FS-UM5, Rabat, Maroc
Youssef ZERAOULI, Univ. Pau PA, France
Izeddine ZORKANI, FSDM, Fès, Maroc

IOP Conf. Series: Earth and Environmental Science **161** (2018) 011004 doi:10.1088/1755-1315/161/1/011004

Sponsoring organizations/funding acknowledgements

ALTEN MOROCCO

National Center for Scientific and Technical Research (CNRST)

Arrondissement Saiss
Arrondissement Agdal
Arrondissement Fès-Médina

German International Cooperation (GIZ Morocco)

Organizers

Sidi Mohamed Ben Abdellah University Fez, Morocco (USMBA)
Faculty of Sciences and Technology Fez, Morocco (FSTF)

Laboratory of Renewable Energies and Intelligent Systems (LERSI)

IOP Conf. Series: Earth and Environmental Science **161** (2018) 011005 doi:10.1088/1755-1315/161/1/011005

Keynote & Invited Speakers (Plenary conferences)

Prof. Abdallah OUGAZZADEN
Director of Georgia Tech Lorraine
Co-President of Lafayette Institute

Plenary conference title: **Wide band gap semiconductors for future generation of solar cells**

Prof. Klaus KUHNKE
Hochschule Osnabrück University, Germany

Plenary conference title: **Energy Storage as a Door-Opener for the Renewable Energies.**

Prof. Ali AHAITOUF
Head of ERSI Laboratory
Sidi Mohammed Ben Abdellah University

Plenary conference title: **CPV in Morocco: History and Situation**

Prof. Mustapha Ayaita
President of the Renewable Energy University Network

Plenary conference title: **Energy transition in Morocco: Context, challenges & perspectives**

IOP Conf. Series: Earth and Environmental Science **161** (2018) 011006 doi:10.1088/1755-1315/161/1/011006

Peer review statement

All papers published in this volume of *IOP Conference Series: Earth and Environmental Science* have been peer reviewed through processes administered by the proceedings Editors. Reviews were conducted by expert referees to the professional and scientific standards expected of a proceedings journal published by IOP Publishing.

Plate flat finned tubes heat exchanger: Heat transfer and pressure drop modeling

M Bououud[1*], O Hachchadi[1], A Mechaqrane[1]

[1]Sidi Mohamed Ben Abdellah University, Faculty of Sciences and Technology Electrical Engineering Department Renewable Energies and Intelligent Systems Laboratory BP 2202 Fez, Morocco

*mahmoud.bououd@usmba.ac.ma

Abstract. Finned tubes heat exchanger is used on an innovative industrial solar dryer which consists of tree main components namely, solar concentrators, thermal room and heat exchanger. In this drying system, the heat transfer fluid coming from solar concentrators field circulates in the fin-and-tube bundles to heat the air moving through this battery to be injected into the thermal room to dry clay bricks. In this paper, we present a modeling and parametric study of the plate flat fin-and-tube bundles considering the most used models in the evaluation of the heat transfer and air flow performances. First, we carry out with a validation of the proposed physical and numerical models based on previous experimental results. In the second part we focus on the variation effects of the different geometric parameters of the tube bundles on the drying temperature and the pressure drop in order to estimate the performance evaluation criterions. Numerical results have shown that big external tube diameter increases the performance evaluation index by 67% when it varies from 20mm to 35mm which promotes the thermo-flow performances of the heat exchanger. However, Small transverse and longitudinal pitches present a higher air velocity at the minimum flow area and a stronger flow disturbance which ensure better air flow and heat transfer characteristics.

Keywords: heat transfer, pressure drop, finned tubes heat exchanger, modeling, performance evaluation criterions, solar drying.

1. Introduction

Many industrial applications include drying in their manufacturing processes such as textiles, phosphate, dairy processing, production of cement, waste water treatment, production of tiles and clay bricks, etc.

Several sources can provide the required energy for this industrial process namely, electricity, fossil fuels, natural gas, wood and solar. For many decades, solar radiation was used for drying agricultural products based on some traditional systems. However, its integration in the industrial sector is very limited due to the low temperatures values reached by the existing systems. Considering the depletion of natural resources in the near future leading the rise of oil prices, the integration of solar drying in industries is expected to become a necessity. Active, inactive, mixed and hybrid solar drying [1,2,3] are the essential modes used in the existing prototypes of solar dryers which have been designed and tested for the drying agricultural products [4, 5, 6, 7, 8].

The cross flow heat exchanger is one of the most effective devices in the heat recovery for its numerous advantages, namely: the large amount of heat which can be transferred from a small exchange area, the simplicity of design and manufacturing, wide operating temperature range, the ability to control high heat flux at different temperature levels [9, 10].

Nomenclature

A	heat transfer surface area (m^2)	S_T	transverse pitch (m)
c_p	specific heat (J/kg K)	S_L	longitudinal pitch (m)
D	external tube diameter (m)	S_D	diagonal pitch (m)
D_i	internal tube diameter (m)	T_w	surface temperature (K)
D_h	hydraulic diameter (m)	T_i	inlet temperature of air (K)
F_s	fin spacing (m)	T_o	outlet temperature of air (K)
f	air side friction factor	ΔT_{lm}	log-mean temperature difference (K)
h	convective heat transfer coefficient (W/m^2. K)	V	air velocity (m/s)
L	total tubes length (m)	U_m	maximum air velocity in the minimum flow area (m/s)
Nu	air side Nusselt number	U_w	water velocity (m/s)
N_T	number of tubes per row		
N_L	number of rows	***Greek symbols***	
N	total tubes number	α	fin Oblique angle (°)
n_f	number of fins	δ	fin thickness (m)
Pr	water side Prandlt number	λ	air thermal conductivity (W/m. K)
Δp	Air Pressure drop (Pa)	ρ	air density (kg/m^2)
Re	air side Reynolds number	μ	average free-stream viscosity, (N.s/m^2)

In order to evaluate the thermo-flow characteristics of flat finned tube bundles, we carry out with a parametric study of different geometric parameters of the heat exchanger which is conducted based on a physical model established according to the most practical models in the performance evaluation of the air flow and heat transfer.

The thermo-flow performance involves many geometric parameters namely fin pitch, fin oblique angle, fin thickness, tube pitches and tube external diameter. The variation effect of the transverse and longitudinal pitches as well as the tube outside diameter will be investigated in this based in order to optimize the architecture of similar heat exchangers for a predefined operating conditions.

2. Methodology

2.1. Air-side flow and heat transfer models

The modeling of thermo-flow characteristics of the flat finned tube heat exchanger (Fig.1) is performed based on the following assumptions:

- Steady-state and incompressible flow conditions
- Unchangeable air properties with the temperature
- Neglected radiation effects.

The heat transfers on the air side of the finned tube bundles in staggered configuration (Fig.2) is modelled based on the following dimensionless parameters [11, 12]:

The Reynolds Number:

$$Re = \rho U_m D / \mu \quad (1)$$

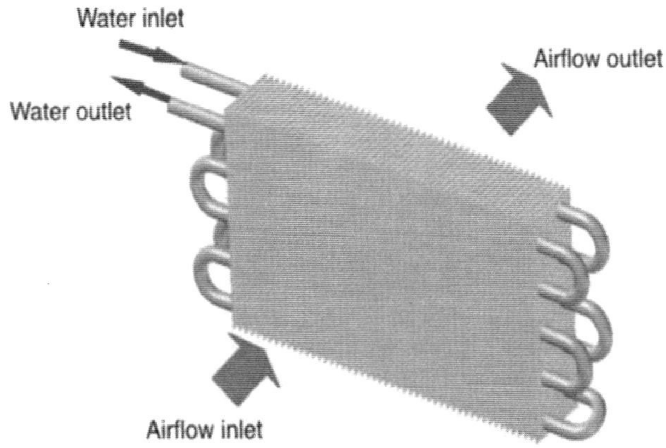

Figure 1: Architecture of the finned tubes heat exchanger

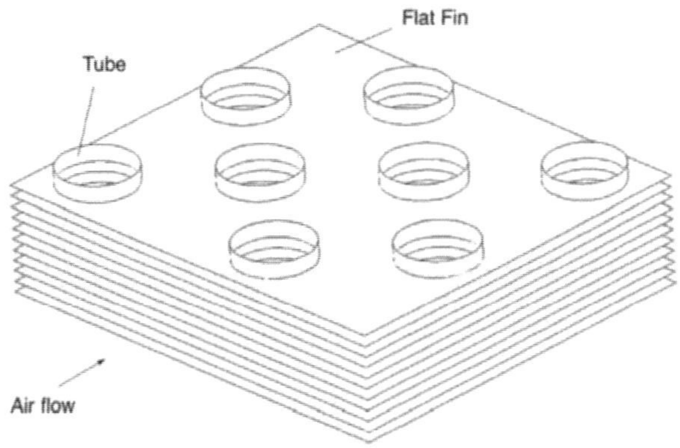

Figure 2: Schematic diagram of flat finned tube bundles in staggered configuration.

with:

$$U_m = \begin{cases} U(S_T/2a) \text{ if } 2a \leq 2b \\ U(S_T/2b) \text{ if } 2a \geq 2b \end{cases} \quad (2)$$

$$2a = (S_T - D) - [(S_T - D)\delta . n_f] \quad (3)$$

$$b = (S_D - D) - [(S_T - D)\delta . n_f] \quad (4)$$

Where, ρ, μ, λ, are the density, dynamic viscosity and thermal conductivity respectively.

The friction factor:

$$f = 2\,\Delta p/\rho U_m^2 \quad (5)$$

Δp is the pressure drop across the finned tube bundles.

The Nusselt Number:

$$Nu = h\,.D/\lambda \quad (6)$$

where h is the average heat transfer coefficient which is defined by:

$$h = \Phi/A\,\Delta T \tag{7}$$

A is the heat transfer surface area and Φ is the total heat flow rate calculated by:

$$\Phi = mC_p(T_o - T_i) \tag{8}$$

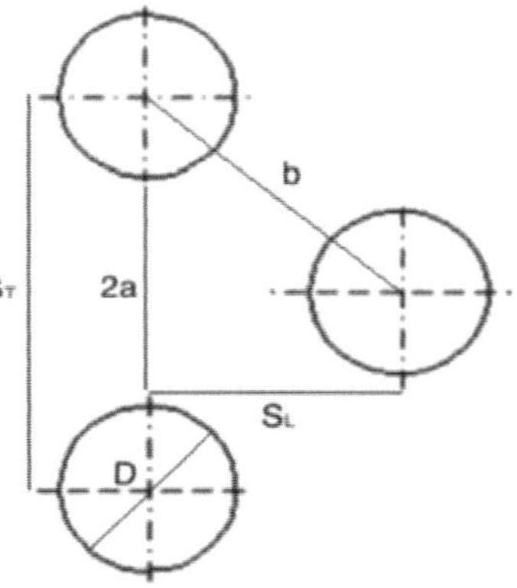

Figure 3: schematic diagram of geometric parameters

The logarithmic mean temperature difference between the tube wall and the air is defined as:

$$\Delta T_{lm} = \frac{T_o - T_i}{ln(T_w - T_i/T_w - T_o)} \tag{9}$$

For different number of rows, several works have been conducted to estimate the average heat transfer coefficient and the pressure drop trough the calculation of the Nusselt number and the friction factor respectively based on the geometric parameters of the finned tube bundles.

For the 4 rows plate flat finned tube bundles Kong et al [13] have established the correlations based on longitudinal and transvers pitches, and tube diameter as follows:

$$f = 662.3561\ Re^{-0.6453}\left(\frac{S_T}{D}\right)^{-0.2801}\left(\frac{S_L}{D}\right)^{-0.3927} \tag{10}$$

$$Nu = 2.6653\ Re^{0.3175}\left(\frac{S_T}{D}\right)^{-0.8732}\left(\frac{S_L}{D}\right)^{-0.5618} \tag{11}$$

with: $1 \le S_T/D \le 1.8\ , 1 \le S_T/D \le 1.8\ , 800 \le Re \le 13000$

Other correlations have been proposed [14], for 6 rows heat exchanger, in function of longitudinal and transverse pitches, tube diameter, hydraulic diameter, fin oblique angle, fin thickness and fin spacing which are given by the following equations:

$$Nu = 0.3313Re^{0.575}(1-\alpha)^{-0.0522}\left(\frac{F_S}{D_h}\right)^{1.3204}\left(\frac{\delta}{D_h}\right)^{0.0351}\left(\frac{D}{D_h}\right)^{-0.1002} \tag{12}$$

$$f = 7.655Re^{-0.4927}(1-\alpha)^{0.1441}\left(\frac{F_S}{D_h}\right)^{1.7383}\left(\frac{\delta}{D_h}\right)^{0.0004}\left(\frac{D}{D_h}\right)^{0.0922} \tag{13}$$

$$D_h = 4A_c/P_c \tag{14}$$

With A_c is the minimum free flow area and $P_c = A/L_2$ is the perimeter at the minimum flow cross-section.
The application ranges of this correlations are:

$$0 \leq \alpha \leq 30°,$$
$$0.5476 \leq \frac{F_s}{D_h} \leq 0.678\,, 0.013 \leq \frac{\delta}{D_h} \leq 0.332\,, 2.228 \leq \frac{D}{D_h} \leq 9.143 \; and \;\; 220 \leq Re \leq 5500$$

For a wider application ranges in term of number of rows from 2 to 7 rows and based on the total number of tubes, fin spacing, tube diameter, and the longitudinal and transverse pitches Xie et al [15] have proposed the following correlations:

$$Nu = 1.565\, Re^{0.3414} \left(N\frac{F_s}{D}\right)^{-0.165} \left(\frac{S_T}{S_L}\right)^{0.0558} \tag{15}$$

$$f = 20.713\, Re^{-0.3489} \left(N\frac{F_s}{D}\right)^{-0.1676} \left(\frac{S_T}{S_L}\right)^{0.6265} \tag{16}$$

With:

$$0.67 \leq V \leq 4m/s,\; 1000 \leq Re \leq 6000,\; 16 \leq D \leq 20mm,\; 38 \leq S_T \leq 46mm,\; 32 \leq S_L \leq 36mm$$

The overall performance estimation of the finned tube bundles is defined by the performance evaluation criterion P_{EC} which is calculated by the following equation:

$$P_{EC} = {}^{Nu}\!\Big/_{f^{1/3}} \tag{17}$$

3. Results and discussion

3.1. Simulations reliability

In order to validate the physical model, we compare our simulations with the experimental results presented in previous works. Table 1 illustrates the used parameters in our simulations. Figure 4 and 5 show that the simulations are reliable for more than 90% and 97% respectively for the Nusselt number and friction factor estimations. Kong et al [13] model was considered due to its wide application ranges.

3.2. variation effects of diameter

The variation effect of the tube diameter was studied between D= 20mm and 35mm for a constant non-dimension pitches of 35 mm and an air velocity of 0.5 m/s in order to fulfil the application ranges of the physical model.
The increase of the tube diameter reduces the cross section of air flow which generates the swirl flows due to the flow separation at the tubes rear part. This disturbed air flow improves the thermo-flow performance, but increases the pressure drop.
Figures 6 and 7 confirm this concept and show that both the drying temperature and the pressure drop increase with increasing the tubes diameter.

Table 1: Parameters used in simulations

Parameters	Values
Ambient temperature T_i	289.15 K
Tube wall temperature T_w	331.72 K
S_T and S_L	0.035m
Air kinematic viscosity	$19.82 x 10^{-6}$
Thermal conductivity of air	0.0229 W/m.K
Air density	1.217 kg/m^2
Specific heat	1009 J/kg K
Number of tubes per row (N_T)	6
Total number of tubes (N)	24
Tube diameter	0.025 m
Air velocity (V)	0.5 m/s

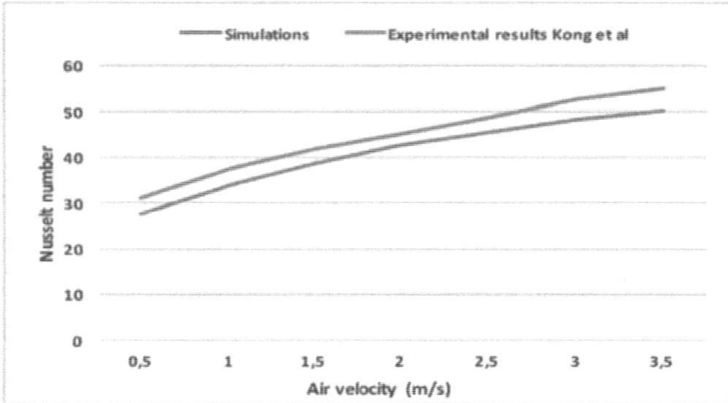

Figure 4: Simulation validation for Nusselt number

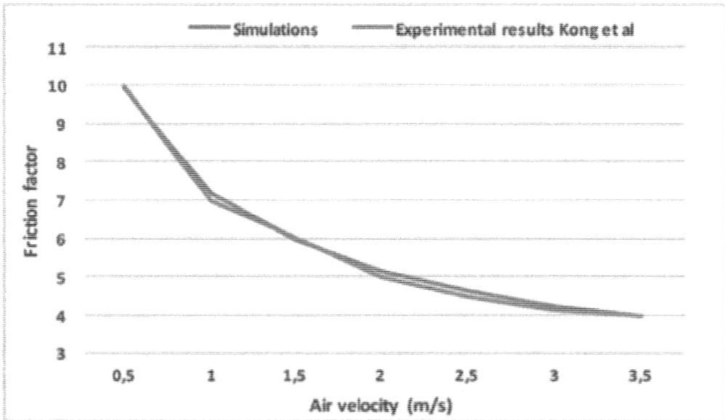

Figure 5: Simulation validation for Friction factor

Figure 8 shows that the performance evaluation index increase with the tube external diameter which means that the heat transfers are more important than the pressure drop, and allows us to conclude that the wide diameter is of benefits to the heat exchanger performances.
It is important to note that the increase of the tube external diameter reduces the fin surface area so it will be necessary to make a techno-economic assessment to determine the tubes diameter.

3.3. Variation effects of transverse pitch
In order to understand how the dimensionless transverse pitch contributes in heat transfer performances, we studied the impact of this parameter on the outlet air temperature, the pressure drop and the performance evaluation factor.

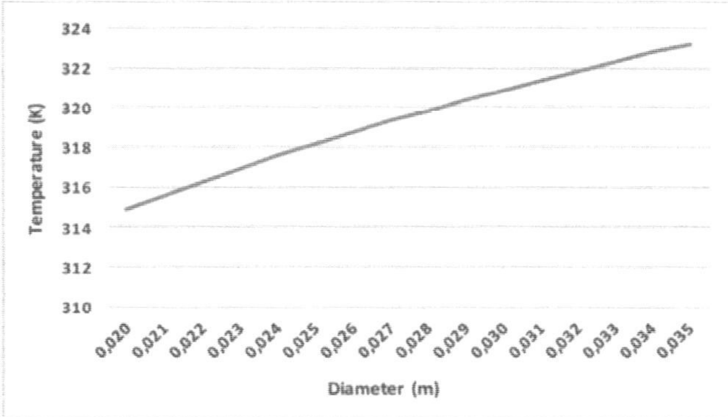

Figure 6: Variation of temperature with diameter

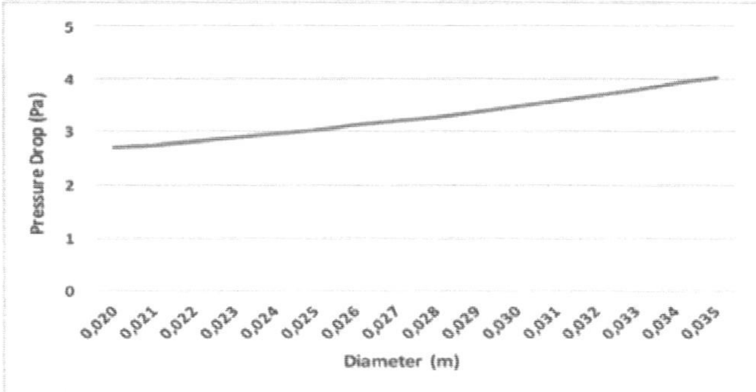

Figure 7: Variation of pressure drop with diameter

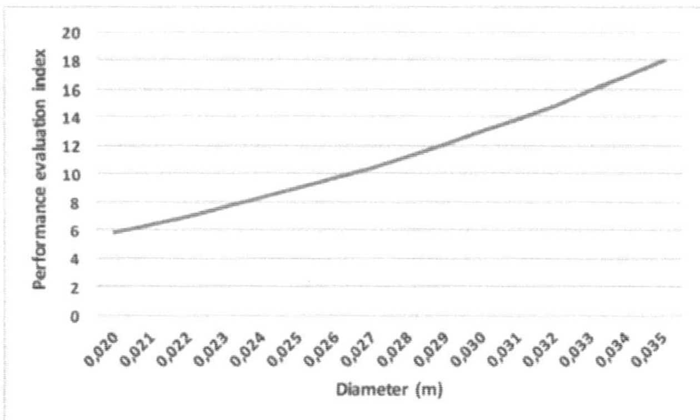

Figure 8: Variation of P_{EC} with diameter

Figures 9 and 10 show that both the outlet air temperature and the pressure drop decrease, respectively, by 3.2% and 31.68% with the transverse pitch increase from 25mm to 45mm.
This can be explained by the reduced air velocity and local air flow at the minimum flow area for a wider transverse pitch which consequently reduce the Reynolds and Nusselt numbers as well as the friction factor.
As a combination of the friction factor and the Nusselt number the performance evaluation index decrease for wider transverse pitch as shown in Figure 11.

3.4. Variation effects of longitudinal pitch

Finally, we investigate the effects of the longitudinal pitch on the heat transfer and the air flow at a constant diameter and transverse pitch. The dimensionless longitudinal pitch is varied from 25mm to 45mm to fulfil the application ranges of the considered model.

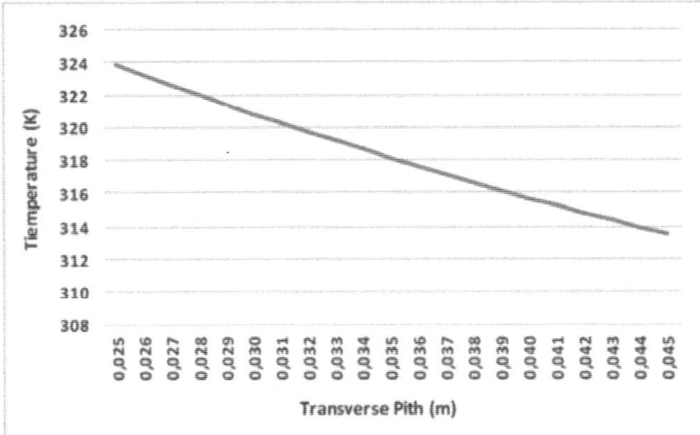

Figure 9: Temperature variation with transverse pitch

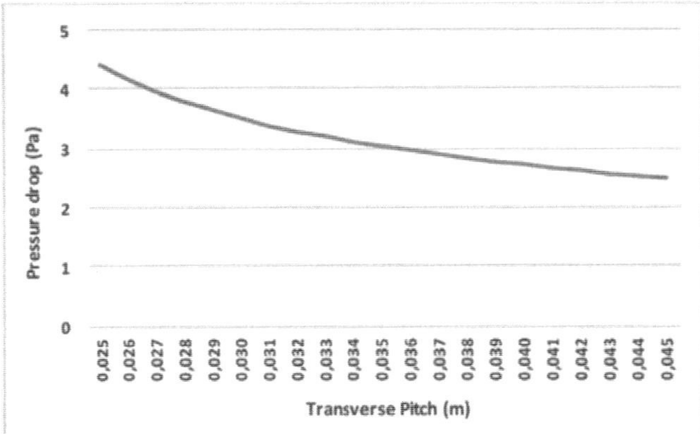

Figure 10: Pressure drop variation with transverse pitch

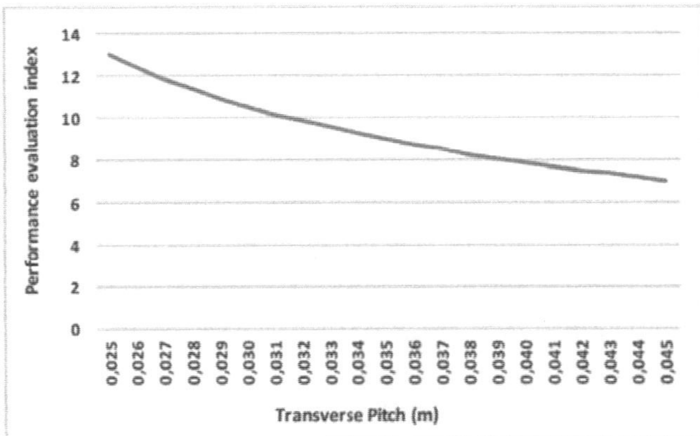

Figure 11: Variation of P_{EC} with transverse pitch

It can be seen from figures 12 and 13 that the drying temperature and the pressure drop decrease for a wider longitudinal pitch due to weakened disturbance of the air flow. Based on the same figure we notice that the air outlet temperature decreases by 1.5% while the pressure drop reduces by 20%.
Figure 14 illustrates that small longitudinal tube pitch is superior to the large one in the overall heat transfer performances of the finned tubes heat exchanger.
According to the performance evaluation index which decreases by 46% with the transverse pitch (Fig.8) and only 22% with the longitudinal pitch (Fig.14) we conclude that the impact of the transverse pitch variation is more significant than that of the longitudinal pitch on the heat exchanger performances.

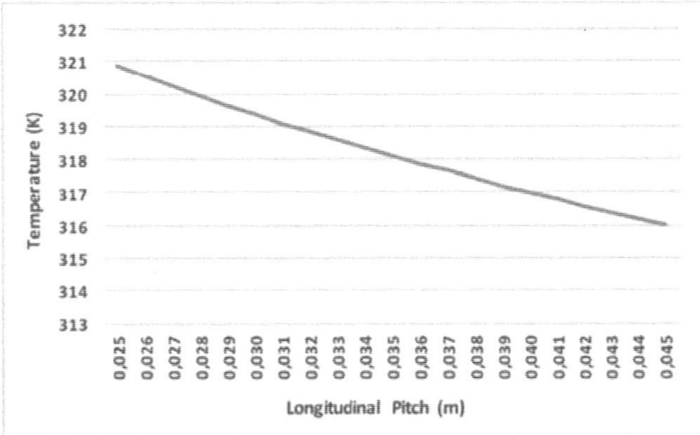

Figure 12: Temperature variation with longitudinal pitch

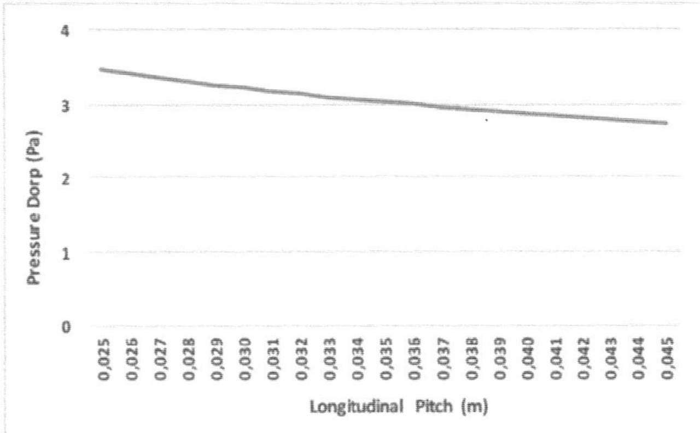

Figure 13: Pressure drop variation with longitudinal pitch

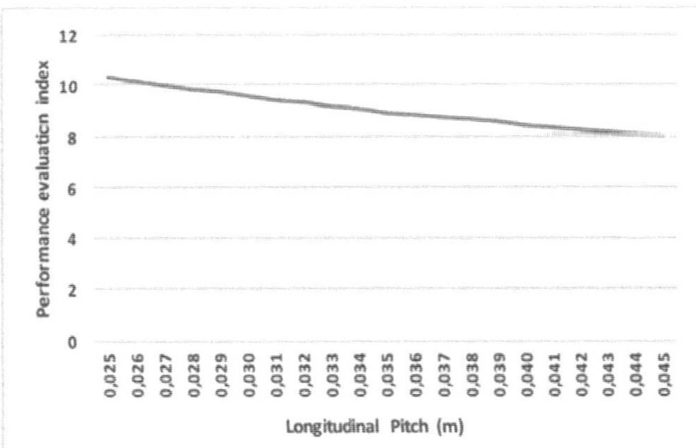

Figure 14: Variation of P_{EC} with longitudinal pitch

4. Conclusion

Parametric study of the thermo-flow characteristics of the flat finned tube bundles is performed at various transverse and longitudinal pitches as well as different tube's external diameter.

Based on the conducted simulations we concluded that a big diameter is of benefits to the air flow and heat transfer performances which is traduced by the increase of the performance evaluation index by 67% when the tube's external diameter varies from 20mm to 35mm.

Small transverse and longitudinal pitches present a higher air velocity at the minimum flow area and a stronger flow disturbance which ensure better thermo-flow characteristics. The proposed physical model and simulations allow us to evaluate the air flow and heat transfer performances in order to optimize the the geometric parameters of the flat finned tubes heat exchanger.
This study can be developed by considering the variations effects of different geometric parameters of fins namely the fin spacing and the fin thickness to achieve a complete analysis of the thermo-flow performances of the finned tube bundles.
Its important to note that all the simulations are conducted at a constant tubes wall temperature which is difficult to ensure when the heat exchanger is coupled with an intermittent heat production system as solar concentrators which require the evaluation of the thermo-flow performances of this heat exchanger at various heat transfer fluid temperatures to get a deeper analysis of the heat transfer performances in order to integrate this solar energy system in the industrial drying process of clay bricks.

5. References

[1] Pardhi C and Bhagoria JL. *Development and performance evaluation of mixed-mode solar dryer with forced convection*. Int J Energy Environ Eng 2013;4(1).
[2] Montero I, Blanco J, Miranda T, Rojas S and Celma AR. *Design, construction and performance testing of a solar dryer for agro-industrial by-products*. Energy Convers Manag 2010;51(7).
[3] Akoy E, Ismail M and El-Fadil AA. *Design and construction of a solar dryer for mango slices*. In: Proceedings of international research on food security; 2006.
[4] Mustayen A, Mekhilef S and Saidur R. *Performance study of different solar dryers: a review*. Renew Sustain Energy Rev 2014; 34:463-470.
[5] Kalogirou SA. Chapter 7, *Industrial process heat, chemistry applications, and solar dryers*. In: Solar energy engineering. 2nd ed. Boston: Academic Press; 2014.
[6] Sevda MS and Rathore NS. *Studies on semi-cylindrical solar tunnel dryer for drying di-basic calcium phosphate*. Agric Eng Int 2007. the CIGR Ejournal. Manuscript EE 07 001. Vol. IX.
[7] Kurt M, Ayşegül A and Sanin F D. *Evaluation of Solar Sludge Drying Alternatives by Costs and Area Requirements*. Water Research 2015; 82: 47-57.
[8] Ferreira AG, Gonçalves LM and Maia CB. *Solar drying of a solid waste from steel wire industry*. Appl Therm Eng 2014;73(1).
[9] Faghri A. *Heat pipe science and technology*. USA: Taylor & Francis; 1995.
[10] Peterson GP. *An introduction to heat pipes*. New York: John Wiley and Sons; 1994.
[11] Wang CC, and Kuan-Yu C. *Heat Transfer and Friction Characteristics of Plain Fin-and-Tube Heat Exchangers, Part I: New Experimental Data*. International Journal of Heat and Mass Transfer 2000; 43, no. 15: 2681–91.
[12] Wang CC, Kuan-Yu C, and Chun-Jung C. *Heat Transfer and Friction Characteristics of Plain Fin-and-Tube Heat Exchangers, Part II: Correlation*. International Journal of Heat and Mass Transfer 2000; 43, no. 15: 2693–2700.
[13] Kong YQ, Yang LJ, Du XZ, and Yang YP. *Air-Side Flow and Heat Transfer Characteristics of Flat and Slotted Finned Tube Bundles with Various Tube Pitches*. International Journal of Heat and Mass Transfer 2016; 99: 357–71.
[14] Kong YQ, Yang LJ, Du XZ, and Yang YP. *Impacts of Geometric Structures on Thermo-Flow Performances of Plate Fin-Tube Bundles*. International Journal of Thermal Sciences 2016; 107: 161–78.
[15] Xie G, Qiuwang W, and Bengt S. *Parametric Study and Multiple Correlations on Air-Side Heat Transfer and Friction Characteristics of Fin-and-Tube Heat Exchangers with Large Number of Large-Diameter Tube Rows*. Applied Thermal Engineering 2009; 29, no. 1: 1–16.

Coupled heat transfers in a three cavity hollow block subjected to an incident solar flux: Effect of heat conductivity

M Najjaoui[1], T Ait-Taleb[1], A Abdelbaki[2], Z Zrikem[2] and H Chaib[1]

[1] ERME, Department of Physics- Chemistry, Polydiscilinary Faculty of Ouarzazate, Ibn Zohr University, P.O. Box 638, Ouarzazate, Morocco.
E-mail: taittaleb@gmail.com, mourad.najjaoui@gmail.com
[2] LMFE, Department of Physics, Cadi Ayyad University, Faculty of Sciences Semlalia, B.P. 2390, Marrakech, Morocco.

Abstract. Hollow blocks or bricks are widely used due to the good thermal insulation, lightweight, and acoustic insulation etc. In this paper, we present a numerical study of heat transfers coupled by natural convection and radiation conduction through a hollow block with three cavities used for the construction of ceilings of buildings. This hollow block is subjected to an incident solar flux and exchanging heat with the air ambient. The governing equations of the mathematical model used for the simulation are discretized by the finite volume method and solved by the SIMPLE algorithm. The main objective of this simulation is to study, in steady state, the effect of the thermal conductivity of the solid partitions on heat flux through this hollow block. The results are presented in terms of streamlines, the isotherms, temperature profiles on the actives surfaces and the obtained global heat flux.

Keywords: Heat transfer; Numerical simulation; Steady state; Thermal conductivity, Hollow block.

1. Introduction

Great and inefficient energy consumption intensifies the energy crisis and causes various environment issues. Buildings energy consumption (residential and tertiary) usually takes up a significant percentage of the total social energy consumption. And more than half of this part energy is eventually consumed for heating and cooling buildings. A significant portion of energy in buildings is consumed due to heat loss or heat gain in the building envelope. Indeed, heat losses through the building envelope are localized at all levels since a building behaves like a container pierced with different types of holes: doors and windows, external walls, ceilings, ventilation, including the envelope of roofing represents 25 to 30% of these thermal losses ... Therefore, the energy efficiency in this sector is very important and there is a great potential to do this. With this in mind, it is important to take this in consideration at the time of design.

Utilization of energy-efficient building envelope may be a good means for improving the building energy efficiency. In recent years, hollow block/brick, a new kind of prefabricated building construction material, is becoming more and more popular. This block has good thermal insulation, lightweight, acoustic insulation etc. (Zhang et al. [1], Gao et al. [2]). But, in general, the heat transfer in the hollow block as the main unit of this envelope is complex and coupled by conduction in solid partitions, natural convection inside the cavities and radiation between the internal surfaces of the cavities.

Consequently, adequate study of the thermal behaviour of such structures must take into account the simultaneous existence of the three modes of heat transfer. Many investigations concerned about the hollow block/brick have been presented. Some of them focus on the thermal properties of this system mainly by using theoretical (Bouchair [3]), numerical (Del Coz Diaz et al. [4]) or experimental studies (Kus et al. [5], Ahmad et al. [6]).

Most of the work carried out in this sense has been devoted to the study of the effects of conduction and/or radiation on natural convection in a differentially heated rectangular block, A part of these works studied the effect of conduction in solid partitions on natural convection [7-8]. Whereas the other part studied the effect of radiation exchange between cavity surfaces on natural convection [9-10]. Recently, Ait-Taleb et al. [11-13] presented a series of numerical studies that take into account the three modes of heat transfer in alveolar blocks heated from below or above, but submitted to the imposed temperatures excitation. These studies play an important role in simulating the thermal performance and optimizing the configurations of the hollow block.

In this paper, we propose a numerical simulation, in steady state, the effect of thermal conductivity on heat transfer in a hollow block with three horizontal rectangular cavities with thick walls subjected to a solar flux imposed, and exchanging heat with ambient air at constant temperatures with surface exchanges coefficients h_e and h_i. The main objective of this simulation is to examine the effect of the thermal conductivity of the solid partitions on the nature of flow and on the overall thermal behaviour of this hollow block.

2. Mathematical formulation

2.1. Studied configuration and governing equations

Figure 1 shows the schematic of a hollow block with one row of air cells. This hollow block has three rectangular cavities surrounded by solid walls. Each cavity has a width l and height h having vertical conductive walls of thickness ex and other horizontal ones of thickness ey. The top horizontal face is subjected to an incident solar flux (G) and exchanges heat with the ambient air at the temperature T_e with a surface exchange coefficient h_e. The lower surface exchanges heat with the indoor air of a room at temperature T_i and with a surface exchange coefficient h_i, while the vertical surfaces are considered adiabatic.

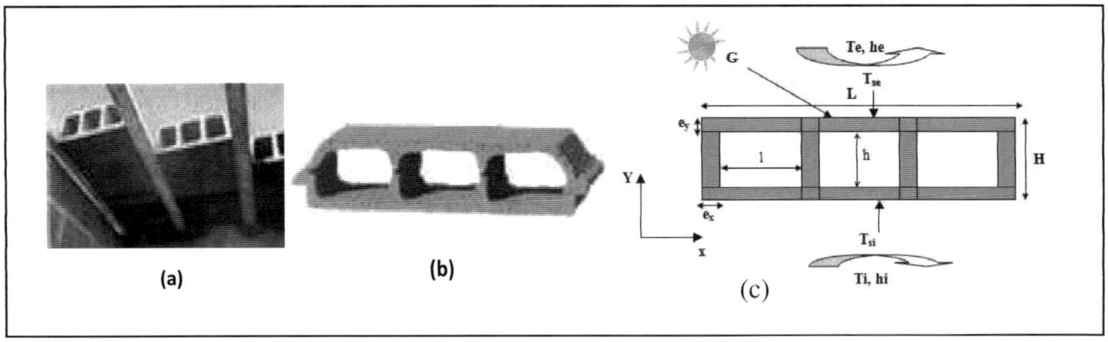

Figure 1. Schematics of hollow blocks: (a) hollow block ceilings; (b) hollow block; (c) cross-section of the hollow block

The heat transfer of the perforations (cavities) of the hollow tile as shown in Figure 1 is complex and coupled with heat conduction, convection and radiation. Therefore, non-linear characteristics exist within the heat transfer process of these perforations. In order to model this problem we adopt some simplifying hypothesis such as the inner surfaces, in contact with the fluid, are assumed to be gray, diffuse emitters and reflectors of radiation with an emissivity ε. The flow is conceived to be laminar, two-dimensional and incompressible with negligible viscous dissipation. The fluid is assumed to be no participating to radiation and the heat transfer is two-dimensional. All the thermophysical properties of the solid and the fluid are assumed constant except the density in the buoyancy term which is assumed to vary linearly with temperature (Boussinesq approximation), such a variation gives rise to the buoyancy forces.

Taking into account the above-mentioned assumptions, the dimensionless equations representing the conservation of the mass, the momentum and the conservation of the energy inside cavities full of air are:

$$\frac{\partial U}{\partial X}+\frac{\partial V}{\partial Y}=0 \tag{1}$$

$$\frac{\partial U}{\partial \tau}+U\frac{\partial U}{\partial X}+V\frac{\partial U}{\partial Y}=-\frac{\partial P}{\partial X}+\Pr\left(\frac{\partial^2 U}{\partial X^2}+\frac{\partial^2 U}{\partial Y^2}\right) \tag{2}$$

$$\frac{\partial V}{\partial \tau}+U\frac{\partial V}{\partial X}+V\frac{\partial V}{\partial Y}=-\frac{\partial P}{\partial Y}+\Pr\left(\frac{\partial^2 V}{\partial X^2}+\frac{\partial^2 V}{\partial Y^2}\right)+\mathrm{Ra}.\Pr.\theta_a \tag{3}$$

$$\frac{\partial \theta_a}{\partial \tau}+U\frac{\partial \theta_a}{\partial X}+V\frac{\partial \theta_a}{\partial Y}=\frac{\partial^2 \theta_a}{\partial X^2}+\frac{\partial^2 \theta_a}{\partial Y^2} \tag{4}$$

Where U, V, τ, θ and P are the dimensionless parameters associated to the primitive quantities u, v, t, T and p, respectively.

The dimensionless variables are obtained from the following expressions:

$$U=\frac{u\,H}{\alpha},\quad V=\frac{v\,H}{\alpha},\quad \theta=\frac{T-T_i}{T_e-T_i},\quad P=\frac{p+\rho_0 g y}{\rho_0(\alpha/H)^2}\ \text{and}\ \ \tau=\frac{t\,\alpha}{H^2} \tag{5}$$

The Rayleigh number (Ra) and the Prandtl number (Pr) are given successively by:

$$\mathrm{Ra}=\frac{g\beta\Delta T H^3}{\alpha_a \nu_a}\quad \text{and}\quad \Pr=\frac{\nu_a}{\alpha_a} \tag{6}$$

The dimensionless equation for heat conduction in solid walls is:

$$\frac{\partial \theta_s}{\partial \tau}=\frac{\alpha_s}{\alpha_a}\left(\frac{\partial^2 \theta_s}{\partial X^2}+\frac{\partial^2 \theta_s}{\partial Y^2}\right) \tag{7}$$

The thermal boundary conditions imposed are:

Adiabaticity conditions :

$$\left(\frac{\partial \theta s}{\partial X}\right)_{X=0}=\left(\frac{\partial \theta s}{\partial X}\right)_{X=1}=0 \qquad 0\leq Y\leq 1 \tag{8}$$

On the external horizontal side:

$$\frac{H}{k_s \Delta T}\alpha G+\frac{he\,H}{k_s}(\theta e-\theta se)=\frac{d\theta_s}{dY} \tag{9}$$

On the inner horizontal side:

$$\frac{hi\,H}{k_s}(\theta si-\theta i)=\frac{d\theta_s}{dY} \tag{10}$$

To these conditions are added those of continuity of the temperature and the heat flux to the fluid-solid interfaces, which given by:

$$\theta_s(X,Y)=\theta_a(X,Y) \tag{11}$$

$$-\frac{\partial\theta_s}{\partial\eta}=-N_k\frac{\partial\theta_a}{\partial\eta}+N_rQr \tag{12}$$

Where η is the dimensionless coordinate normal to the considered interface. The expression for calculating the dimensionless net radiative flux is given by the radiosity method of Siegel and Howell [14]:

$$Q_r(r_k)=\varepsilon_k\left(1-\frac{1}{G}\right)^4\left(\theta_k(r_k)+\frac{1}{G-1}\right)^4-\varepsilon_k\sum_{j=1}^{4}\int_{S_j}J_j(r_j)dF_{ds_k-ds_j} \tag{13}$$

The dimensionless average heat flux across the structure is given by:

$$Q_a=-\frac{H}{L}\int_0^{\frac{L}{H}}\frac{d\theta_S}{dY}\Bigg)_{Y=0}dX=-\frac{H}{L}\int_0^{\frac{L}{H}}\frac{d\theta_S}{dY}\Bigg)_{Y=1}dX \tag{14}$$

The dimensional average heat flux in ($W.m^{-2}$) is calculated by this equation:

$$Q=(Q_a\times k_s\times\Delta T)/H \tag{15}$$

2.2. Method of solution

Governing equations are discretised using the finites differences method based on the control volumes approach (Patankar [15]) with a power law scheme and are solved by the SIMPLE algorithm. The resulting system of algebraic equations is solved by the Tri-Diagonal-Matrix-Algorithm.
To accelerate the convergence of solutions, the governing equations are solved in their instationary form. To realize a compromise between accuracy and computation time, a study on the effects of both grid spacing and time step on the simulation results has been conducted. This study leads to the non-uniform grid size of 80 × 40. This number of mesh point used is sufficient to modeling accurately the heat transfer and fluid flow inside the hollow tile. The dimensionless time used is 10^{-6}. The convergence criterion is 10^{-4} which is based on the relative change in the variables U, V, P, and Q_r at different nodes of the calculation domain:

$$\left|\frac{f^{n+1}(i,j)-f^{n}(i,j)}{f^{n(i,j)}}\right|\le 10^{-4} \tag{16}$$

Where $f^n(i,j)$ is the variable $f(f=U,V,P,\theta,Q_r)$ value at node (i, j) calculated at iteration n.

2.3. Parameters of simulation

The results presented in this study are obtained for a hollow block representing a type of hollow bricks. This block is formed by three rectangular cavities each one of the geometrical dimensions such as the width l = 13cm and the height h = 7cm (type of 50 × 11) and having vertical conductive walls of thickness ex_i = 2.5cm. And other horizontal ones of thickness ey_j = 2cm. The thermal conductivity k_s varies between 0.5 and 1W/mK, and its thermal diffusivity is $\alpha_s = 4.25\times10^{-7}m^2/s$. The cavities are full by air of thermal conductivity k_a and thermal diffusivity α_a are equal to 0.0262W/mK and $1.57\times10^{-5}m^2/s$ respectively. The Prandtl number is Pr = 0.71. The emissivity of the internal sides of the cavities is equal 0.9. In accordance to real conditions, the average surface exchange coefficients were set at the values, $h_e = 17W/m^2°C$ and $h_i = 8.3W/m^2°C$. The indoor air temperature is maintained at temperature T_i = 20°C, while the ambient temperature is set to T_e = 40°C. The incident solar flux G on the outside face of the structure is fixed at the value $1000W/m^2$.

3. Results and discussion

3.1. Streamlines and isotherms

Figure 2. illustrates the streamlines (at the left) and the isotherms (at the right) obtained for the considered three thermal conductivity values k_s=0,5 ; k_s=0,8 and k_s=1W/m.K.We are in a situation of heating from the top, due to the positions of the hot and cold facades. As expected, the flow within the three cavities is characterized by the existence of four small cells rotating in opposite directions. These small recirculation cells (at the streamlines) are due to low intensity flows ($\Psi_{max} = 0.86$) resulting from the temperature gradients created by conduction and by radiation at the corners of the various cavities. Concerning the effect of the thermal conductivity k_s, examination of the current lines shows that this does not affect the nature of the flow. Indeed, this nature of flow is similar in the various cavities of the block and so for the three values of k_s considered: The air particles in direct contact with the upper surface of the various cavities, heated by conduction, are the lightest, And consequently their movements, remain limited in the vicinity of the active walls. Temperature field analysis shows that heat transfer is mainly through conduction and radiation. Indeed, the isotherms are almost parallel lines in the cavities (horizontally stratified temperature fields) with small distortions of the isotherms at the level of the vertical solid walls which are due to the difference between the thermal conductivities of the two solid and fluid environments.

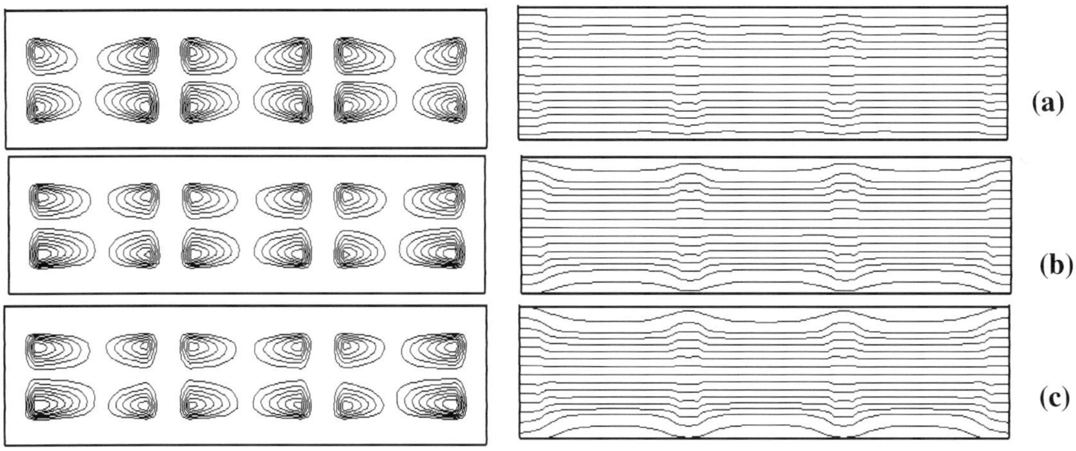

Figure 2. Streamlines (left) and isotherms (right) obtained for G=1000W/m^2 and different values of conductivities: (a) k_s=0,5W/m.K ; (b) k_s=0,8W/m.K and (c) k_s=1W/m.K.

3.2. Temperature Profiles

Figure 3 shows the temperature profiles of Ts(X) on both high and low horizontal surfaces of the studied hollow bloc. Through this figure, it is shown that the conductive thermal transfer is more important in solid partitions than via fluid environment. Furthermore, we notice that the temperature of the higher wall decreases noticeably as the conductivity k_s increases. This situation is reversed in the higher horizontal side where the wall temperature is lower in the environment, and increases as k_s increases but in more significant way. All this can be explained by the radiation transfer who is less important in the lower side than in the higher. The increase/(decrease) of temperature in the higher side/(lower) as k_s increases can be explained simply by the increase of heat conduction in accordance with Fourier's law, in the isolation characteristics of materiel that increases when k_s decreases more precisely. We also notice that in the higher side, the temperature is low in the lower corner and increases in the function of X. It reaches its maximum inside every cavity as we decrease and move the higher corner. This behaviour matches the statements made while we analyse the isotherms, in which we noticed small distortions of isotherms at the level of solid vertical walls. This is due to the differences between thermal conductivities in both solid and fluid environments.

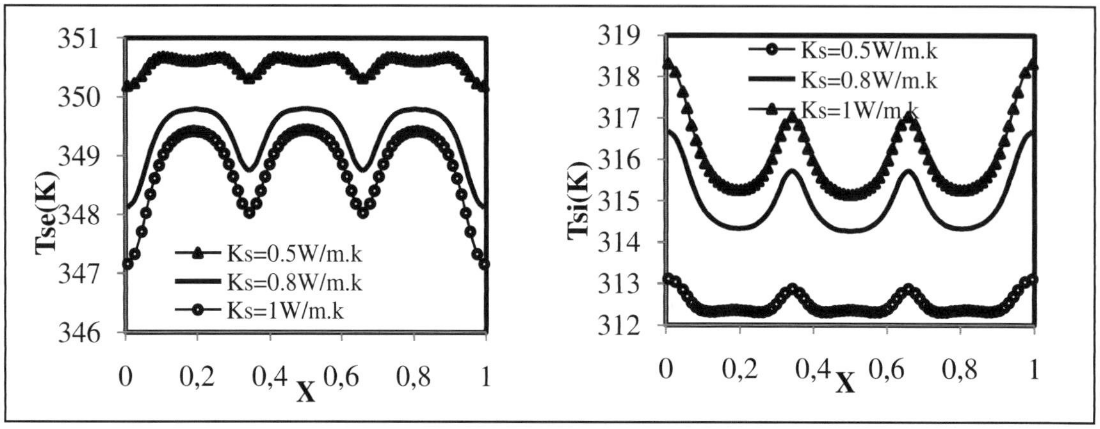

Figure 3. Temperature profiles (left) upper horizontal surface, (right) lower horizontal surface.

3.3. Global heat transfer

With respect to the total heat transfer in the vertical direction, figure 4 for G = 1000W/m², represents the distribution as a function of the width X, of the overall heat flux passing through the cold (Y = 0) and hot (Y = 1) hollow block, respectively for the values of the thermal conductivity k_s = 0.5, 0.8 and 1.0 W/mK.

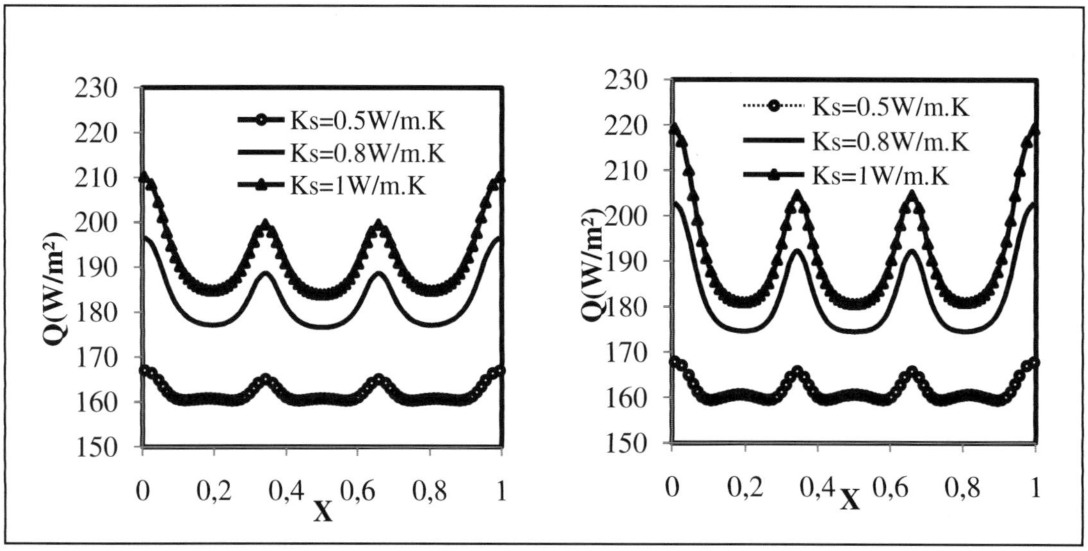

Figure 4. Variation of the global heat flux crossing the upper (left) and lower (right) surfaces of the structure.

From this figure it can be seen, as expected, that as the thermal conductivity k_s increases there is a considerable increase in the heat flux, especially when k_s increases from 0.5 to 1.0 W/m.K. Indeed, the average value of the total flux increases from value 161.56W/m² for k_s = 0.5 W/m.K to 191.32 W/m² for k_s = 1.0W/m.K. This corresponds to an increase of about 18%. It is also noted that the heat flow is maximum at the solid partitions separating the three cavities due to the thermal conductivity of the solid k_s. Inside the cavities there is a moderation of the heat transfer, marking the absence of convection (because it is a heating from above) whatever the value of k_s. The comparison of the distributions of the total flux obtained on the two hot and cold sides of the block show that the range of variation of the overall flux on the hot side (upper face) is more extensive than that on the cold side (lower face) regardless of the value of the conductivity k_s considered. However, and as expected, the average heat flow is retained.

4. Conclusion

The coupled heat transfers by natural convection, conduction and radiation in a hollow block with three cavities which is used in the construction the ceilings of buildings, subjected to an incident solar flux and exchanging heat with the air ambient, have been studied numerically. The simulation results show that heat transfer through this hollow block heated from above by the incident solar flux and exchanging heat with the air ambient makes mainly by conduction and radiation. The intensity of flow within the different cavities is very weak confirm that there is a very low convective heat transfer. It has been found that thermal transfer through this system depends highly on the thermal conductivity of solid partitions k_s. This parameter changes the structure of flow inside the cavity and contributes in the global heat transfer growth in a significant way.

References

[1] Zhang Y, et al. 2014 Impact factors analysis on the thermal performance of hollow block wall, Energy and Buildings, 75, 330-341.

[2] Gao Y, et al. 2004 Reduced linear state model of hollow blocks walls, validation using hot box measurements, Energy and Buildings, 36 (11), 1107-1115.

[3] Bouchair A 2008 Steady state theoretical model of fired clay hollow bricks for enhanced external wall thermal insulation, *Building and Environment*, **43**(10), 1603-1618.

[4] Del Coz Díaz J.J. et al. 2007 Analysis and optimization of the heat-insulating light concrete hollow brick walls design by the finite element method, *Applied Thermal Engineering*, **27**, 1445-1456.

[5] Kus H, et al. 2013 Hot box measurements of pumice aggregate concrete hollow block walls, *Construction and Building Materials*, **38**, 837-845.

[6] Ahmad A, et al. 2014 In situ measurement of thermal transmittance and thermal resistance of hollow reinforced precast concrete walls, *Energy and Buildings*, **84**, 132-141.

[7] Kim D.M, R Viskanta R 1984 Study of the effects of wall conductance on natural convection in differentiallyoriented square cavities, *J. Fluid. Mech.* **144**, 153-176,

[8] Zhang W, Zhang C.H, Xi G 2010, Conjugate conduction-natural convection in an enclosure with time-periodicsidewalltemperature and inclination, *Int. Journal of Heat and Fluid Flow*, **32**, 52–64,

[9] Balaji C, Venkateshan S.P 1993 Interaction of surface radiation with free convection in a square cavity, *Int. Journal of Heat and Fluid Flow*, **14**, 260-267,

[10] Wang H, Xin S, et Le Quéré P 2006 Etude numérique du couplage de la convection naturelle avec le rayonnement de surfaces en cavité carrée remplie d'air, *C. R. Mécanique*, **334**, 48-57,

[11] Ait-Taleb T, Abdelbaki A, Zrikem Z 2008 Numerical simulation of coupled heat transfers by conduction, natural convection and radiation in hollow structures heated from below or above. *Int. J. of Thermal Sciences* **47**, (4), 378-387,

[12] Ait-Taleb T, Abdelbaki A, Zrikem Z 2008 Transfer function coefficients for time varying coupled heat transfers. Application to hollow concrete brick'. *Building Simulation Journal*, **1**, 303-310.

[13] Ait-Taleb T, Abdelbaki A, and Zrikem Z 2014, "Simulation of coupled heat transfers in a hollowtile with two vertical and three horizontal uniform rectangular cavities heated from below or above". *Energy and Buildings*, **84**, 628–632.

[14] Siegel R, Howell JR 1981 Thermal radiation heat transfer. 2nd ed. New York: McGraw-Hill.

[15] Patankar S.V 1980 Numerical Heat Transfer and Fluid Flow, *Hemisphere*, New York,

IOP Conf. Series: Earth and Environmental Science **161** (2018) 012003 doi:10.1088/1755-1315/161/1/012003

Deep-level traps and recombination centers effects on photovoltaic conversion efficiency of GaAs based crystalline solar cell

H El Achouby, M Zaimi, A Ibral and E M Assaid

Optics and Electronics of Semiconductor Nanostructures Team, Laboratory of Instrumentation, Measurements and Control, Department of Physics, Faculty of Sciences, Chouaib Doukkali University, P. O. Box 20 El Jadida, Morocco

E-mail:eassaid@yahoo.fr

Abstract. In this work, we proceed to analytical and numerical resolutions, using finite difference method, of semiconductor four equations system governing the variations of electrostatic potential, electrons and holes densities, density profile of occupied levels associated to deep-level traps as well as concentration profile of occupied levels related to recombination centers, in one-dimensional thin-film crystalline solar cell based on Gallium Arsenide (GaAs). In addition to physical quantities mentioned above, the problem resolution, achieved using Maple computer algebra software, enables us to retrieve electrostatic field, electrons and holes current densities as well as energy band diagram throughout the device. Ultimately, particular attention is devoted to specific phenomena related to semiconductors such as the presence of deep-level traps or recombination centers whose levels are in the middle of the gap. In this way, we show that the presence of deep-level traps or recombination centers leads, via Shockley-Read-Hall mechanisms, to a decrease of solar cell photovoltaic conversion efficiency. We also show that these effects are dependent on energy levels of traps and defects, on traps and defects concentrations as well as their electron and hole capture cross sections.

1. Introduction

Nowadays, global energy needs are growing exponentially. This is essentially due to increased energy demand of newly industrialized Asian countries such as India and China, and also of emerging countries all around the world such as Indonesia, Brazil, Turkey and South Africa. Now, all climate specialists agree on the fact that excessive consumption of fossil fuels such as oil, coal or natural gas leads to an increase of carbon dioxide, methane and nitrous oxide rates in the atmosphere. Unfortunately for us and for our planet Earth, these gas emissions have two major negative environmental effects, they lead to an increase of greenhouse gases rates in the atmosphere and make air pollution more pronounced which brings about climatic changes that result in acute droughts or catastrophic floods. Nowadays, there is only one alternative that satisfy growing energy needs while contributing to the protection of the environment: renewable energies. The main source of renewable energy is the Sun. With its two components solar thermal and solar photovoltaic, the Sun represents an inexhaustible source of energy which is able to fulfil all energy needs of humanity. Here, we are interested in converting incoming solar radiation into energy via crystalline photovoltaic solar cells. Now, the main component in solar photovoltaics is silicon solar cell technology but thin film technology is gaining importance since it offers high efficiency and in some cases low cost solar cells. In the present work, we perform the simulation of a thin-film crystalline semiconductor solar cell based on Gallium Arsenide (GaAs). We are particularly interested in the effects of deep-level traps and recombination centers, whose energy levels are in the middle of the gap, on solar cell photovoltaic conversion efficiency. The present paper is divided into six sections. After the present general introduction, we present, in the second section, the four semiconductor fundamental equations driving electrostatic potential, densities of charge carriers, density of occupied levels associated with deep-level traps as well as density of occupied levels associated with recombination centers, in the case of a one-dimensional crystalline solar cell. In the third section, we present the photonic part of our work. We particularize to the case of Gallium Arsenide based solar cell and give refractive index and extinction coefficient of GaAs. Then, we calculate the reflectance and the transmittance of air/GaAs interface as well as the absorption coefficient of the material. Next, we calculate the generation rate of electron-hole pairs in GaAs under real solar illumination corresponding to an air mass equal to 1.5 (AM1.5). In the fourth section, we perform the normalization of semiconductor fundamental equations. Then, we proceed to numerical and analytical resolutions of three coupled differential equations system obtained in the stationary case. In the fifth section, we present the results of numerical and analytic calculations. We show that the presence of deep-level traps or recombination centers leads to a decrease of open-circuit voltage, maximum power point coordinates and photovoltaic conversion efficiency of thin-film crystalline semiconductor solar cell. Finally, we end our article with a general conclusion.

2. Theoretical framework - Electronics background

2.1. Semiconductor fundamental equations formulated as functions of natural variables V(x), n(x), p(x)

The semiconductor fundamental equations as they have been established by W. Van Roosbroeck for a one-dimensional electronic device are [1]:

$$\frac{d^2V(x)}{dx^2}+\frac{q}{\varepsilon_0\varepsilon_r}\left(p(x)-n(x)+N_D^+(x)-N_A^-(x)\right)=0 \tag{2.1}$$

$$\frac{dn(x)}{dt}=\frac{1}{q}\frac{dJ_n(x)}{dx}-R_n(x)+G_n(x) \tag{2.2}$$

$$\frac{dp(x)}{dt}=-\frac{1}{q}\frac{dJ_p(x)}{dx}-R_p(x)+G_p(x) \tag{2.3}$$

$$\frac{dN_{trap,defect}(x)}{dt}=R_n(x)-R_p(x) \tag{2.4}$$

q is the elementary charge, ε_0 is the vacuum dielectric permittivity and ε_r is the semiconductor relative permittivity. The first equation is Poisson equation. It expresses the variations of electrostatic potential *V(x)* and electrostatic field *E(x)* as functions of total charge volume density $\rho(x)$. $N_D^+(x)$ and $N_A^-(x)$ are respectively donor-type and acceptor-type impurity density profiles. The second and the third equations are respectively electrons and holes continuity equations. They describe the conservation of electrons charge $-qn(x)$ and holes charge $qp(x)$ through the device. $R_n(x)$ and $R_p(x)$ are respectively the terms describing electrons and holes recombination. $G_n(x)$ and $G_p(x)$ are respectively electrons and holes generation rates. The fourth equation represents the continuity equation of deep-level traps or defects. $N_{trap}(x)$ ($N_{defect}(x)$) is deep-level traps (defects) density profile. $J_n(x)$ and $J_p(x)$ are current densities due to electrons and holes flows:

$$J_n(x)=-q\mu_n n(x)\frac{dV(x)}{dx}+qD_n\frac{dn(x)}{dx} \tag{2.5}$$

$$J_p(x)=-q\mu_p p(x)\frac{dV(x)}{dx}-qD_p\frac{dp(x)}{dx} \tag{2.6}$$

D_n and D_p are electrons and holes diffusion coefficients. They are related to mobilities of electrons μ_n and holes μ_p by Einstein's relations:

$$\frac{D_{n,p}}{\mu_{n,p}}=\frac{K_BT}{q} \tag{2.7}$$

In the stationary case and for normal operating conditions, all impurities are ionized. So, semiconductor fundamental equations as functions of natural variables become:

$$\frac{d^2V(x)}{dx^2}+\frac{q}{\varepsilon_0\varepsilon_r}\left(p(x)-n(x)+N_D(x)-N_A(x)\right)=0 \tag{2.8}$$

$$D_n\frac{d^2n(x)}{dx^2}+\mu_nE(x)\frac{dn(x)}{dx}+\mu_nn(x)\frac{dE(x)}{dx}-R_n(x)+G_n(x)=0 \tag{2.9}$$

$$D_p\frac{d^2p(x)}{dx^2}-\mu_pE(x)\frac{dp(x)}{dx}-\mu_pp(x)\frac{dE(x)}{dx}-R_p(x)+G_p(x)=0 \tag{2.10}$$

$$R_n(x)-R_p(x)=0 \tag{2.11}$$

2.2. *Recombination-generation mechanisms in semiconductors*

In the present study, we take into account the following recombination-generation mechanisms:

2.2.1. *Direct band-to-band or bimolecular recombination*

Direct band-to-band recombination is a radiative recombination involving one electron from conduction band and one hole from valence band. Its rate is given by [2]:

$$R_{rad}=B_r\left(n.p-n_i^2\right) \tag{2.12}$$

Where B_r is band-to-band recombination coefficient [3, 4] and n_i is the intrinsic concentration of charge carriers.

2.2.2. *Indirect Shockley-Read-Hall (SRH) recombination of shallow impurities*

Indirect Shockley-Read-Hall recombination of shallow impurities is a non-radiative recombination involving one electron and a donor-type hydrogenic impurity or one hole and an acceptor-type hydrogenic impurity. Its rate is given by [2]:

$$R_{SRH}^{Shallow}=\frac{n.p-n_i^2}{\tau_p\left(n+n_1\right)+\tau_n\left(p+p_1\right)} \tag{2.13}$$

τ_n (τ_p) is the electron (hole) lifetime. $n_1=n_i\exp\left((E_D-E_{F_i})/k_BT\right)$ and $p_1=n_i\exp\left(-(E_A-E_{F_i})/k_BT\right)$. E_D and E_A are, respectively, energy levels of donors and acceptors. E_{Fi} is the intrinsic Fermi level.

2.2.3. *Indirect Shockley-Read-Hall (SRH) recombination of deep-level traps or recombination centers*

Indirect Shockley-Read-Hall recombination of deep-level traps or recombination centers is a non-radiative recombination involving one charge carrier and a deep-level trap or a recombination center in the middle of the gap. Its rate is given by [2]:

$$R_{SRH}^{Deep}=\frac{\sigma_n\sigma_pv_e^{th}v_h^{th}N_{t,r}\left(n.p-n_i^2\right)}{\sigma_nv_e^{th}\left(n+n_{t,r}\right)+\sigma_pv_h^{th}\left(p+p_{t,r}\right)} \tag{2.14}$$

N_t (N_r) is deep-level traps (recombination centers) concentration. σ_n (σ_p) is electron (hole) capture cross-section by deep-level traps or recombination centers. $n_{t,r}=n_i\exp\left((E_{t,r}-E_{F_i})/k_BT\right)$ and $p_{t,r}=n_i\exp\left(-(E_{t,r}-E_{F_i})/k_BT\right)$. E_t (E_r) is the energy of deep-level traps (recombination centers). $v_e^{th}=\sqrt{3k_BT/m_e}$ $\left(v_h^{th}=\sqrt{3k_BT/m_h}\right)$ is the electron (hole) velocity due to thermal agitation.

3. Theoretical framework - Photonics background

3.1. Electron-hole pairs generation rate

The generation rate $G(x)$ of electron-hole pairs in an illuminated semiconductor at a distance x from air/semiconductor interface receiving incoming light is written as [2]:

$$G(x)=\int_0^{\lambda_c}(1-R(\lambda))\alpha(\lambda)\frac{\lambda}{hc_0}I(\lambda)e^{(-\alpha(\lambda)x)}d\lambda \tag{3.1}$$

$I(\lambda)$ is the real spectral solar irradiance. $R(\lambda)$ is the reflectance of illuminated air/semiconductor interface. $\alpha(\lambda)$ is the semiconductor absorption coefficient. λ_c which is the critical value of wavelength beyond which semiconductor material is transparent, is linked to semiconductor energy band-gap by the expression $\lambda_c = hc/E_g$.

3.2. Real solar spectrum

In the present study, we carry out the simulation of a GaAs based thin-film crystalline solar cell exposed to a real illumination corresponding to an air mass AM1.5, which means an incident solar radiation making an angle $\theta = 48.2°$ with the zenith [5] and whose incident power P_{in} is equal to 1000.37 $\mathrm{W\,m^{-2}}$ (see figure 1).

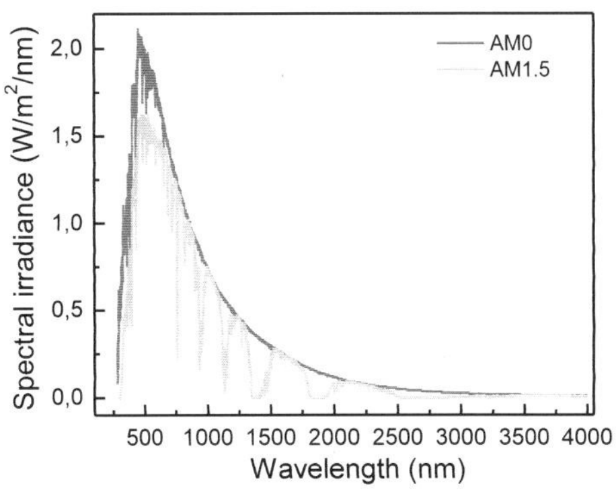

Figure 1. Spectral irradiance $I(\lambda)$ for an extra-terrestrial collector (AM0) and a tilted terrestrial collector making an angle $\theta = 48.2°$ with the zenith (AM1.5) [5].

3.3. Optical and dielectric properties of the semiconductor

The reflectance and the transmittance of air/semiconductor interface are expressed as [6]:

$$R(\lambda)=\frac{\left|1-n^*(\lambda)\right|^2}{\left|1+n^*(\lambda)\right|^2} \tag{3.2}$$

$$T(\lambda)=1-R(\lambda) \tag{3.3}$$

$n^*(\lambda)$ is the semiconductor complex refractive index. It may be expressed as:

$$n^*(\lambda)=n(\lambda)-ik(\lambda) \tag{3.4}$$

Where $n(\lambda)$ and $k(\lambda)$ are respectively refractive index and extinction coefficient of semiconductor. The absorption coefficient $\alpha(\lambda)$ of semiconductor is related to the extinction coefficient $k(\lambda)$ by the following equation:

$$\alpha(\lambda)=4\pi\frac{k(\lambda)}{\lambda} \tag{3.5}$$

IOP Conf. Series: Earth and Environmental Science **161** (2018) 012003 doi:10.1088/1755-1315/161/1/012003

In figure 2, we present the variations of refractive index $n(\lambda)$ and extinction coefficient $k(\lambda)$ of Gallium Arsenide as functions of incoming photon wavelength [7, 8].

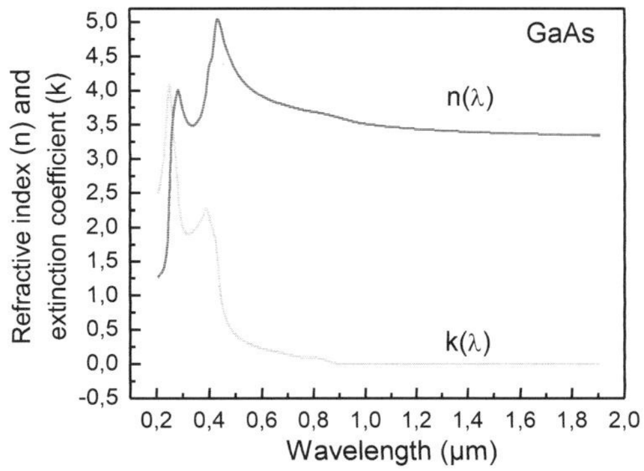

Figure 2. Real $n(\lambda)$ and imaginary $k(\lambda)$ parts of Gallium Arsenide complex refractive index $n^*(\lambda)$ as functions of incident photon wavelength [7, 8].

In figure 3, we plot the variations of air/Gallium Arsenide interface reflectance $R(\lambda)$ and transmittance $T(\lambda)$ as functions of incoming photon wavelength.

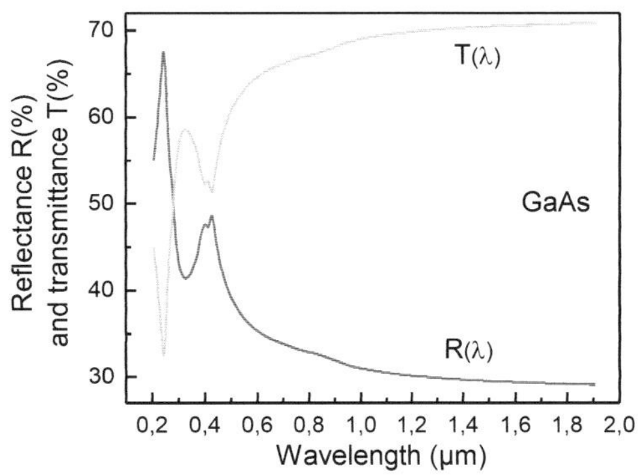

Figure 3. Reflectance $R(\lambda)$ and transmittance $T(\lambda)$ profiles of air/Gallium Arsenide interface as functions of incident photon wavelength.

Figure 4 presents the variations of Gallium Arsenide absorption coefficient $\alpha(\lambda)$ versus incoming photon wavelength [7, 8]. The reader can easily remark that for photons of wavelength $\lambda > \lambda_c(GaAs) = hc/E_g(GaAs) = 873.128\,\text{nm}$, the absorption coefficient $\alpha(\lambda)$ vanishes.

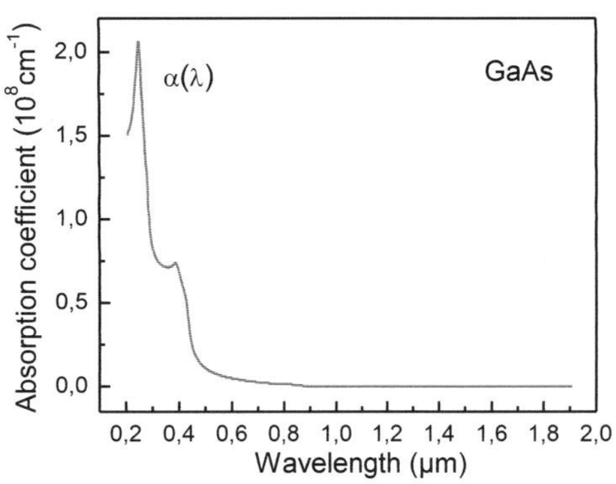

Figure 4. Variation of Gallium Arsenide absorption coefficient $\alpha(\lambda)$ as a function of photon wavelength [7, 8].

In figure 5, the variations of electron-hole pairs generation rate $G(x)$ (*cf.* eq 3.1) are plotted against the depth x in GaAs material. After a careful analysis of generation rate profile, we show that the process of photons absorption and electron-hole generation occurring in Gallium Arsenide at a distance x from illuminated air/semiconductor interface may be described by the following equation $G(x) = Ae^{-ax} + Be^{-bx}$ where $A = 0.9574\,10^{28}$ m^{-3}, $B = 0.2345\,10^{28}$ m^{-3}, $a = 24.0456\,10^{6}$ m^{-1} and $b = 2.6758\,10^{6}$ m^{-1}.

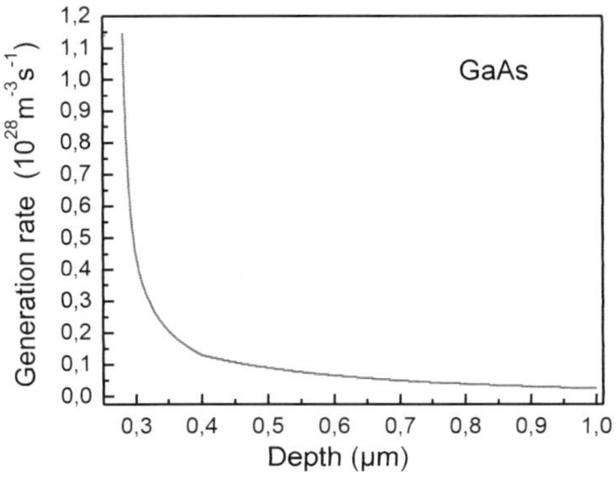

Figure 5. Electron-hole pairs generation rate $G(x)$ profile drawn versus the distance x to air/GaAs illuminated interface.

4. Theoretical framework - Normalization and scaling of semiconductor equations

4.1. Normalization and scaling of semiconductor equations

Equations (2.8) to (2.10) are coupled non-linear non-homogeneous second order differential equations. In order to make easier numerical handling and computational treatment of the system composed by these equations via Maple computer algebra software [9], and also in order to avoid precision loss due to the large difference between orders of magnitude of *x*, *V(x)*, *n(x)* and *p(x)*, we proceed to a normalization of all physical quantities involved by using the parameters given in table 1.

Table 1. Physical quantities and normalization parameters of semiconductor equations [10, 11].

Symbol with dimension	Symbol dimensionless	Meaning	Normalization parameters
x	X	Length	$L = w_{p0} + w_{n0}$
V, φ_n, φ_p	U, F_n, F_p	Electrostatic potential, Electrons and holes quasi-Fermi levels	$k_B T / q$
$n(x)$, $p(x)$	$n'(X)$, $p'(X)$	Electrons and holes densities	n_i
$J_n(x)$, $J_p(x)$	$J'_n(X)$, $J'_p(X)$	Electrons and holes current densities	$qn_i D_{n,p} / L$

In the table above, w_{p0} and w_{n0} denote penetration depths of the quasi-depletion layer, i.e. the space charge zone respectively into P-type and N-type regions at thermodynamic equilibrium, in the dark and without polarization.

4.2. Normalized semiconductor equations system

After normalization, the basic equations of one-dimensional semiconductor device are expressed via dimensionless potential $U(X)=qV(X)/k_BT$, electrons density $n'(X)=n(X)/n_i$ and holes density $p'(X)=p(X)/n_i$, where dimensionless variable $X=x/L$:

$$\frac{d^2U(X)}{dX^2} - \frac{L^2}{L_{Di}^2}\left(\frac{n'(X)-p'(X)}{2} - \left(\frac{N_D(X) - N_A(X)}{2n_i}\right)\right) = 0 \tag{4.1}$$

$$\frac{d^2n'(X)}{dX^2} - \frac{dU(X)}{dX}\frac{dn'(X)}{dX} - n'(X)\frac{d^2U(X)}{dX^2} - \frac{R_n(X)}{D_n n_i}L^2 + \frac{G_n(X)}{D_n n_i}L^2 = 0 \tag{4.2}$$

$$\frac{d^2p'(X)}{dX^2} + \frac{dU(X)}{dX}\frac{dp'(X)}{dX} + p'(X)\frac{d^2U(X)}{dX^2} - \frac{R_p(X)}{D_p n_i}L^2 + \frac{G_p(X)}{D_p n_i}L^2 = 0 \tag{4.3}$$

Where L_{Di} is the Debye length of charge carriers in intrinsic semiconductor:

$$L_{Di} = \sqrt{\left(\varepsilon_0\varepsilon_r k_B T / 2n_i q^2\right)} \tag{4.4}$$

4.3. Analytical and numerical resolutions

The system of three coupled differential equations (4.1) to (4.3) is solved in the case of the thin-film GaAs crystalline solar cell presented in figure 6. The resolution of this system is carried out both analytically and numerically in P-type quasi-neutral region $\left(-X_p < X < -W_p\right)$ and N-type quasi-neutral region $\left(W_n < X < X_n\right)$. The resolution is performed numerically via finite difference approach in the space charge zone $\left(-W_p < X < W_n\right)$. In all regions, analytical and numerical calculations are achieved using Maple computer algebra software [12].

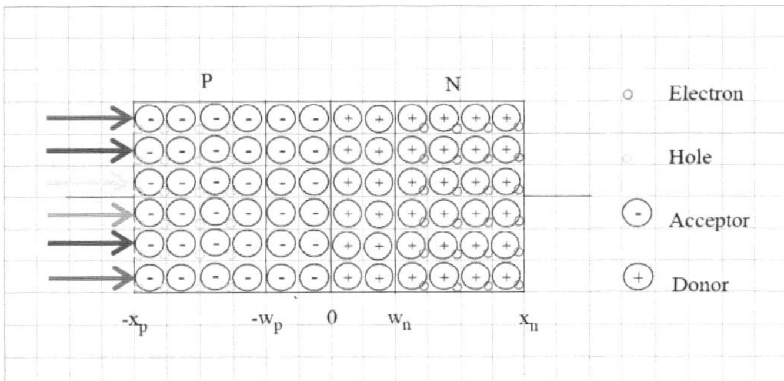

Figure 6. Sketch of thin-film crystalline solar cell under polychromatic illumination.

5. Results and discussions

In the present section, we investigate the variations of total current density $J=J_n+J_p$ as well as power density $P = J(V).V$ delivered by the solar cell as functions of applied voltage V. We also calculate the form factor $FF=J_{max}V_{max}/J_{sc}V_{oc}$ and solar cell photovoltaic conversion efficiency $\eta = P_{\max} / P_{in}$.

In figure 7, we present the variations of Gallium Arsenide based crystalline thin-film solar cell characteristics $J=f(V)$, in the dark and under an illumination corresponding to AM1.5 air mass. The solar cell emitter length x_p and base length x_n are both equal to $50.50\mu\text{m}$. Acceptors doping profile $N_A(x)$ in P-type region and donors doping profile $N_D(x)$ in N-type region are uniform and respectively equal to 5.10^{15}cm^{-3} and 10^{16}cm^{-3}. When the solar cell is illuminated with real solar spectrum, we get a short circuit current density $J_{sc} = -20.445\text{ mAcm}^{-2}$, an open-circuit voltage $V_{oc} = 0.92\text{ V}$, a maximum power point whose current density $J_{\max} = -19.96\text{ mAcm}^{-2}$ and voltage $V_{\max} = 0.833\text{ V}$ (see figure 8). One can easily deduce that the form factor $FF = 88.39\,\%$ and solar cell photovoltaic conversion efficiency $\eta = 16.62\,\%$. The results obtained above show that, for similar acceptors and donors doping profiles and concentrations, efficiency of GaAs based solar cells is clearly higher than efficiency of Si based solar cells. We also remark that these results are in very good agreement with those given by solar cells modelling and simulation software used in photovoltaic field such as AMPS1D [2], PC1D [13, 14] and SCAPS3300 [15].

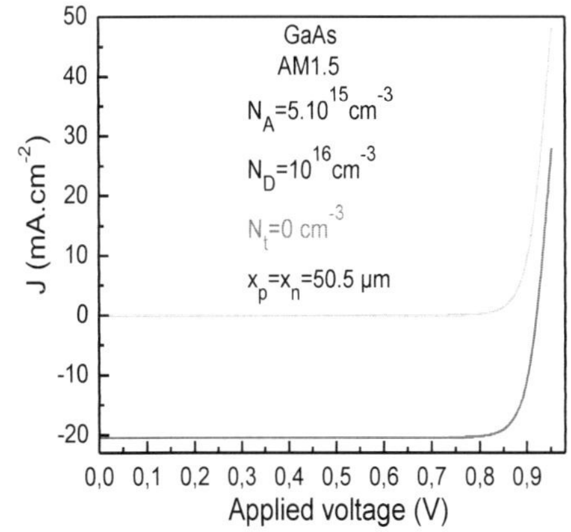

Figure 7. GaAs based thin-film crystalline solar cell characteristics *J=f(V)* in the dark and under real solar spectrum illumination corresponding to AM1.5 air mass.

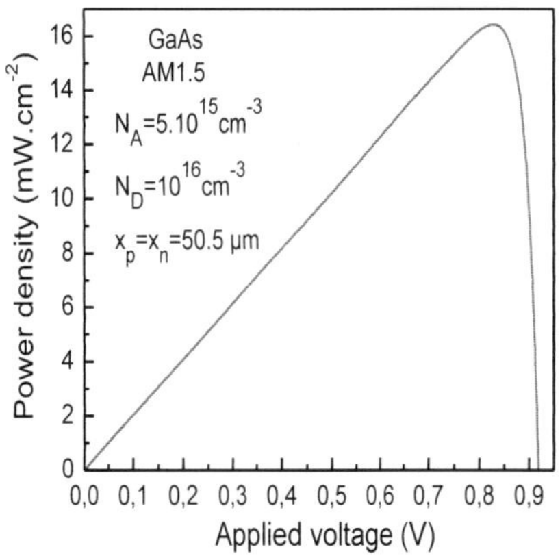

Figure 8. GaAs based thin-film crystalline solar cell power density characteristics *P=f(V)* under real solar spectrum illumination corresponding to AM1.5 air mass.

In figure 9, GaAs based solar cell characteristics under illumination *J=f(V)* are drawn versus applied voltage for different values of uniform deep-level traps concentration: $N_t = 0, 5.10^{12}, 10^{13}, 10^{14}$ and 5.10^{14}cm^{-3}. The energy of deep-level traps $E_t = 0.46\,\text{eV}$ [16]. Electrons capture cross-section $\sigma_n = 10^{-12}\text{cm}^2$ [16]. Holes capture cross-section $\sigma_p = 10^{-14}\text{cm}^2$ [16]. One can notice that deep-level traps concentration has a small influence on short-circuit current density J_{sc}. On the other hand, increasing deep-level traps concentration leads to an appreciable decrease of open-circuit voltage and also implies a significant decrease of maximum power point coordinates (see figure 10).

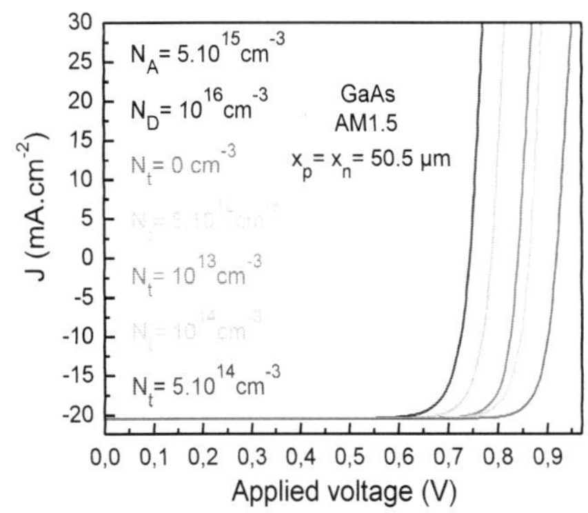

Figure 9. GaAs based thin-film crystalline solar cell characteristics *J=f(V)* under real solar spectrum illumination corresponding to AM1.5 air mass for different concentrations of deep-level traps.

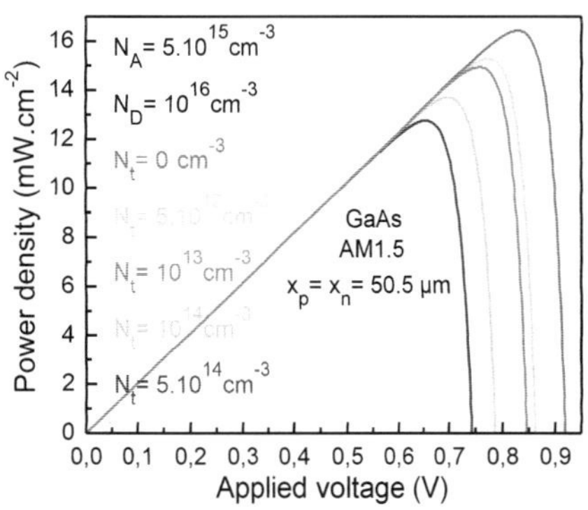

Figure 10. GaAs based thin-film crystalline solar cell power density characteristics *P=f(V)* under real solar spectrum illumination corresponding to AM1.5 air mass for different concentrations of deep-level traps.

Figure 11 presents GaAs based crystalline solar cell characteristics under illumination *J=f(V)* against output voltage for different values of uniform recombination centers concentration: $N_r = 0, 5.10^{12}, 10^{13}, 10^{14}$ and 5.10^{14}cm^{-3}. The energy of mid-gap recombination centers is $E_r = 0.71\,\text{eV}$ [17]. Electrons capture cross-section $\sigma_n = 10^{-14}\text{cm}^2$ [17]. Holes capture cross-section $\sigma_p = 10^{-14}\text{cm}^2$ [17]. Here also, we remark that recombination centers concentration has a minor effect on short-circuit current density J_{sc}. However, increasing recombination centers concentration implies an important decrease of open-circuit voltage and also causes an appreciable reduction of maximum power point coordinates (see figure 12).

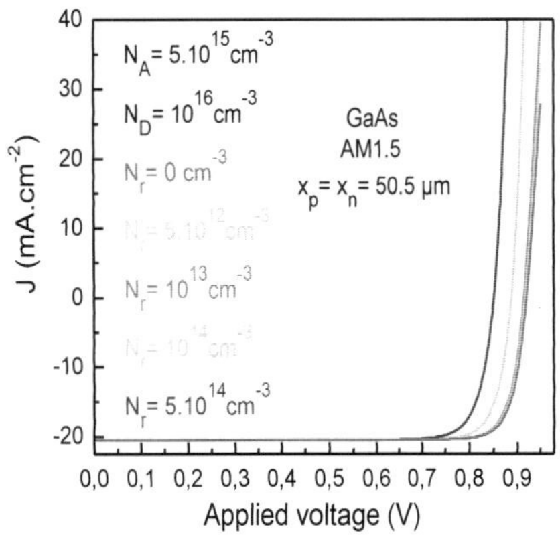

Figure 11. GaAs based thin-film crystalline solar cell characteristics *J=f(V)* under real solar spectrum illumination corresponding to AM1.5 air mass for different concentrations of recombination centers.

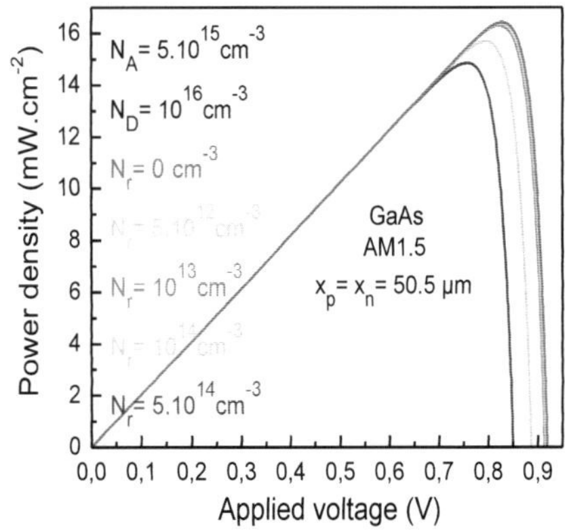

Figure 12. GaAs based thin-film crystalline solar cell power density characteristics *P=f(V)* under real solar spectrum illumination corresponding to AM1.5 air mass for different concentrations of recombination centers.

6. Conclusion

In conclusion, analytical and numerical resolutions of second order non-linear non-homogeneous coupled equations governing the variations of electrostatic potential, electrons and holes densities as well as occupied deep-level traps density (or recombination centers density) throughout one-dimensional GaAs based crystalline thin-film solar cell, was performed according to finite difference approach using Maple computer algebra software. Special attention was given to the effects due to the existence of deep-levels in the middle of the gap. In this way, we have shown that the presence of deep-level traps or recombination centers has a minor influence on short-circuit current density. However, the existence of these traps (or centers) leads, via Shockley-Read-Hall process, to an important decrease of open circuit voltage and to a noticeable drop of maximum power point coordinates as well as solar cell photovoltaic conversion efficiency.

Acknowledgments

The authors H El Achouby and M Zaimi are grateful to Kingdom of Morocco Ministry of Higher Education and to Chouaib Doukkali University for their financial supports.

References

[1] Van Roosbroeck W 1950 Theory of the flow of electrons and holes in Germanium and other semiconductors *Bell System Tech. J* **29** 560-607

[2] Fonash S J 1981 Solar cell device physics **2** Second Edition *Academic Press* (Cambridge, USA)

[3] Nelson J 2003 The Physics of Solar Cells *Imperial college Press* (London)

[4] Williams E W and Hall R 1978 Luminescence and the light emitting diode *Pergamon press* (Oxford)
[5] Gueymard C, Myers D and Emery K 2002 Proposed reference irradiance spectra for solar energy systems testing *Solar Energy* **73** (6), 443-467
[6] Yu P Y and Cardona M 1996 Fundamentals of Semiconductors, Physics and Materials Properties *Springer-Verlag* (Berlin, Germany)
[7] Schubert E F 2004 Refractive index and extinction coefficient of materials https://www.ecse.rpi.edu/~schubert/Educational-resources/Materials-Refractive-index-and-extinction-coefficient.pdf
[8] Aspnes D and Studna A 1983 Dielectric functions and optical parameters of Si, Ge, GaP, GaAs, GaSb, InP, InAs, and InSb from 1.5 to 6.0 eV *Phys. Rev. B* **27** (2) 985–1009
[9] Maple V Release 2015 Copyright 1981-2017 Waterloo Maple Inc
[10] Slotboom J W 1969 Iterative scheme for 1 and 2 dimensional dc transistor simulation *Electronics Letters* **5** 677-678
[11] De Mari A 1968 An accurate numerical steady-state one-dimensional solution of the p-n junction *Solid-State Electron.* **11** 33-58
[12] El Achouby H, Zaimi M, Ibral A and Assaid E M 2017 Physics of Silicon based P-N junction *Maple Application Centre* http://fr.maplesoft.com/applications/view.aspx?SID=154248
[13] Wang X, Khan M R, Gray J L, Ashraful Alam M and Lundstrom M S 2013 Design of GaAs solar cells operating close to the shockley–queisser limit *IEEE Journal of Photovoltaics* **3** (2) 737-744
[14] Khellaf S and Ounissi A 2014 Optimizing the efficiency of solar cells based on GaAs *International Journal of Advanced Science and Technology* **69** 57-64
[15] Khelifi S, Ayat L and Belghachi A 2008 Effects of temperature and series resistance on GaAs concentrator solar cell *Eur. Phys. J. Appl. Phys.* **41** 115–119
[16] Li S S and Loo R Y 1991 Deep-level defects and numerical simulation of radiation damage in GaAs solar cells *Solar Cells* **31** 349-377
[17] Pons D and Bourgoin J C 1985 Irradiation-induced defects in GaAs *J. Phys. C: Solid State Phys.* **18** 3839–3871

Direct Torque Control Strategy Based on Fuzzy Logic Controller for a Doubly Fed Induction Motor

N El Ouanjli[1], A Derouich[1], A El Ghzizal[1], A Chebabhi[2], M Taoussi[3], B Bossoufi[3,4]

[1] Laboratory of Production Engineering, Energy and Sustainable Development, the Higher School of Technology, Sidi Mohamed Ben Abdellah University Fez, Morocco.
[2] Faculty of Sciences and Technology, University of Bordj Bou Arreridj, Algeria.
[3] Laboratory of Systems Integration and Advanced Technologies, Faculty of Sciences Dhar El Mahraz, Sidi Mohamed Ben Abdellah University Fez, Morocco.
[4] Laboratory of Electrical Engineering and Maintenance, ESTO School of Technology, University Mohammed I Oujda, Morocco

E-mail: najib.elouanjli@usmba.ac.ma

Abstract. In this paper, we are interested in the study and improvement of the performances of a Doubly Fed Induction Machine (DFIM) functioning in motor mode using the strategy of Direct Control Torque (DTC) based on fuzzy logic. At first, we are focusing in the study of the functioning principle and the modeling of the DFIM. Secondly, we applied the fuzzy approach to the DTC control. Finally, the simulation results of the control are validated on the Matlab/Simulink environment followed by a detailed analysis.

1. Introduction

For few years, various control strategies have been developed in order to carry out a decoupled control of the doubly fed induction machine. These methods called "vector controls" provide dynamic performance equivalent to that obtained by the direct current machine [1] [2]. Today, the development of new digital signal processing techniques has led to much more sophisticated control strategies. The most recent steps in this direction are those grouped under the term direct torque control [3].
The direct torque control, which is based on the direct regulation of the flux and the electromagnetic torque of the machine [4], has been the most important research and best adapted to industrial requirements, it is characterized by good stability, precision, a good torque response, robustness and less complexity than other controls [5].
Our objective consists in improving the DTC control, by reducing the ripples of the flows and torque by an intelligent technique based on fuzzy logic, the algorithm of this control is based on the use of two fuzzy tables which replace the hysteresis comparators and the conventional tables [6] [7]. we have checked this control strategy by simulation in MATlab / simulink, it remains the validation by the tests PIL as the works [8]. This article is organized as follows: The first part will be devoted to the modeling of the doubly fed induction machine and the two-level voltage inverter. The second part concerns with the fuzzy DTC control of this machine. In a third part and in order to validate our model, simulation results curves using the Matlab/Simulink software will be presented and analysed.

2. Modeling of the system

The power supply of the doubly fed induction machine operating in motor mode and variable speed is assured by two voltage converters which are reciprocally connected to the stator and rotor windings [9].

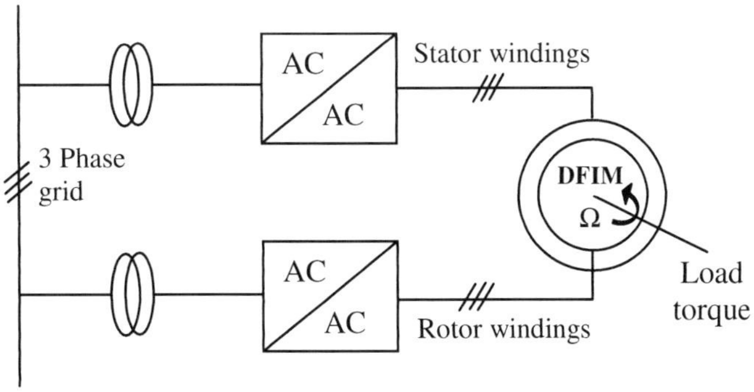

Figure 1. Overall scheme of the system studied

2.1. *Modeling of the DFIM*

Before doing the modeling of a doubly fed induction machine, it is necessary to represent a two-phase model (d, q) given by Park transformation [10].

The model of the doubly fed induction machine after the Park transformation is defined by the electrical, magnetic and mechanical following equations [11].

2.1.1. Electrical equations

The equations of the stator and rotor voltages $V_{s,r\,(d,q)}$ in the reference (d,q), are expressed by:

$$\begin{cases} V_{sd} = R_s.i_{sd} + \frac{d\psi_{sd}}{dt} - \omega_s.\psi_{sq} \\ V_{sq} = R_s.i_{sq} + \frac{d\psi_{sq}}{dt} + \omega_s.\psi_{sd} \\ V_{rd} = R_r.i_{rd} + \frac{d\psi_{rd}}{dt} - \omega_r.\psi_{rq} \\ V_{rq} = R_r.i_{rq} + \frac{d\psi_{rq}}{dt} + \omega_r.\psi_{rd} \end{cases} \quad (1)$$

2.1.2. Magnetic equations

The stator and rotor flux are expressed by:

$$\begin{cases} \psi_{sd} = L_s.i_{sd} + M_{sr}.i_{rd} \\ \psi_{sq} = L_s.i_{sq} + M_{sr}.i_{rq} \\ \psi_{rd} = L_r.i_{rd} + M_{sr}.i_{sd} \\ \psi_{rq} = L_r.i_{rq} + M_{sr}.i_{sq} \end{cases} \quad (2)$$

2.1.3. Mechanical equation

The mechanical equation is given by:

$$T_{em} = T_r + J.\frac{d\Omega}{dt} + f.\Omega \quad (3)$$

The expression of electromagnetic torque based on rotor flux and currents is given, by:

$$T_{em} = p.\left(\psi_{rq}.i_{rd} - \psi_{rd}.i_{rq}\right) \quad (4)$$

With:

$V_{s(d,q)}, V_{r(d,q)}$: Stator and rotor voltages in the Park reference.
$I_{s(d,q)}, I_{r(d,q)}$:Stator and rotor currents in the Park reference.
$\psi_{s(d,q)}, \psi_{r(d,q)}$: Stator and rotor flux in the Park reference.
R_s, R_r : Stator and rotor resistances.
L_s, L_r, M : Stator, rotor and mutual inductances.
ω_s, ω_r : Stator and rotor pulsations.
T_r : Load and electromagnetic torque.
T_{em} : Electromagnetic torque.
P : Pole pair number.
Ω : Rotation speed of the machine.
J : Moment of inertia.
f : Coefficient of viscous friction.

2.2. Modeling of the inverter

The mathematical model of the two-level inverter is represented by the following matrix [12]:

$$\begin{pmatrix} V_a \\ V_b \\ V_c \end{pmatrix} = \frac{1}{3}U_c \cdot \begin{pmatrix} 2 & -1 & -1 \\ -1 & 2 & -1 \\ -1 & -1 & 2 \end{pmatrix} \cdot \begin{pmatrix} S_a \\ S_b \\ S_c \end{pmatrix} \tag{5}$$

The state of switches, assumed to be perfect, can be represented by three Boolean magnitude of control S_i(i=a, b, c), such as:

- S_i=1 if T_j (i= 1,2,3) is closed and T_k (j=4,5,6) is open.
- S_i=0 if T_j (i= 1,2,3) is open and T_k (j=4,5,6) is closed.

3. Direct Torque Control based on the Fuzzy Logic

3.1. Principle of the direct torque control

The direct torque control consists in directly controlling the flows and the electromagnetic torque of the doubly fed induction machine by the use of two switching tables whose functions are the direct determination of the control sequences applied to the switches of the voltage inverters [13].
The expression of the inverter voltage vector can be given in the form below:

$$V = \sqrt{\frac{2}{3}}U_c(S_a + S_b e^{\frac{j2\pi}{3}} + S_c e^{\frac{j4\pi}{3}}) \tag{6}$$

With:

S_a , S_b, S_c : Switching logic states (0 or 1).
U_c : DC bus voltage.

The set of voltage vectors delivered by each inverter of voltage at two levels are represented in the following figure:

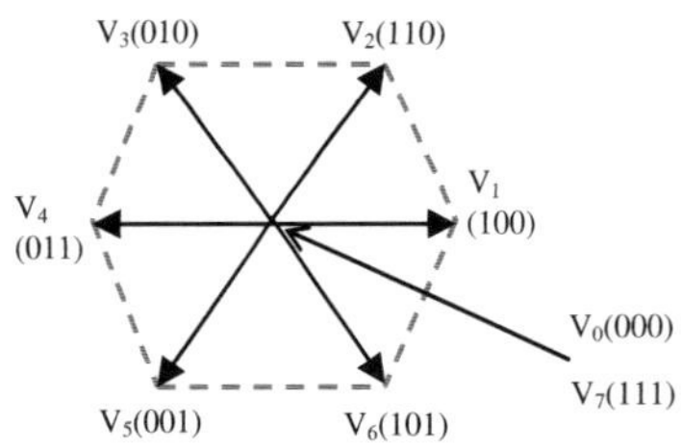

Figure 2. Voltage vectors delivered by the voltage inverter

Direct torque control is also based on the estimation of the stator flux, rotor flux and electromagnetic torque of the machine [14], the module of the stator and rotor flux is written:

$$\hat{\psi}_s = \sqrt{\hat{\psi}_{s\alpha}^{\ 2} + \hat{\psi}_{s\beta}^{\ 2}} \text{ and } \hat{\psi}_r = \sqrt{\hat{\psi}_{r\alpha}^{\ 2} + \hat{\psi}_{r\beta}^{\ 2}} \tag{7}$$

With: $\psi_{s\alpha}$, $\psi_{s\beta}$, $\psi_{r\alpha}$ and $\psi_{r\beta}$ are of the stator and rotor flux in the reference frame (α,β).
The position of the flux vectors calculated as follows:

$$\theta_s = arctg\,(\frac{\hat{\psi}_{s\beta}}{\hat{\psi}_{s\alpha}}) \text{ and } \theta_r = arctg\,(\frac{\hat{\psi}_{r\beta}}{\hat{\psi}_{r\alpha}}) \tag{8}$$

The electromagnetic torque is estimated from the measured rotor currents:

$$\hat{T}_{em} = p.\left(\hat{\psi}_{r\alpha}.i_{r\beta} - \hat{\psi}_{r\beta}.i_{r\alpha}\right) \tag{9}$$

3.2. DTC control based on the fuzzy logic
To improve the performances of the system and to minimize ripples in the electromagnetic torque and flux, two fuzzy logic controllers presented to replace the controllers of hysteresis and the conventional switching tables [15]. In this system there are three inputs for the fuzzy logic controller, which are flux angle, torque error and flux error [16].

$$\theta_s = arctg\,(\frac{\hat{\psi}_{s\beta}}{\hat{\psi}_{s\alpha}}) \text{ and } \theta_r = arctg\,(\frac{\hat{\psi}_{r\beta}}{\hat{\psi}_{r\alpha}}) \tag{10}$$

$$\varepsilon_{T_{em}} = (T_{em})_{ref} - \hat{T}_{em} = \Delta T_{em} \tag{11}$$

$$\varepsilon_{\psi_s} = (\psi_s)_{ref} - \hat{\psi}_s = \Delta\psi_s \tag{12}$$

$$\varepsilon_{\psi_r} = (\psi_r)_{ref} - \hat{\psi}_r = \Delta\psi_r \tag{13}$$

The figure 3 bellow gives the block diagram adopted for fuzzy-DTC control of the doubly fed induction machine:

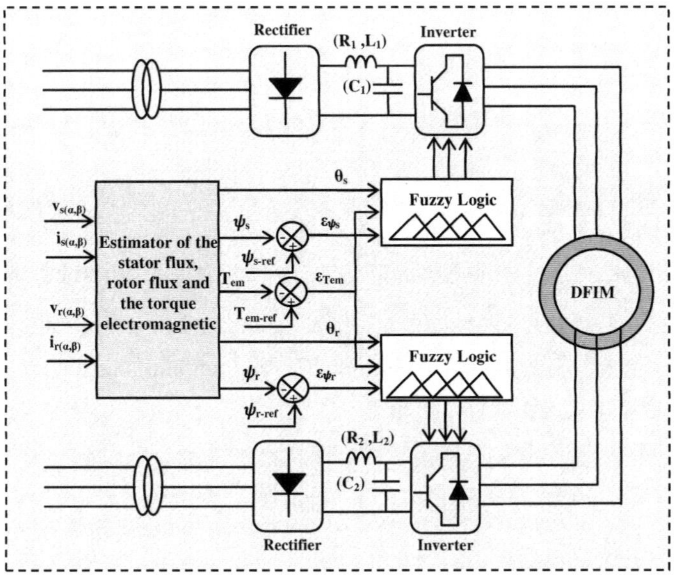

Figure 3. Synoptic schema of the fuzzy DTC applied to DFIM

The switching table at the stator side by the fuzzy logic is illustrated in the following figure 4, (the same reasoning for the rotor side):

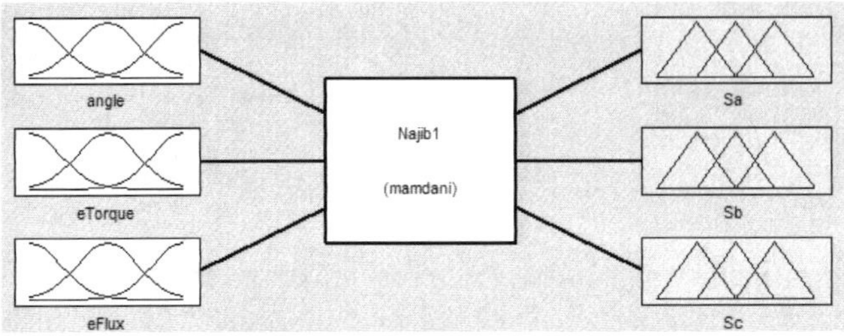

Figure 4. switching table by fuzzy logic

The universe of discourse of the flux position (stator flux or rotor flow) is divided into six fuzzy sets (θ_1 to θ_6). We choose the triangular membership function for all angles θ_i.

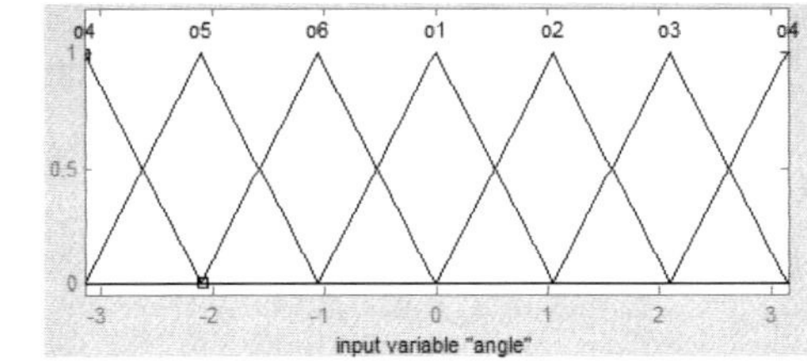

Figure 5. Membership function for flux position

The universe of discourse of the electromagnetic torque error is divided into three fuzzy sets (torque error is positive (P); torque error is zero (Z) and torque error is negative (N)).We choose the triangular membership function for the fuzzy set (Z). And the trapezoid membership functions for the fuzzy sets (P) and (N).

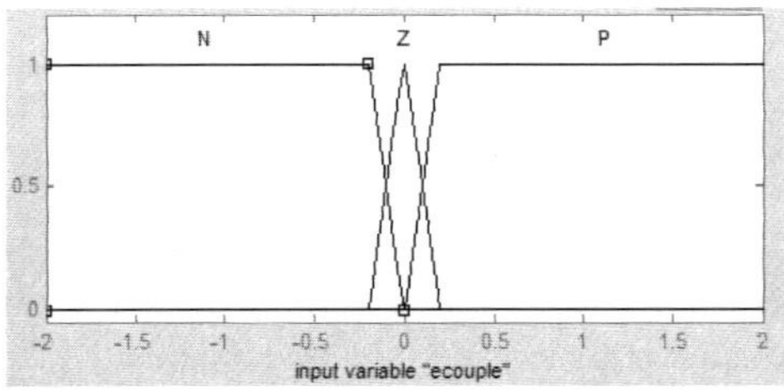

Figure 6. Membership function for error of the electromagnetic torque

The universe of discourse of the flux error is divided into two fuzzy sets (flux error is positive (P); flux error is negative (N)). We choose the trapezoid membership functions for the two fuzzy sets.

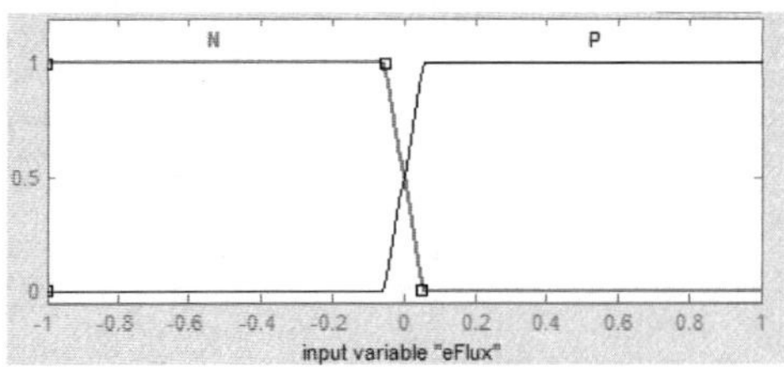

Figure 7. Membership function for flux error

The output variable is decomposed into three sub-outputs representing three switching magnitudes (S_a, S_b S_c) of switches of the inverter at two levels, the universe of discourse of each output is divided into two fuzzy sets (zero and one) whose membership functions are chosen by trapezoid forms.

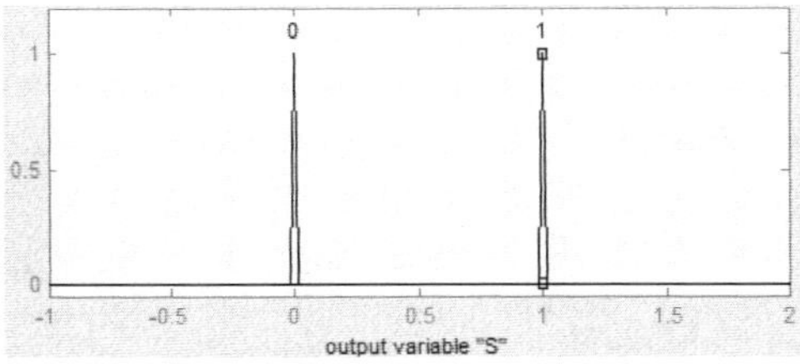

Figure 8. Membership functions for output variables

The fuzzy rules base with 36 rules can be obtained from the following table 1.

Table 1. Set of fuzzy rules

e_ψ	e_{Tem}	θ_1	θ_2	θ_3	θ_4	θ_5	θ_6
N	N	v_6	v_1	v_2	v_3	v_4	v_5
N	Z	v_7	v_0	v_7	v_0	v_7	v_0
N	P	v_2	v_3	v_4	v_5	v_6	v_1
P	N	v_5	v_6	v_1	v_2	v_3	v_4
P	Z	v_0	v_7	v_0	v_7	v_0	v_7
P	P	v_3	v_4	v_5	v_6	v_1	v_2

4. Simulation results

In order to show the performance and robustness of the fuzzy DTC control of DFIM, we simulated the dynamic behavior of DFIM in the Matlab/Simulink environment.

The following figures show the behavior of the structure of the direct torque control by the fuzzy logic applied to the doubly fed induction machine of 1,5 kW.

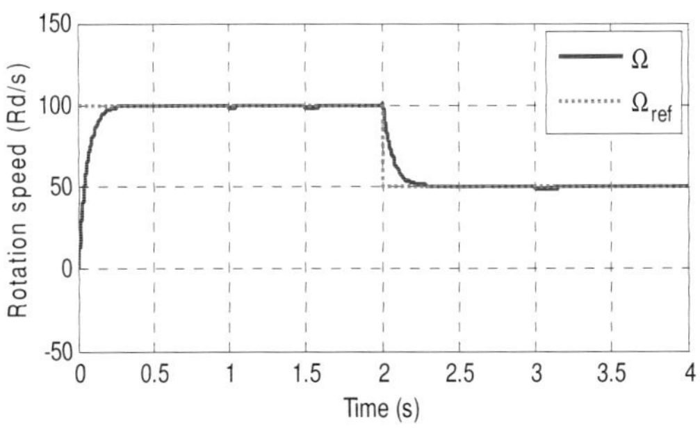

Figure 9. Rotation speed

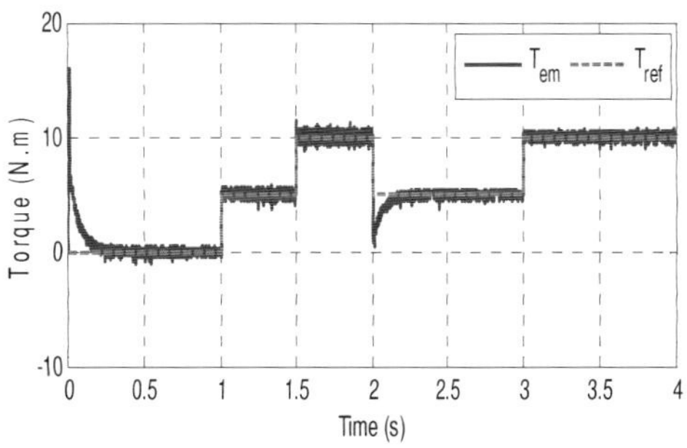

Figure 10. Electromagnetic torque

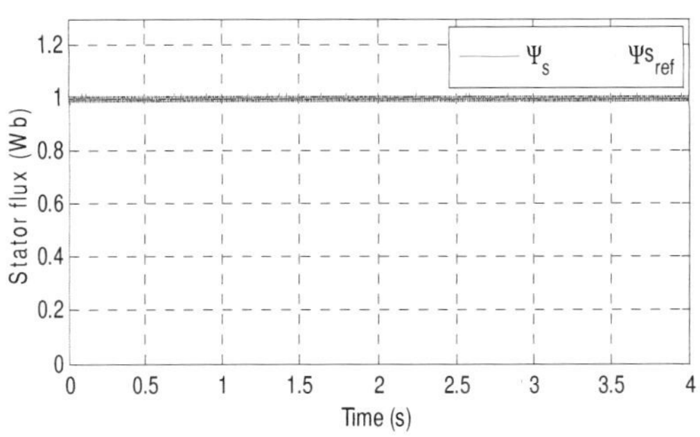

Figure 11. Stator flux

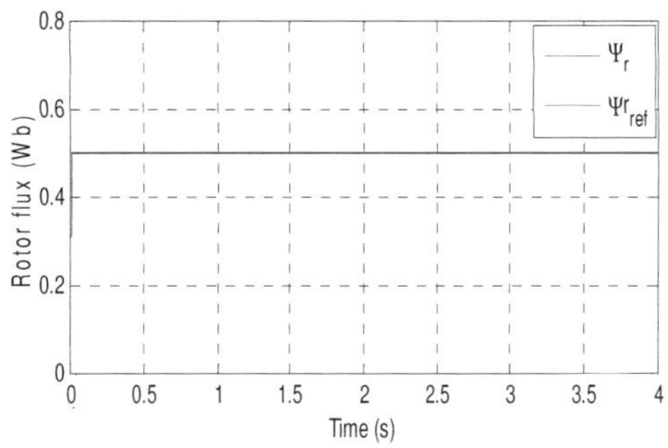

Figure 12. Rotor flux

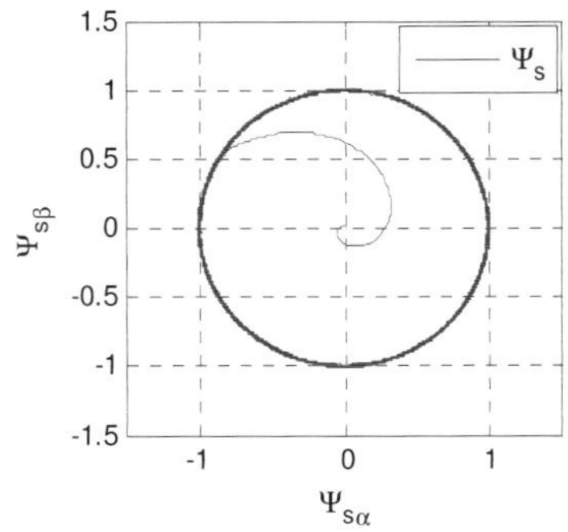

Figure 13. Evolution of the stator flux

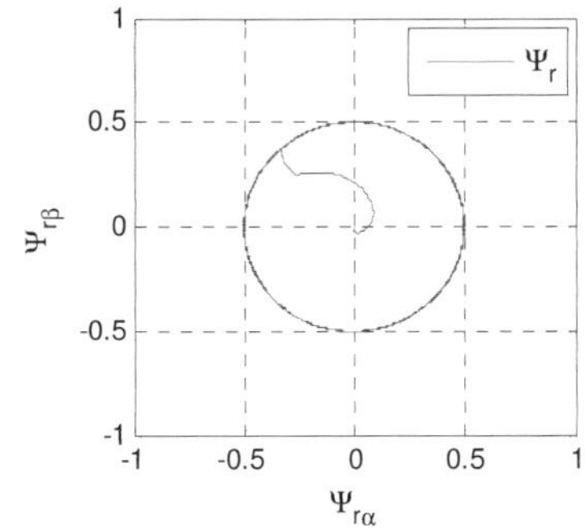

Figure 14. Evolution of the rotor flux

According to the simulation results, it is found that the speed and the electromagnetic torque reached its reference values with a fast response time (t_r=0.3 s). The stator and rotor flux remain constant at their reference values whatever the applied load which shows that the torque and the flux are decoupled. The evolution of these flows in the reference frame (α, β) is perfectly circular. We also note that the electromagnetic torque and flows of the machine present less undulation compared to other classic commands.
The results obtained show efficiency of the chosen algorithm for improving performances and for minimizing ripples of the torque and flux of the machine.

5. Conclusion

The objective of this work consists in the modeling, the development and the simulation of the fuzzy DTC control for a doubly fed induction machine powered by two voltage converters. The simulation results clearly show that the system perfectly follows the reference values and that the control approach studied provides a very interesting solution to the problems encountered in to other classic controls. To continue and to complete this work, we will test and implement this control law by using a test bench based on an FPGA target in the laboratory.

Appendix

Table 2. Parameters of the DFIM

Variable	Symbol	Value (unit)
Nominal power	P_m	1.5 kW
Stator nominal voltage	V_{sn}	230 V
Rototor nominal voltage	V_{rn}	130 V
Frequency	f	50 Hz
Pair pole number	P	2
Stator self inductance	Ls	0.295 H
Rotor self inductance	Lr	0.104 H
Maximum of mutual inductance	M	0.165 H
Stator resistance	R_s	1.75 Ω
Rotor resistance	R_r	1.68 Ω
Total viscous frictions	f	0.0027 Kg.m²/s
Total inertia	J	0.01 Kg.m²

References

[1] Taoussi M, Karim M, Bossoufi B, Hammoumi D and Lagrioui A 2015 Speed Backstepping control of the double-fed induction machine drive *J. Theo. Appl. Info. Tech.* 74 189-199

[2] El Ouanjli N, Derouich A, El ghzizal A, Taoussi M 2016 Contribution à l'optimisation des performances d'une Machine Asynchrone à Double Alimentation (MADA) fonctionnant en mode moteur *Int. conf.TISPI* Oujda

[3] El Ouanjli N, Derouich A, El ghzizal A, Chebabhi A and Taoussi M 2017 A comparative study between FOC and DTC controls of the Doubly Fed Induction Motor (DFIM) *IEEE Int. Conf. on Electrical and Information Technologies.*

[4] Takahashi I and Noguchi T 1988 A new quick-response and high-efficiency control strategy of an induction motor *IEEE Trans.Ind.*Appl 34 473-479

[5] El Ouanjli N, Derouich A, El Ghzizal A, El Mourabit Y, Bossoufi B and Taoussi M 2017 Contribution to the Performance Improvement of Doubly Fed Induction Machine Functioning in Motor Mode by the DTC Control *Int. J. Pow. Electron. Dri. Sys.* 8 1117-1127

[6] Romeral L, et al. 2003 "Novel Direct Torque Control (DTC) Scheme with Fuzzy Adaptive Torque-Ripple Reduction," *IEEE Trans. Ind. Electron.* 50 487-492

[7] Chebabhi A and A. Meroufel A 2010 Commande directe du couple d'une MAS alimentée par un onduleur de tension à trois niveaux à base de la logique floue *JIEMCEM, ENSET-Oran*

[8] Motahhir S, El Ghzizal, Sebti S, and Derouich A 2017 MIL and SIL and PIL tests for MPPT algorithm *Cogent Engineering* 4, 1378475

[9] Taoussi M, Karim M, Bossoufi B, Hmmoumi D, Lagrioui A and Derouich A 2016 Speed Variable Adaptive Backstepping Control of the Double-Fed Induction Machine Drive *Int. J. Autom. Contr.*10 12-33

[10] Taoussi M, Karim M , Hammoumi D, El Bekkali C, Bossoufi B and El Ouanjli N 2017 Comparative study between backstepping adaptive and field-oriented control of the DFIG applied to wind turbines *Int. Conf. Adv. Tech. Sig. Ima. Proc*

[11] El Ouanjli N, Derouich A, El ghzizal A, Errouha M and Taoussi M 2017 Direct Torque Control of the Doubly Fed Induction Motor (DFIM) *Int. conf. COFMER'2*

[12] Taoussi M, Karim M, Bossoufi B, Hmmoumi D, Bakkali C, Derouich A and El Ouanjli N 2017 Low-Speed Sensorless Control for Wind Turbine System *WSEAS Transactions on Systems and Control* 12 405-417

[13] El Mourabet Y, Derouich A, El ghzizal, El Ouanjli N and zamzoum O 2017 DTC-SVM Control for Permanent Magnet Synchronous Generator based Variable Speed Wind Turbine *Int. J. Pow. Electron. Dri. Sys.* 8 1732-1743

[14] Telford D, Dunnigan M.W and Williams B.W 2001 A Novel Torque-Ripple Reduction Strategy for Direct Torque Control *IEEE Trans. Ind. Electron.* 48 867-870.

[15] Arias A, Romeral J.L, Aldabas E, and Jayne M.G 2000 Fuzzy logic direct torque control *Proc. IEEE Int. Symp. Ind. Electron.* 253–258.

[16] Beerten J, Verveckken J and Driesen J 2010 Predictive Direct Torque Control for Flux and Torque Ripple Reduction *IEEE Tran. Indus. Electron.* 57 404- 412

IOP Conf. Series: Earth and Environmental Science **161** (2018) 012005 doi:10.1088/1755-1315/161/1/012005

Daily direct, diffuse and global irradiances modelling for an accurate estimate of daily solar radiation intensity at the city of El Jadida (Morocco)

M Zaimi, H El Achouby, A Ibral and E M Assaid

Optics and Electronics of Semiconductor Nanostructures Team, Laboratory of Instrumentation, Measurements and Control, Department of Physics, Faculty of Sciences, Chouaib Doukkali University, P. O. Box 20 El Jadida, Morocco

E-mail:eassaid@yahoo.fr

Abstract. In this work, we derive analytic expressions and calculate numerical values of incident daily solar radiation on horizontal, fixed tilted, single-axis and double-axis tracking apertures as functions of day number, collector latitude, tilt and azimuth angles. The effect of non-uniform atmospheric absorption of incident radiation throughout a day is taken into account and new models of direct normal B_n and diffuse horizontal D_h components of global horizontal (tilted) irradiance G_h (G_t) are proposed, tested and validated on local experimental measurements. This enables us to extract global irradiance model parameters for average day of each month and to predict, accurately, daily solar radiation (energy per unit area) effectively received by any collector placed on earth during the whole year.

1. Introduction

Today, global energy demand is increasing exponentially. This is mainly due to increase of energy needs of newly industrialized or emerging countries such as China, India, Brazil, Indonesia, Mexico, South Africa or Turkey [1]. Meanwhile, major oil and natural gas discoveries throughout the world are scarce and world fossil energy reserves increase slightly. In addition, excessive consumption of coal and fossil fuels increases carbon levels in the atmosphere and generates air pollution and climate perturbation resulting in floods followed by long periods of drought [2]. The inadequacy between increase in demand and that in energy reserves as well as the climate change and its disastrous consequences has impelled political and economic decision-makers to turn to renewable energy sources which are more abundant and more respectful for environment. Solar energy, with its two components: Concentrated Solar Power (CSP) and Photovoltaic (PV), is the flagship renewable energy, it is at the center of interest of any government that seeks to diversify its bouquet of energy. The government of Morocco has established an energy plan aiming to reduce carbon footprint of the country and to increase the share of renewable energies to 42% of the global capacity by 2020 before to achieve 52% share of renewable energies by 2030 [3-5]. The high point of this plan was the inauguration, in February 2016, of the first part of solar power plant Noor Ouarzazate (NOORo I) [6]. NOORo I uses solar thermal technology with parabolic cylindrical troughs to capture sunlight. By 2020, NOORo I, II, III and IV will be able to generate 580 MW of electric power. They will also allow an annual saving of greenhouse effect gases equivalent to 3.7 million tones of CO_2.

Before opting for such technology, a preliminary study of incident solar radiation throughout the year as well as the knowledge of number of sunny days per year are required. In the present work, we will attempt to address these two issues. More precisely, we will investigate the solar potential in the city of El Jadida (33° 15' 00'' N, 8° 30' 00'' W) which is situated on Moroccan Atlantic coast. To our knowledge, there is only one complete study which was devoted to characterization of solar potential at the site of El Jadida [7]. In this thesis, the author measured hourly, daily, monthly and yearly direct, diffuse and global solar irradiance falling on a horizontal collector during two years. In the present paper we will give analytical models of daily diffuse, direct and global irradiance and proceed to extraction of model parameters. We use this models to predict energy densities received by horizontal and fixed tilted collectors during a given day. We compare these densities to local experimental values. The present paper may be broken into five sections. In the second section, physics and astronomy backgrounds necessary for the calculation of incident irradiance (energy density per unit of time and area) on a given aperture at higher atmosphere boundary are presented. In the third section, details of calculations of daily solar radiation (energy density per unit area) radiated on horizontal, fixed tilted, double-axis tracking, north to south oriented single-axis tracking and east to west oriented single-axis tracking apertures placed at higher atmosphere border are presented. In the fourth section, effects of non-uniform atmospheric absorption throughout a day are taken into account and models of direct normal B_n and diffuse horizontal D_h components of daily global horizontal (tilted) irradiance G_h (G_t) are proposed, tested and validated using local irradiance experimental data. Then, the extracted model physical parameters are used to accurately predict, the daily solar energy per unit area (radiation) effectively received by all types of collectors placed on earth for each day of the year. A comparison between measured and predicted daily global horizontal irradiances over one year is done and acceptable agreement is found.

2. Physics background and theoretical framework

2.1. The non-uniform extra-atmospheric spectral irradiance

The spectral irradiance is the power per unit area in a very short spectral range $d\lambda$, its unit is $\mathrm{Wm^{-2}\mu m^{-1}}$. Considering isotropic solar emission, the non-uniform extra-atmospheric spectral irradiance received by a point at the top of terrestrial atmosphere is [8]:

$$I_\lambda = S_{sun}\left(E_\lambda \Big/ \int_0^\infty E_\lambda d\lambda\right) \quad (1)$$

S_{sun} is the solar ''constant'' [8], it represents the total power per unit area radiated to a square meter of an extra-atmospheric surface normal to the sun's rays and placed at a distance equal to one astronomical unit $\left(1\mathrm{AU} = 1.496\times10^{11}\mathrm{m}\right)$. S_{sun} should be normalized to take into account the dependence on the day number n. E_λ is the spectral emissivity, it stands for the spectral power radiated per unit of area and wavelength, its unit is $\mathrm{W\mu m^{-1}m^{-2}}$ [8]. The extra-atmospheric spectral irradiance depends on the earth-sun factor which also depends on the day number [9]. This factor accounts for the change in intensity that occurs as the distance between earth and sun varies during one year [10]. Earth–sun factor is unity when earth-sun distance is at its average value equal to $1\mathrm{A.U}$ as occurs around the 3rd of April and the 1st of October [10]. For a given photon wavelength, I_λ is maximal at perihelion where earth to sun distance is minimal and I_λ is minimal at aphelion where earth to sun distance is maximal (see figure 1) [10].

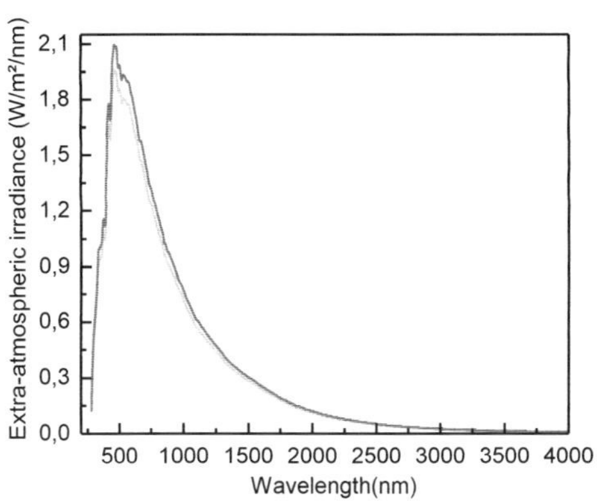

Figure 1. Extra-atmospheric spectral irradiance at perihelion (January the 2nd 2016) (red line) and aphelion (july the 4th 2016) (green line) as a function of photon wavelength.

2.2. *The non-uniform daily extra-atmospheric solar "constant"*

The extra-atmospheric solar "constant" S_{sun} is the continuous sum i.e. integral of extra-atmospheric spectral irradiance I_λ over the whole electromagnetic spectrum:

$$S_{sun} = \int_0^\infty I_\lambda d\lambda \tag{2}$$

For a given point in the nearest neighbourhood of terrestrial atmosphere, the extra-atmospheric solar "constant" S_{sun} depends on the day number n as [11]:

$$S_{sun}(n) = 1347.4466\left(1+0.034\cos\left(\frac{360(n-2)}{365.25}\frac{\pi}{180}\right)\right) \tag{3}$$

For the leap year 2016, eq. (3) shows that S_{sun} is maximal $\left(1393.25\ \mathrm{Wm}^{-2}\right)$ at perihelion which occurs on January the 2nd, minimal $\left(1301.63\ \mathrm{Wm}^{-2}\right)$ at aphelion which occurs on july the 4th and equal to 1347.44 Wm^{-2} on April the 2nd (see figure 2). This latter value is obtained by integrating numerically extra-atmospheric spectral irradiance corresponding to april the 2nd between 280 nm and 4000 nm [11]. It's a little bit lighter than 1360.8 Wm^{-2} which is the most recent value found in the literature [12, 13].

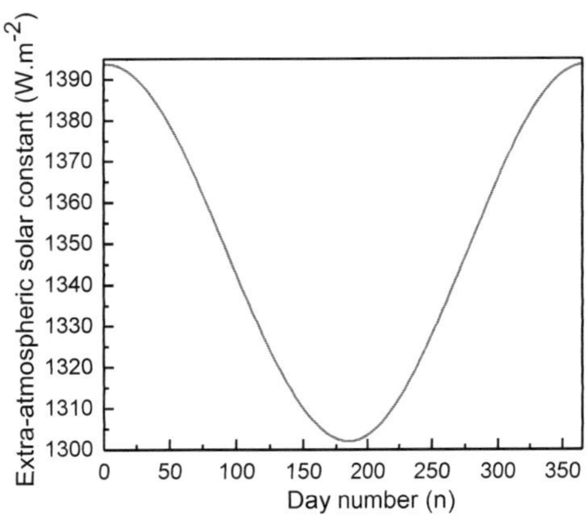

Figure 2. Extra-atmospheric solar constant as a function of the day number n during the leap year 2016.

2.3. *The declination of the sun*

The declination of the sun $\delta(n)$ is the angle between the ecliptic i.e. the plane containing the earth's orbit and the celestial equator. The declination of the sun $\delta(n)$ depends on the day number n as [14]:

$$\delta(n) = 0.39795\cos\left(\frac{0.98563(n-173)\pi}{180}\right) \tag{4}$$

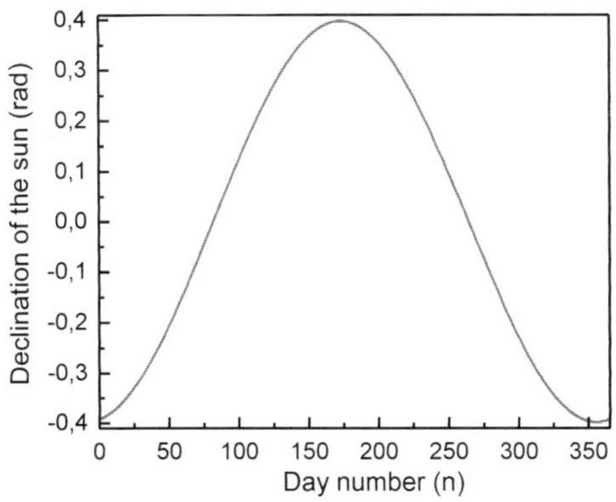

Figure 3. Declination of the Sun i.e. the angle between ecliptic and celestial equator plotted versus day number n during the leap year 2016.

2.4. Horizontal coordinates of the sun

The sun horizontal coordinates according to a given observation site on earth are:

The solar elevation angle or the altitude of the sun. At a given observation site, it's the angle between the centre of the sun's disc and the horizon. It's given by [14]:

$$\sin(h) = \sin(\varphi)\sin(\delta) + \cos(\varphi)\cos(\delta)\cos(\omega) \tag{5}$$

φ is the observation site latitude, δ is the sun declination and ω is the hour angle.

The solar zenith angle θ_z. At a given site, it's the angle between the centre of the sun's disc and the zenith. θ_z is the complementary angle to h, it's expressed as follows [14]:

$$\cos(\theta_z) = \sin(h) \tag{6}$$

The sun azimuth angle a_s, it's the angle between a reference vector pointing towards the north and the perpendicular projection of the sun down onto the horizon. The expression giving a_s is [14]:

$$\sin(a_s) = \frac{\cos(\delta)\sin(\omega)}{\cos(h)} \tag{7}$$

In figure 4, the variations of solar elevation angle h (green line), zenith angle θ_z (red line) and azimuth angle a_s measured in radians clockwise from the north (blue line), are drawn against time during the 346th day of the leap year 2016 for the site of El Jadida (Morocco) (33° 15' 00'' N and 8° 30' 00'' W). One can remark that at sunrise and sunset times, $\theta_z = \pi/2$ and $h = 0$. At solar midday, θ_z is minimal and h is maximal. For the solar azimuth angle, one can notice that it increases monotonically from 2.09 rad at sunrise to 4.22 rad at sunset, passing by π at solar midday.

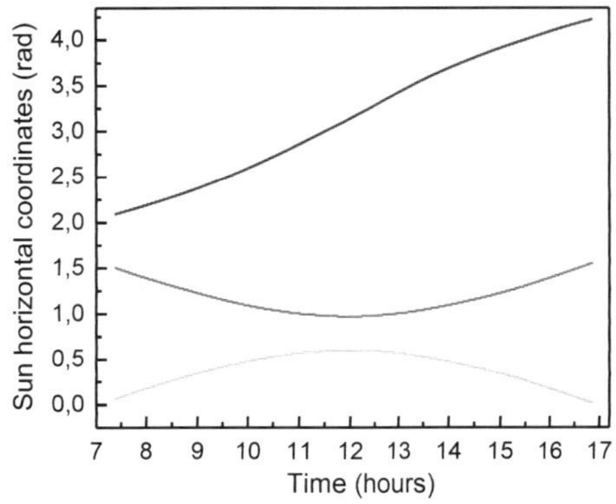

Figure 4. Sun elevation angle (green line), sun zenith angle (red line) and sun azimuth (blue line) at the site of El Jadida (Morocco) plotted against time during the 346th day (December 11th) of the leap year 2016.

2.5. Angle of incidence on a tilted surface

The angle of sunlight incidence on a sloped surface is the angle between sun direction and normal direction to this surface [14]:

$$\cos(\theta) = \cos(\theta_z)\cos(\beta) + \sin(\theta_z)\sin(\beta)\cos(a_s - a) \tag{8}$$

β and a are, respectively, the surface tilt and azimuth angles. For a horizontal surface $\beta = 0$ and $\theta = \theta_z$, for a normal surface to sunlight $\beta = \theta_z$, $a = a_s$ and $\theta = 0$.

2.6. The sunrise and sunset hour angles

In the particular case of a horizontal collector, the sunrise and sunset angles ($-\omega_s$ and ω_s) are obtained by setting $h = 0$ in eq (5) which leads to the following expression:

$$\omega_s(n, \varphi) = \arccos(-\tan(\varphi)\tan(\delta)) \tag{9}$$

In a general case, the tilted collector sunrise and sunset angles (ω_{ss} and ω_{ps}) are solutions of the equation $\theta = \pi/2$ (or $\cos(\theta) = 0$). ω_{ss} and ω_{ps} depend on n, φ, β and a. Taking into account the fact that tilted collector receives sunlight if and only if $-\omega_s \le \omega \le \omega_s$, then the effective tilted collector sunrise and sunset angles are respectively equal to $\max(-\omega_s, \omega_{ss})$ and $\min(\omega_s, \omega_{ps})$.

According to eq (8), one can deduce calculated sunshine duration in hours for a horizontal collector as the difference between times corresponding to sunset and sunrise at a given site:

$$\Delta t_h = \frac{24}{\pi}\arccos(-\tan(\varphi)\tan(\delta)) \tag{10}$$

The calculated sunshine duration in hours for tilted collector at a given site is the difference between times corresponding to effective sunset and sunrise:

$$\Delta t_t = \frac{24}{2\pi}\left(\min(\omega_s, \omega_{ps}) - \max(-\omega_s, \omega_{ss})\right) \tag{11}$$

Figure 5, gives the variations of calculated sunshine duration for horizontal collector (red line) and south-facing tilted collector ($\beta = (33.25)°$, $a = 180°$) (green line) at the site of El Jadida (Morocco) corresponding to the latitude $\varphi = (33.25)°$ as functions of the day number throughout the year.

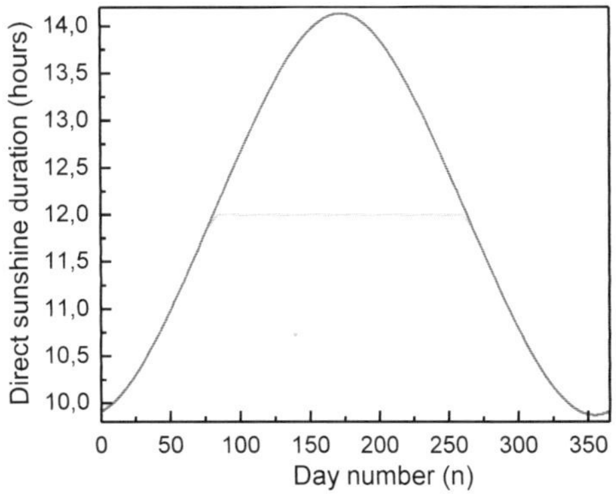

Figure 5. Calculated sunshine duration at the site of El Jadida (Morocco) for horizontal (red line) and tilted (green line) collectors drawn versus day number n during the leap year 2016.

3. Estimated solar irradiance emitted towards the site of El Jadida (Morocco)

3.1. Irradiance emitted towards a tilted surface. Case of a fixed aperture oriented arbitrarily

At a given time t in a fixed day n, the spectral irradiance $I_\lambda(n)$ is the energy density radiated per unit of time and area in a very short spectral range $d\lambda$. According to eq (2), the irradiance $I(n)$ of the whole spectrum for a fixed day n is equal to $S_{sun}(n)$. The elementary energy density per unit of time and area radiated towards a tilted collector, oriented arbitrarily and placed at a given site, during an elementary period of time dt corresponding to an elementary rotation of hour angle $d\omega$ such as $d\omega = 2\pi dt/24$, is equal to $S_{sun}(n)\cos(\theta)d\omega$.

The energy density per unit area radiated towards the tilted surface during an entire day is expressed as follows:

$$H_t(n,\varphi,\beta,a) = \frac{24\times 3600}{2\pi}\int_{\max(\omega_{ss},-\omega_s)}^{\min(\omega_{ps},\omega_s)} I(n)\cos(\theta)d\omega \quad (12)$$

where $\max(-\omega_s,\omega_{ss})$ and $\min(\omega_s,\omega_{ps})$ are respectively effective tilted collector sunrise and sunset angles.

3.2. Irradiance emitted towards a horizontal surface.

The elementary energy density per unit of time and area radiated towards a perpendicular surface to zenith axis placed at a given site during an elementary rotation of hour angle $d\omega$ is equal to $S_{sun}(n)\cos(\theta_z)\,d\omega$.

The energy density radiated per unit area towards a perpendicular surface to zenith axis during one day is expressed as follows:

$$H_h(n,\varphi) = \frac{24\times 3600}{2\pi}\int_{-\omega_s}^{\omega_s} I(n)\left(\sin(\varphi)\sin(\delta)+\cos(\varphi)\cos(\delta)\cos(\omega)\right)d\omega \quad (13)$$

where $-\omega_s$ and ω_s are respectively horizontal collector sunrise and sunset angles.

3.3. Irradiance emitted towards a perpendicular surface to incident radiation. Case of a two-axis tracking aperture

The energy density per unit area radiated towards a perpendicular surface to incident radiation placed at a given extra-atmospheric observation site during one day reads as follows:

$$H_p(n,\varphi) = \frac{24 \times 3600}{2\pi} \int_{-\omega_s}^{\omega_s} I(n)\, d\omega \tag{14}$$

3.4. Irradiance emitted towards a tilted surface. Case of a single-axis tracking aperture

When the tracking axis is oriented in north to south direction, eq (8) reduces to [14]:

$$\cos(\theta) = \sqrt{1 - \left(\sin(\varphi)\sin(\delta) + \cos(\varphi)\cos(\delta)\cos(\omega)\right)^2} \tag{15}$$

The energy density per unit area radiated towards this mobile tilted surface during one day writes as follows:

$$H_{s-a,n-s}(n,\varphi) = \frac{24 \times 3600}{2\pi} \int_{-\omega_s}^{\omega_s} I(n) \sqrt{1 - \left(\sin(\varphi)\sin(\delta) + \cos(\varphi)\cos(\delta)\cos(\omega)\right)^2}\, d\omega \tag{16}$$

When the tracking axis is oriented in east to west direction, eq (8) reduces to [14]:

$$\cos(\theta) = \sqrt{1 - \left(\cos(\delta)\sin(\omega)\right)^2} \tag{17}$$

The energy density per unit area emitted towards this mobile sloped surface during one day writes as follows:

$$H_{s-a,e-w}(n,\varphi) = \frac{24 \times 3600}{2\pi} \int_{-\omega_s}^{\omega_s} I(n) \sqrt{1 - \left(\cos(\delta)\sin(\omega)\right)^2}\, d\omega \tag{18}$$

3.5. Annual irradiance mean value

In the particular case where the sun inclination is considered as constant during the day, the annual mean value of energy density per unit area $H_i(\varphi,\beta,a)$ for a given observation site writes:

$$\langle H_i \rangle(\varphi,\beta,a) = \frac{1}{366} \sum_{n=1}^{366} H_{i,n}(\varphi,\beta,a) \tag{19.1}$$

In all cases, the annual mean value of energy density per unit area $H_i(n,\varphi,\beta,a)$ for a given observation site writes:

$$\langle H_i \rangle(\varphi,\beta,a) = \frac{1}{365.25} \int_0^{365.25} H_i(n,\varphi,\beta,a)\, dn \tag{19.2}$$

where i is identified with *t, h, p, (s-a, n-s)* and *(s-a, e-w)* respectively for tilted, horizontal, perpendicular, single-axis oriented in east-west direction and single-axis oriented in north-south direction collectors.

In sections 3.1 to 3.5, Maple computer algebra software is used to derive analytic expressions for ω_{ss} and ω_{ps} and to achieve symbolic and numeric calculation of $H_i(n,\varphi,\beta,a)$ and $\langle H_i \rangle(\varphi,\beta,a)$. Due to excessive lengths of analytic expressions of $H_i(n,\varphi,\beta,a)$, we simply present their variations as functions of the day number n.

In figure 6, daily energy densities (solar radiation or solar insolation) radiated towards horizontal collector (red line), south-facing fixed tilted collector ($\beta = (33.25)°$, $a = 180°$) (green line), north-south direction oriented single-axis collector (cyan line), east-west direction oriented single-axis collector (magenta line) and perpendicular collector to sunlight (blue line) are drawn as functions of the day number n for the observation site of El Jadida (Morocco). Dashed lines correspond to annual mean values of energy densities.

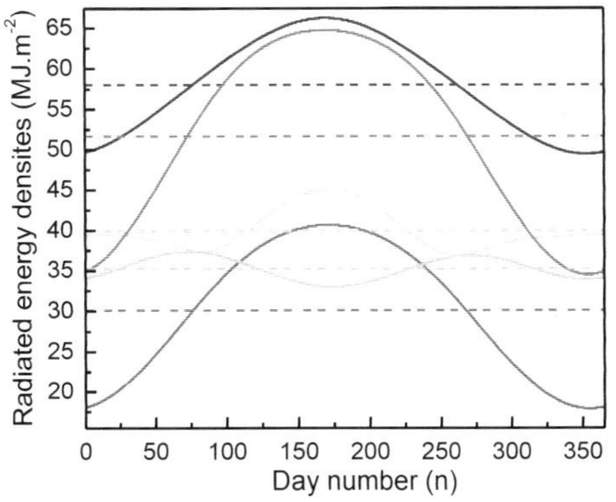

Figure 6. Daily energy densities (solid lines) and annual mean values of energy densities (dashed lines) radiated towards horizontal (red line), south facing fixed tilted (green line) north-south direction oriented single-axis (cyan line) east-west direction oriented single-axis (magenta line) and perpendicular (blue line) collectors at the site of El Jadida (Morocco) plotted versus the day number n.

4. Experimental measures of daily irradiance received at the site of El Jadida (Morocco)

4.1. Tracking global irradiance received by fixed tilted and horizontal collectors at the site of El Jadida

In the present work, we use two datasets of experimental values coming from two stations. The first dataset gathers experimental values collected during two years by A Rerhrhaye and M S El Maliki [7, 15, 16] in the Solar Energy Laboratory which is equipped with two Kipp and Zonen CM11 pyranometers. The first pyranometer measures the global horizontal irradiation. The second pyranometer, equipped with a sunscreen band, measures horizontal diffuse irradiation. These two instruments are linked to a Kipp and Zonen CC12 solar integrator which records the hourly values of these two quantities. The duration of insolation is measured by a Campelle-Stockes heliograph. The experimental setup allows tracking of global horizontal irradiance and diffuse horizontal irradiance every five minutes. The second dataset gathers experimental values corresponding to twelve days. They are collected by M Khaidar and L Birba [15, 17, 18] in Condensed Matter Laboratory which is equipped with two photovoltaic receivers allowing indirect measurement of horizontal and tilted irradiances from the output power. The first photovoltaic receiver is horizontal, the second one is south facing and tilted with an inclination angle equal to the latitude of El Jadida. This experimental setup allows tracking of global horizontal and tilted irradiances every five minutes. To achieve a comparison between estimates based on numerical and symbolic calculations performed in section 3, and energy density effectively received on the earth, we proceed to a careful analysis of experimental global irradiance received by tilted and horizontal apertures from metrological and photovoltaic stations at El Jadida University (Morocco). We also use simulated values of direct, diffuse and global irradiances from Climate-SAE PVGIS databank [19, 20].

4.2. The Concept of Air Mass (AM)

The Air Mass coefficient (AM) is the direct optical path length through earth's atmosphere to zenith optical path length ratio. In this paper, we work with the most accurate model of Air Mass which considers the earth as a sphere and the atmosphere as a spherical layer around it:

$$AM(\theta_z)=\sqrt{\left(\left(r\cos(\theta_z)\right)^2+2r+1\right)}-r\cos(\theta_z) \tag{20}$$

where $r=796$ is the earth's radius to atmosphere thickness ratio [21].

4.3. Modelling direct normal, diffuse horizontal and global irradiances

The solar radiation reaching a horizontal collector on earth is composed by direct horizontal component B_h and diffuse horizontal component D_h. The sum of these components makes the global radiation: $G_h=B_h+D_h$.

Taking into account atmospheric absorption depicted by the concept of Air Mass, direct normal irradiance may be described by the following model:

$$B_n = I(n)\left(A + B\exp\left(-C\left(AM(\theta_z)\right)\right)\right) \quad (21)$$

The direct horizontal irradiance B_h and direct tilted irradiance B_t may be deduced from eq (21) by: $B_h = B_n\cos(\theta_z)$ and $B_t = B_n\cos(\theta)$.

The diffuse horizontal component D_h may be depicted by the following model:

$$D_h = I(n)\cos(\theta_z)\left(D + EB\left(1 - \exp\left(-CAM(\theta_z)\right)\right)\right) \quad (22)$$

Concerning diffuse tilted irradiance, it may be found by applying Liu and Jordan equation [22]: $D_t = D_h(1+\cos(\beta))/2$.

The global horizontal irradiance G_h collected locally is then modelled by the following equation:

$$G_h = I(n)\cos(\theta_z)\left(A + D + EB + B(1-E)\left(\exp\left(-CAM(\theta_z)\right)\right)\right) \quad (23)$$

The models proposed of direct normal irradiance eq (21), diffuse horizontal irradiance eq (22) and global horizontal irradiance eq (23) are different from other models in the literature [23, 24]. *A*, *B*, *C*, *D* and *E* are real parameters which are varied numerically to fit simulated data from PVGIS and experimental data from local station to models of eq (21), (22) and (23) by minimizing the sum of squared errors. These parameters are determined for twelve average days of the year i.e. one average day for each month. In tables 1 and 2, we respectively give parameters values corresponding to the models fitting well with data from Climate-SAE PVGIS and experimental irradiance measured at the site of El Jadida [15, 17, 18].

Table 1. Numerical values of parameters corresponding to models fitting well with data from Climate-SAE PVGIS for twelve average days of the year.

Date	A (Wm^{-2})		B (Wm^{-2})		C		D (Wm^{-2})		E	
	PVGIS normal sky	PVGIS clear sky	PVGIS normal sky	PVGIS clear sky	PVGIS normal sky	PVGIS clear sky	PVGIS normal sky	PVGIS clear sky	PVGIS normal sky	PVGIS clear sky
10 Jan	0.115	0.156	0.499	0.677	0.212	0.212	0.087	0.062	0.580	0.304
15 Feb	0.109	0.140	0.552	0.711	0.268	0.269	0.080	0.062	0.508	0.303
16 Mar	0.130	0.180	0.479	0.709	0.247	0.199	0.090	0.035	0.870	0.310
15 Apr	0.151	0.187	0.575	0.712	0.349	0.349	0.045	0.032	0.590	0.355
16May	0.205	0.260	0.540	0.682	0.408	0.409	0.024	0.013	0.644	0.355
16 Jun	0.215	0.263	0.558	0.685	0.441	0.441	0.018	0.013	0.615	0.356
17 Jul	0.199	0.219	0.646	0.712	0.460	0.460	0.024	0.020	0.478	0.366
16Aug	0.166	0.178	0.671	0.725	0.412	0.410	0.038	0.035	0.434	0.355
16 Sep	0.140	0.163	0.619	0.721	0.341	0.341	0.055	0.044	0.490	0.339
16 Oct	0.136	0.162	0.595	0.707	0.317	0.318	0.072	0.055	0.482	0.322
15 Nev	0.111	0.144	0.545	0.701	0.249	0.249	0.083	0.067	0.467	0.280
11 Dec	0.116	0.172	0.463	0.688	0.181	0.181	0.088	0.052	0.610	0.249

Table 2. Numerical values of parameters corresponding to models fitting well with experimental measures for twelve average days of the year.

Date	A (Wm^{-2})	B (Wm^{-2})	C	D (Wm^{-2})	E
10 Jan	0.063	0.792	0.249	0.077	0.177
15 Feb	0.120	0.586	0.328	0.239	0.153
16 Mar	0.167	0.643	0.220	0.084	0.090
15 Apr	0.064	0.875	0.294	0.021	0.115
16 May	0.071	0.878	0.442	0.128	0.011
16 Jun	0.123	0.783	0.401	0.107	0.010
17 Jul	0.153	0.692	0.285	0.030	0.011
16Aug	0.120	0.822	0.526	0.011	0.103
16 Sep	0.089	0.737	0.475	0.021	0.010
16 Oct	0.112	0.701	0.203	0.089	0.077
15 Nev	0.143	0.715	0.263	0.095	0.069
11 Dec	0.050	0.830	0.288	0.027	0.167

In this study, we define clear and normal states of the sky as the situations for which diffuse transmission function at the zenith is respectively equal to 0.05 and 0.22 [7].

In figure 7, the variations of direct normal irradiance B_n values from PVGIS databank (dashed lines) and their numerical models (solid lines) for clear (red lines) and normal (green lines) skies are plotted as functions of time during december 11th 2016. As expected, B_n vanishes for sunrise and sunset times and is maximal for solar midday. In addition, one can remark that the model proposed in eq (21) fits perfectly with simulated data from PVGIS.

In figure 8, variations of diffuse horizontal irradiance D_h values from PVGIS databank (dashed lines) as well as its numerical models (solid lines) for clear (red lines) and normal (green lines) skies are drawn versus time throughout december 11th 2016. Here also, one can easily notice that the model proposed in eq (22) gives a good description of diffuse horizontal irradiance in a clear contrast with some models given in the litterature [25, 26].

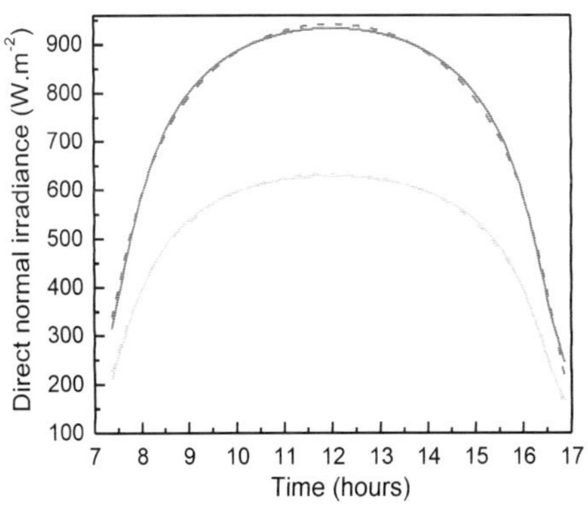

Figure 7. Direct normal irradiance B_n data from Climate-SAE PVGIS (dashed lines) and its numerical models (solid lines) for clear (red lines) and normal (green lines) skies drawn against time during December 11th 2016.

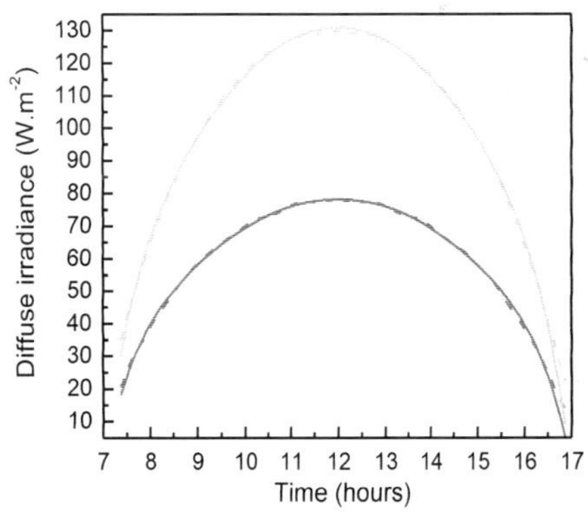

Figure 8. Diffuse horizontal irradiance D_h values from Climate-SAE PVGIS (dashed lines) and its numerical models (solid lines) for clear (red lines) and normal (green lines) skies drawn against time during December 11th 2016.

Figure 9 presents experimental measurements collected locally and simulated values from PVGIS databank of global irradiance G_h (dashed lines) at El Jadida (Morocco) as well as their numerical models (solid lines) for clear (red lines) and normal (green lines) skies as functions of time during december 11th 2016. Blue lines correspond to experimental data and its numerical model. We can remark that experimental measurements are lying between clear and normal skies data, we also notice that maximal value of D_h reaches 138.82 Wm^{-2} at midday which is consistent with cloudy days of December.

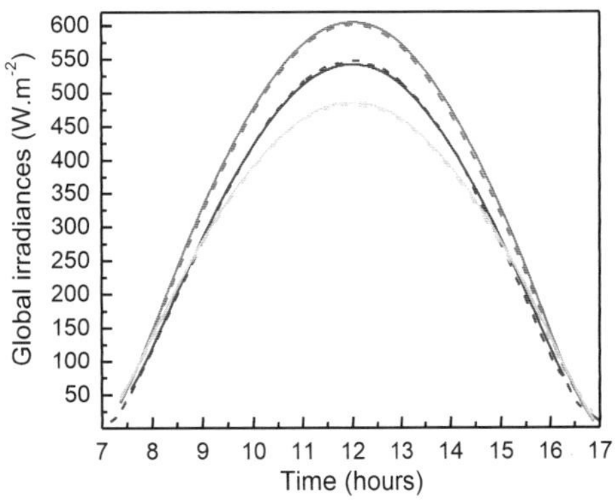

Figure 9. Experimental measurements and simulated values from Climate-SAE PVGIS of global irradiance G_h (dashed lines) and their numerical models (solid lines) for clear (red lines) and normal (green lines) skies plotted versus time during December 11th 2016. Experimental data correspond to blue lines.

4.4. Prediction of daily energy density received at the site of El Jadida (Morocco):
The daily energy density received at the site of El Jadida (Morocco) by an arbitrarily oriented surface during one day is expressed as follows:

$$H_i(n,\beta,a) = 3600\sum_{j=1}^{N}\left(G_{i,j+1}(n,\beta,a)+G_{i,j}(n,\beta,a)\right)\frac{\left(t_{j+1}-t_j\right)}{2} \tag{24}$$

where i is identified with t, h, p, $(s\text{-}a,\ n\text{-}s)$ and $(s\text{-}a,\ e\text{-}w)$ respectively for tilted, horizontal, perpendicular, single-axis oriented in east-west direction and single-axis oriented in north-south direction collectors.
In table 3, the values of incident and received energy per unit area for horizontal, normal and tilted fixed surfaces with $\beta=(33.25)°$ and $a=180°$ for the average day of December 2016 are gathered.

Table 3. Daily energy densities radiated and received at El Jadida (Morocco) by horizontal, normal and tilted fixed apertures ($\beta=(33.25)°$, $a=180°$).

	Daily energy densities radiated and received at El Jadida (Morocco) (MJm^{-2})		
n = 346	Horizontal aperture	Two-axis aperture	Tilted fixed aperture
Radiated extra-atmospheric	18.06	49.57	34.00
Global for clear sky	13.16	28.89	22.10
Global for normal sky	10.91	20.73	16.62
Global experimental	11.56	-	20.58

As mentioned above, the values of models parameters are determined for 12 average days, one average day for each month, from experimental measurements and simulated data from PVGIS. Then, numerical interpolations are performed to generate the values of models parameters for each day of the year and to calculate daily direct, diffuse and global irradiances using the equation:

$$H_i(n,\varphi,\beta,a) = \frac{24\times3600}{2\pi}\int_{\max(\omega_{ss},-\omega_s)}^{\min(\omega_{ps},\omega_s)} G_i(n)d\omega \quad (25)$$

where G_i is collector global irradiance given in paragraph 4.3 which is dependent on interpolated values of A, B, C, D and E for normal sky. i is identified with t, h and p, respectively for tilted, horizontal and perpendicular to sun rays apertures. This allows us to predict, with a good degree of accuracy, energy densities received by horizontal, normal and fixed tilted apertures (see figure 10).

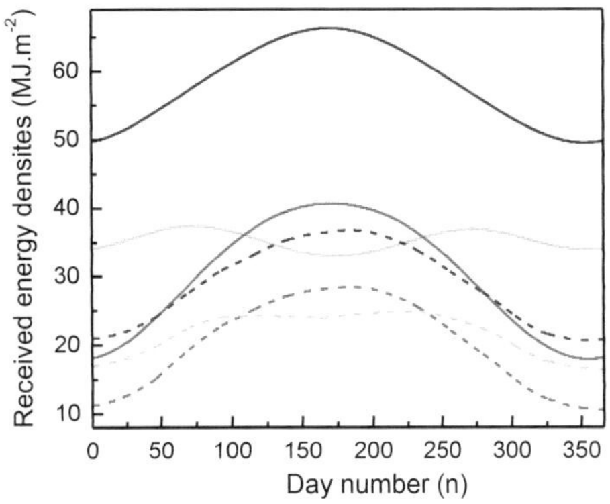

Figure 10. Calculated daily energy densities radiated (solid lines) and effectively received (dashed lines) by normal to sun rays (blue lines), horizontal (red lines) and fixed tilted apertures with $\beta=(33.25)°$, $a=180°$ (green lines) as functions of the day number.

In figure 11, we present daily energy densities measured locally and calculated according eq (25) and using the model given in paragraph 4.3 for a horizontal collector at the site of El Jadida. The reader can notice that measured and calculated data are in good agreement.

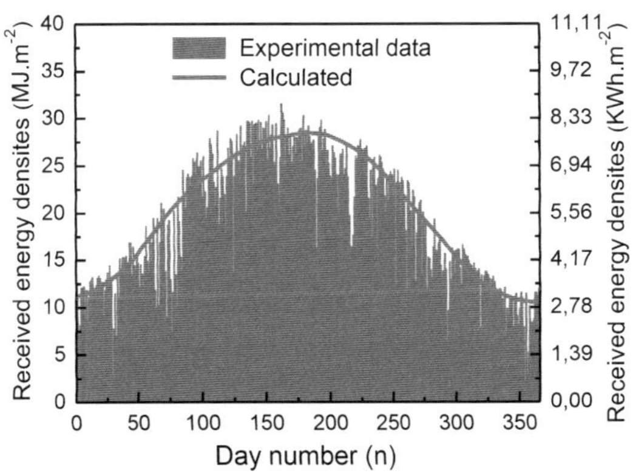

Figure 11. Daily energy densities measured and calculated for a horizontal collector at the site of El Jadida as functions of the day number n.

5. Conclusion

In this paper, Maple computer algebra software is used to determine analytic expressions and to calculate numerical values of daily solar radiation directed towards horizontal, fixed tilted, single-axis and double-axis tracking collectors placed outside atmosphere as functions of the day number, collector latitude, collector tilt and azimuth angles. Effects of non-uniform atmospheric absorption of incident solar radiation throughout a day, from sunrise to sunset, are taken into account and new models of direct normal and diffuse horizontal components of daily global horizontal and tilted irradiances are proposed, tested and validated on local experimental data. This enables us to extract global irradiance model parameters for average day of each month and to predict, accurately, daily solar radiation (energy per unit area) effectively received by any collector placed on earth during the whole year.

Acknowledgments
The authors H El Achouby and M Zaimi are grateful to Kingdom of Morocco Ministry of Higher Education and to Chouaïb Doukkali University for their financial supports.

References

[1] http://data.worldbank.org/topic/energy-and-mining
[2] http://data.worldbank.org/topic/climate-change
[3] Loi 13-09 relative aux énergies renouvelables March 18th 2010 *Bulletin Officiel du Royaume du Maroc* N° 5822 pp 229
[4] Loi 16-09 relative à l'Agence Nationale pour le Développement des Energies Renouvelables et de l'Efficacité Energétiques March 18th 2010 *Bulletin Officiel du Royaume du Maroc* N° 5822, pp 235
[5] Loi 57-09 portant création de la société ''Moroccan Agency for Solar Energy'' March 18th 2010 *Bulletin Officiel du Royaume du Maroc* N° 5822 pp 237
[6] http://noorouarzazate.com/
[7] Rerhrhaye A 1995 Caractérisation du gisement solaire du site d'El Jadida Thèse de Doctorat d'Etat en Physique Université Chouaïb Doukkali Faculté des Sciences (El Jadida)
[8] Castaner L and Silvestre S 2002 Modelling Photovoltaic Systems Using PSpice *John Wiley and Sons* (Chichester, England)
[9] Gueymard C 2004 The sun's total and spectral irradiance for solar energy applications and solar radiation models *Solar Energy* **76** (2004) 423–453
[10] https://www.pvlighthouse.com.au
[11] Gueymard C 1995 SMARTS2, A simple model of the atmospheric radiative transfer of sunshine: Algorithms and performance assessment *Florida Solar Energy Center*
[12] Sengupta M, Habte A, Gueymard C, Wilbert S and Renné D 2017 Best Practices Handbook for the Collection and Use of Solar Resource Data for Solar Energy Applications: Second Edition NREL Technical Report http:/www.osti.gov/scitech
[13] Kopp G and Lean J L 2011 A new lower value of total solar irradiance: evidence and climate significance Geophys Res Lett **38**:L01706
[14] William B S and Michael G 2001 Power From The Sun http://www.powerfromthesun.net/book.html
[15] El Maliki M S 2017 Modèles d'estimation des composantes du rayonnement solaire & Productivité photovoltaïque d'installations connectées au réseau Habilitation Universitaire en Energétique Université Chouaïb Doukkali Faculté des Sciences (El Jadida)
[16] Rerhrhaye A, El Maliki M S and Zehaf M 2002 Estimation des valeurs quotidiennes du diffus sur surface horizontale Phys. Chem. News **6** 18-22
[17] Birba L, Khaidar M, Zradba A, Monkade M, El Maliki M S, Zehaf M, Rerhrhaye A, Aarich N, Raoufi M and Bennouna A 2016 The behaviour of a grid connected PV plant under a semiarid and maritime climates in summer 1st International Conference on Solar Energy and Materials (Marrakesh)
[18] Birba L, Khaidar M, Louardi A, Monkade M, Zradba A, El Maliki M S and Zehaf M 2017 Modeling of photovoltaic systems taking into account the thermal inertia of PV materials The International conference on Advanced Materials for Photonics, Sensing and Energy Applications (Agadir)
[19] Suri M, Huld T and Dunlop E 2005 PVGIS, a Web Based Solar Radiation Database for the Calculation of PV Potential in Europe *Int. J. of Sustainable Energy* **24** (2) 55-67
[20] http://re.jrc.ec.europa.eu/pvgis/apps4/pvest.php?map=africa&lang=fr
[21] Bruneton E and Neyret F 2008 Precomputed atmospheric scattering *Comput. Graph. Forum.* **27** (4) Special Issue: Proceedings of the 19th Eurographics Symposium on Rendering 1079–1086
[22] Liu B Y H and Jordan R C 1961 Daily insolation on surfaces tilted towards the equator *ASHRAE Journal* **3** (0) 53–59
[23] El Mghouchi Y, El Bouardi A, Choulli Z and Ajzoul T 2016 Models for obtaining the daily direct, diffuse and global solar radiations Renewable and Sustainable Energy Reviews **56** 87-99

[24] El Mghouchi Y, Ajzoul T and El Bouardi A 2016 Prediction of daily solar radiation intensity by day of the year in twenty-four cities of Morocco Renewable and Sustainable Energy Reviews **53** 823-831
[25] Perrin de Brichambaut C 1975 Estimation des Ressources Energétiques en France *Cahiers de l'A. FEDES, N°1*
[26] Capderou M 1985 Editor of Atlas solaire de l'Algérie énergétique Tome 2. **2** 140–399 (Alger)

Economic Sizing of a Grid-Connected Photovoltaic System: Case of GISER research project in Morocco

M Khanfara[1], R EL Bachtiri[2], M Boussetta[3] and K EL Hammoumi[4]

[1,2,4]CED STI, FST, PERE Laboratory EST-USMBA, Fez, Morocco
[3]FSDM, LESSI Laboratory, REPEER Group EST-USMBA, Fez, Morocco

maha.khanfara@usmba.ac.ma

Abstract. This paper is within the framework of the GISER (Gestion Intelligent des Systèmes d'Energies Renouvelables). It's a project about 3kVA hybrid energy management in PERE Laboratory (Procedés des Energies Renouvelables) of EST fez Morocco. We present a sizing study and a cost analysis of a photovoltaic system using simulation results with Sunny Design software. The goal of our work is to compute and optimize the dimensioning of a photovoltaic installation in order to find the best configuration of the system. We have considered a PV installation connected to the grid with 2.5kWp through an inverter. We have taken into account meteorological conditions in the installation site (radiation and temperature). Our optimization approach considers the operation of each part of the installation, including energy production, energy consumption and the overall quality/price of the installation.

1. Introduction

The rapid depletion of fossil-fuel resources on a worldwide basis has necessitated an urgent search for alternative energy sources. Photovoltaic and wind energy are considered as promising toward meeting the continually increasing demand for energy. The solar energy is inexhaustible and pollution-free.

With the growing world population, electrical network become nonoperational due to redundant and overstressed infrastructure. Micro-grids are a promising solution to provide local energy generation and management [1].

Generated energy from PV systems depends mainly on solar energy available at selected site. Geographical location, ambient temperature, clearness index, tilt angle and orientation of PV panel. In order to efficiently and economically use of PV array system sizing and economic analysis is necessary [2].

Several researches, deal with system cost optimization of either grid-connected or stand-alone using HOMER optimization software which is based on the analytical simulation and storage system with anaerobic digestion biogas Power Plants [3]- [5]. Others [6][7] have developed heuristic methodologies, relying on evolutionary algorithms to optimize micro grids or HRESs size according to cost, environmental, or reliability parameters. However, few of them
deal with solar-, wind-, and biomass-resource hybridization in a grid-connected scheme [8] – [11].

In [12] the authors showed that the Moroccan climate is very favorable for the installation of grid-connected microgrid in several regions of the country. In [13] a grid coupling and injection into isolated grid networks with a proposed method to eliminate the first harmonics using a multilevel inverter. The photovoltaic systems, particularly those linked to the grid, have been developed in reply to an increasing demand of a renewable and clean energy [14], the aim is that the system must be economically attractive, the components for PV installation have a great influence on the system cost and its lifetime, and thus each part of the system must be well optimized and configured [15] [16].

The goal of this work is to find the optimal configuration of the 2.5kWp PV installation connected to grid for a research project in Morocco. The main contribution of this paper is to discuss an economic study and optimize the dimensioning of a photovoltaic installation compared to simulation results.
The optimal PV sizing depends on climatic data based in Fez. We have identified the optimal configuration between all the possible structures of a microgrid based on the power peak of the system which defines our project, thus on real climatic data of the region of Fez where we intend to realize our system.
We have found as a best configuration: Trina Solar Energy TSM-250PC05A for the PV panel. SB 2.5-1VL_40 for the inverter which have been chosen for its best performance and its compatibility with the PV module. We have also computed cable section, economic sizing and compared the results with Sunny Design Software simulations. We have found that the total investment including PV modules, inverter and lines is equal to 28,049 MAD.

2. Studied system

The studied system is a 3 kVA hybrid system developed within the framework of the CNRST-funded GISER project about the management of renewable energies consisting 2.5 kW peak PV module, 1.5 kVA wind turbine, 1 kW battery pack in order to ensure the electrical energy needs during the night and 1.5 kVA diesel generator as a backup power supply in the absence of PV-wind generation and if the batteries are discharged as shown in figure 1.

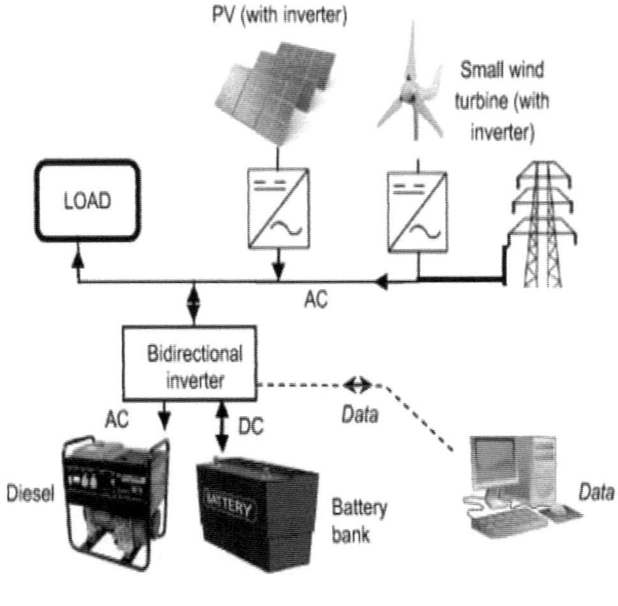

Figure 1. Hybrid Power System

2.1 Location of the site

The performance of a photovoltaic panels depends on climatic data which are principally solar radiation and temperature. To cater for a given load profile, the system size increases when climate resources are low [14]. Information required for this study was collected based on a project in Fez city of Morocco.

2.2 Meteorological DATA

Global irradiation of the location is 1.982 kWh/m² per year. The performance of PV modules is strongly depending on the sun light condition and temperature [15].
Figure. 2 and 3 give data of global irradiation and global temperature per month for Fez site.

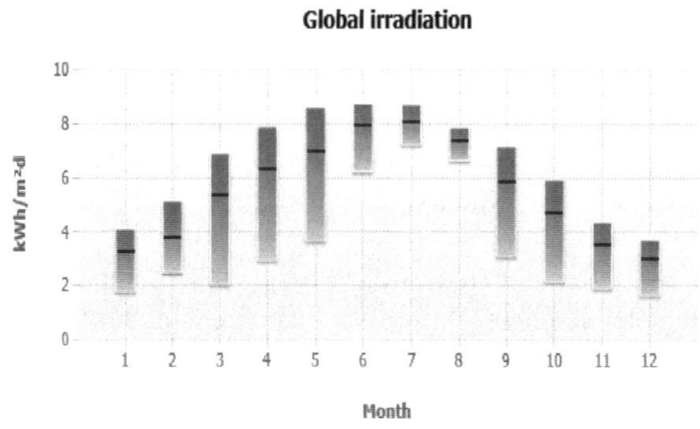

Figure 2. Global irradiation per month

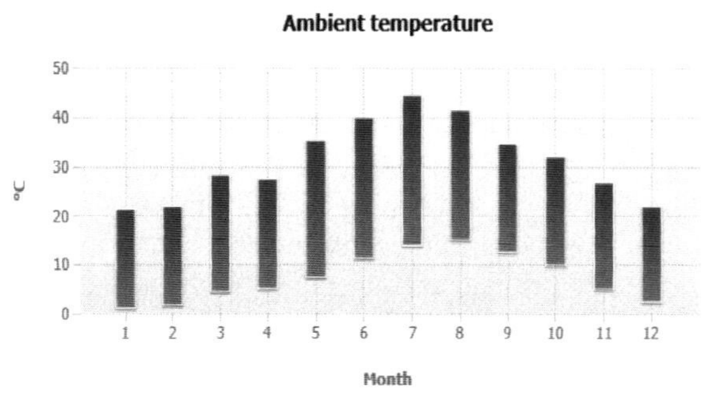

Figure 3. Global temperature per month

3. System Configuration

After the data input of the project (location and information on the PV power installation (Peak power, Orientation/Mounting type) the software proposes the following configuration:

3.1 Photovoltaic design data

Figure 4. Shows the information design of the PV array

10 x Trina Solar Energy TSM-250PC05A (35mm) , Azimuth angle: 0 °, Tilt angle: 25 °, Mounting type: Ground mount	
Number of strings:	1
PV modules per string:	10
Peak power (input):	2.50 kWp
Typical PV voltage:	280 V
Min. PV voltage:	254 V
Min. DC voltage (Grid voltage 230 V):	50 V
Max. PV voltage:	410 V
Max. DC voltage:	600 V
Max. MPP current of PV array:	8.3 A
Max. operating input current per MPPT:	10 A
Max. input short-circuit current per MPPT:	18 A
Photovoltaic Output Circuit Current:	8.8 A

Figure 4. PV array configuration

The important technical requirement which characterize the PV is the efficiency:

$$\mu = \frac{P}{E * S} \tag{1}$$

P= Power peak of the PV
E= Irradiation
S= Area of the PV
$\mu = \frac{250}{1000*1.6368}$=15.27%

3.2 Inverter

The sizing of the inverters will impose the way of cabling the modules between them. figure 5. Shows the inverter configuration proposed.

1 x SB 2.5-1VL-40 (PV system section 1)	
Peak power:	2.50 kWp
Total number of PV modules:	10
Number of PV inverters:	1
Max. DC power (cos φ = 1):	2.65 kW
Max. AC active power (cos φ = 1):	2.50 kW
Grid voltage:	230V (230V / 400V)
Nominal power ratio:	106 %
Dimensioning factor:	100 %
Displacement power factor cos φ:	1

Figure 5. Inverter configuration

The Sunny Boy is a transformerless photovoltaic inverter that converts the direct current of the photovoltaic generator into grid-compliant alternating current and feeds it into the grid.
Figure.6. shows the performance and the compatibility between the inverter and PV modules.
The choice of inverters is based on three requirements: Power Compatibility, Voltage Compatibility and Current Compatibility.

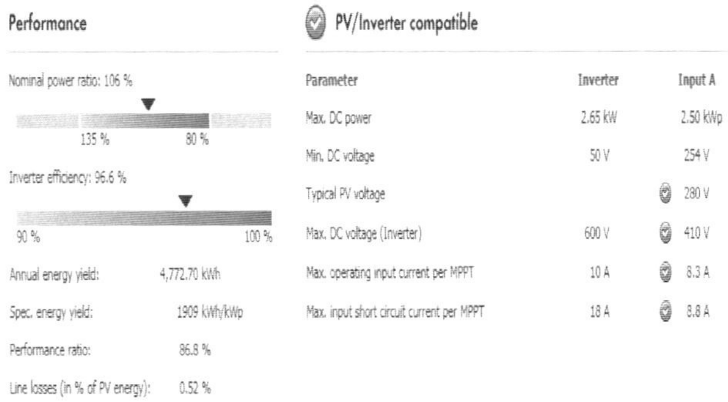

Figure 6. Performance and PV Inverter compatibility

3.3 Lines

The choice of the cables section on DC side is carried out according two important requirements:
The permissible current maximal in the cable $I_{max}$and the voltage drop in line. In [17] the permissible current through the lines must be equal to or greater than 1.25 times of the PV string short-circuit current.

$$I_{max} \geq 1.25 * Icc \tag{2}$$

Voltage drop in the cable should be less than 3% and ideally to equal to 1%.

$$S = \frac{2*\rho*L*I}{\epsilon * U} \tag{3}$$

S: Cable section
ρ: Resistivity of the semiconductor = 0.02314Ωmm²/m
L: Cable length = 15 m
I: Current flowing in the cable = $I_{mpp} = 8.27\ A$

ϵ: Voltage drop
U: Voltage at the origin of the cable = $U_{mpp} = 30.3\ V$

$$S = \frac{2 * 0.02314 * 15 * 8.27}{0.03 * 10 * 30.3} = 0.63\ mm^2 \cong 1.5mm^2$$

In [17] l'UTC C15-712-1 contains a table giving the value of the permissible current as a function of cable cross-section, installation mode, the temperature.
The value of the maximal permissible currentI_{max}at a given ambient temperature is defined by the following formula:

$$I_{max} = I_{120} * k$$

(4)

I_{120}: permissible current in an ambient temperature

So $I_{max} = 22 * 0.82 = 18.04$ A

We must verify if the permissible current of the lines are equal to or greater than 1.25 times of the PV string short-circuit current.

$$I_{max} \geq 1.25 * I_{CC}$$

we notice that $I_{max} \geq 10.987$
The cable section on AC side as follow:

$$S = \frac{b * \rho * L * I * cos_{\varphi}}{\epsilon * V_n} \quad (5)$$

S: Cable section
b: Coefficient equal to 1 for one phase and 2 for 3 phases
ρ: Resistivity of the semiconductor = 0.02314Ωmm²/m
L: Cable length = 25 m
I: Reverse Current = $I = 1.25 * 8.27\ A$
ϵ: Voltage drop
V_n: Nominal Voltage = $V_n = 220\ V$

$$S = \frac{2 * 0.02314 * 25 * 8.79 * 1}{0.03 * 220} = 1.54\ mm^2$$

Figure.7. shows wire sizing
We notice that the section of cables that have been proposed are equal to that calculated.

	DC	LV	Total
Power loss at nominal operation	17.27 W	27.10 W	44.36 W
Rel. power loss at rated nominal operation	0.65 %	1.08 %	1.74 %
Total cable length	20.00 m	10.00 m	30.00 m
Cable cross-sections	1,5 mm²	1,5 mm²	1,5 mm²

Figure 7. Line section

The energy yield contains two major elements: the target yield is the theoretical annual energy production (on the DC side of the module), only taking into account the energy of the incoming light and the module's nominal efficiency and The Performance Ratio is the ratio between actual yield (i.e. annual production of electricity delivered at AC) and the target yield, it takes into account all pre-conversion losses, inverter losses, thermal losses and conduction losses.
figure 8 shows that the energy yield and the performance ratio of the PV system.

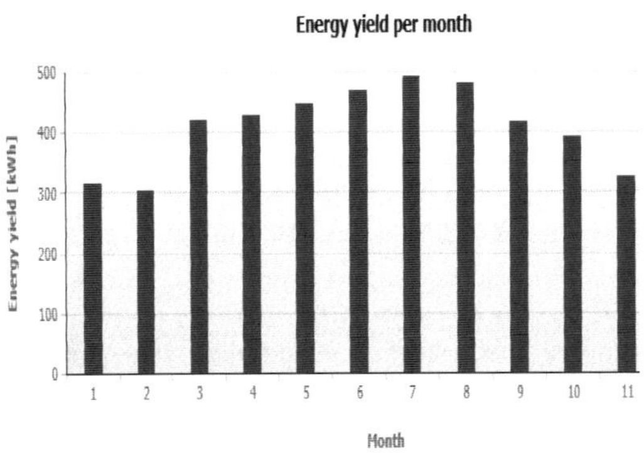

Month	Energy yield [kWh]	Performanc
1	313 (6.6 %)	89 %
2	303 (6.3 %)	89 %
3	417 (8.7 %)	88 %
4	426 (8.9 %)	88 %
5	445 (9.3 %)	87 %
6	467 (9.8 %)	85 %
7	491 (10.3 %)	85 %
8	480 (10.1 %)	85 %
9	415 (8.7 %)	86 %
10	392 (8.2 %)	87 %
11	325 (6.8 %)	88 %
12	299 (6.3 %)	89 %

Figure 8. PV energy yield and the performance ratio per month

3.4 Profitability analysis

This section shows the economic analysis for the project.

	The total costs
PV modules`	2,143.80 USD
Inverter	808.00 USD
Lines	19.75 USD
Investment	2,971.5528 USD
Specific investment (CapEx / kWp)	1,180.72 USD/kWp

4. Conclusion

This paper presents a technical economic study and a dimensioning of a PV system Trina. The dimensioning was realized with two methods, theoretically and by using the sunny design software, the results shows that we have found the best configuration of the system which consist: 10 of Trina Solar Energy TSM-250PC05A and one inverter SB 2.5-1VL_40. The Total investment is 2,971.5528 USD, and the total revenue from grid feed-in after 20 years is 11687 USD.

Acknowledgments
This research is within the framework of the GISER project funded by the CNRST (for the MESRSFC, Morocco)

References

[1] Gungor V C, Lu B, and Hancke G P 2010 *Opportunities and Challenges of Wireless Sensor Networks in Smart Grid* (*IEEE Transactions on Industrial Electronics* vol 57)pp. 3557–3564
[2] Zhou W, Lou C, Li Z, Lu L, and Yang H 2010 *Current status of research on optimum sizing of stand-alone hybrid solar–wind power generation systems* (*Applied Energy* vol 87) no. 2, pp. 380–389
[3] Montuori L, Alcázar-Ortega M, Álvarez-Bel C, and Domijan A 2014 *Integration of renewable energy in microgrids coordinated with demand response resources: Economic evaluation of a biomass gasification plant by Homer Simulator* (*Applied Energy* vol 132) pp. 15–22, Nov. 2014.
[4] Ma T, Yang H, and Lu L 2014 *A feasibility study of a stand-alone hybrid solar–wind–battery system for a remote island* (*Applied Energy* vol 121)pp. 149–158
[5] Lai C S and McCulloch M D 2017 *Sizing of Stand-Alone Solar PV and Storage System With Anaerobic Digestion Biogas Power Plants*(*IEEE Transactions on Industrial Electronics* vol 64) pp. 2112–2121
[6] Zhao B, Zhang X, Li P, Wang K, Xue M, and Wang C 2014 *Optimal sizing, operating strategy and operational experience of a stand-alone micro grid on Dongfushan Island* (*Applied Energy* vol 113)pp. 1656–1666
[7] Ma T, Yang H, Lu L, and Peng J 2015 *Pumped storage-based standalone photovoltaic power generation system: Modeling and techno-economic optimization* (*Applied Energy* vol. 137)pp. 649–659
[8] González A, Riba J R, and Rius A 2015 *Optimal Sizing of a Hybrid Grid-Connected Photovoltaic–Wind–Biomass Power System (Sustainability* vol 7)pp. 12787–12806
[9] Yazdani A, Bhuiyan F A, and Primak S L 2015 *Optimal sizing approach for islanded microgrids* (*IET Renewable Power Generation* vol 9)pp. 166–175
[10] Tazi K, Abbou F M, Bannour Chaka A, and Abdi F 2017 *Modeling and simulation of a residential microgrid supplied with PV/batteries in connected/disconnected modes—Case of Morocco* (*Journal of Renewable and Sustainable Energy* vol 9)p. 025503
[11] Mondol J D, Yohanis Y G, and Norton B 2006 *Optimal sizing of array and inverter for grid-connected photovoltaic systems* (*Solar Energy* vol 80) pp. 1517–1539
[12] Boussett M, El Bachtiri R, Khanfara M, and El Hammoumi K 2017 *Assessing the potential of hybrid PV–Wind systems to cover public facilities loads under different Moroccan climate conditions* (Sustainable Energy Technologies and Assessments vol 22)pp 74–82
[13] Khanfara M, Elbachtiri R, El hammoumi K, Boussetta M 2016 *A Multicarrier PWM Technique for Five Level*
*Inverter connected to the grid (*International Conference on Information Technology for Organization
Development (IT40D)) 978-1-4673-7689-1/16$31.00
[14] Nordin N D and Rahman H A 2014 *Design and economic analysis in stand-alone photovoltaic system* (in Energy Conversion CENCON IEEE Conference) pp. 152–157
[15] Kolhe M 2009 *Techno-Economic Optimum Sizing of a Stand-Alone Solar Photovoltaic System* (*IEEE Transactions on Energy Conversion* vol. 24) pp. 511–519
[16] Boussetta M, Elbachtiri R, El hammoumi K, and Khanfara M 2016 *optimal sizing of grid-connected pv-wind system case study: agricultural farm in morocco (Journal of Theoretical and Applied Information Technology* vol 86) p. 196
[17] GUIDE PRATIQUE *Installations photovoltaïques*, UTC C15-712

IOP Conf. Series: Earth and Environmental Science **161** (2018) 012007 doi:10.1088/1755-1315/161/1/012007

Thin layer solar drying of Moroccan carob pulp (*Ceratonia Siliqua L.*)

Z Tagnamas, Y Bahammou, M Kouhila, A Lamharrar and A Idlimam

Team of Solar Energy & Aromatic and Medicinal Plants, ENS Marrakesh, Physics Department, Marrakesh, Morocco

zakariatagnamas@gmail.com, younesbahammou3@gmail.com, kouhilamounir@gmail.com, alamharrar@yahoo.fr, aidlimam@gmail.com

Abstract. Most biological products are highly perishable, so the issue of their conservation during large transportation distances and storage is a sensitive problem. Solar drying is the most useful method of keeping food staff safe in terms of both energy and time. Our work aims to contribute to the optimization of solar convective drying at the pulp of carob widely produced in Morocco by an experimental approach in an indirect forced convection solar dryer. In this paper, the effect of air temperature on the drying kinetics of carob pulp was investigated in convective solar drying. The relative humidity of air was varied from 15% to 45.5%. This kinetic measurement is carried out at four temperature values of drying air (50, 60, 70 and 80 °C) and also for one value of drying air flow rate D_v (0.0833 ± 0.002 m^3 s^{-1}). Four models have been used to adjust experimental curves. In conclusion, the experimental study shows that solar drying is adapted for the valorisation of carob pulp. Middili & Kucuk model is the most appropriate to describe drying curves in various drying conditions.

1. Introduction

Carob tree (*Ceratonia siliqua, L.*) is native vegetation in the Mediterranean area; it is mainly in human and animal nutrition. This tree is mainly grown in Italy, Spain, Portugal, Greece and Morocco, considered as the principal producers of carob in the world [1]. Traditionally, the seeds of the pods are used to obtain the carob gum and the pulp is quite rich in sugars [2].

Nowadays, the carob constituents are used for many applications in the food, pharmaceutical and cosmetics industries, with particular interest in locust bean gum made from seeds. Carob pulp contains high levels of sugars, particularly low molecular weight carbohydrates, such as sucrose, therefore is considered as a good source of potential energy for animal feed. In addition, the carob pulp is characterized by a protein content and very low fat content. However, carob pulp has a favorable fatty acid composition due to the presence of essential fatty acids and may be a natural source of acids [3].

The drying of foods is a very important method for the food industry and offers possibilities for novel products to consumers. In recent years, there have been many advances in food drying technology, including pre-treatments, techniques, equipment, energy efficiency and quality.

Drying is among the most ancient method of food preservation. It is aimed at lowering the water content of foodstuff and is used predominately for foods such as fruits, vegetables, spices and other products with a high moisture content, and ones which are considered highly perishable [4]. Dried foods offer multiple benefits including: extended product shelf-life, reduced packaging, storage, handling and transportation.

Energy consumption, in drying industry, is a major problem, because over 85 % of industrial dryers are convective type with either hot air or combustion gases [5]. Being a complex phenomenon involving simultaneous heat and mass transfer, the process of drying is energy intensive and accounts for roughly 12 to 20 % of energy consumed in the industry sector [6].

Dried product nutritional quality is another major concern. With conventional drying methods using conductive and convective modes of heat transfer, final product preserves its quality and the probability of product contamination is low [4]. For this reasons, over recent years, there has been significant technological advancement in food drying in terms of drying pre-treatments, techniques, equipment and quality [7,8]. These works address the growing need to find an improved solar drying technique to preserve the quality of the end product at improved utilization of energy.

According to many studies [9] which focused on drying of biological products, it has been noted that it is difficult to find a mathematical model governing the evolution of the products during drying. For this reason, only an experimental study allows to determine their drying kinetics. Therefore, it seems useful to study variations of moisture content and those of the drying rate versus time, for different controllable drying parameters, using an indirect solar dryer operating in forced convection. This study is a contribution in the valorization of carob pulp, by using solar energy which is abundant in Morocco.

The aims of this experimental study were to:

- Study the drying kinetics of carob pulp in a convective solar dryer;
- Determine the effects of drying air temperature on the drying kinetics of carob pulp;
- Determine the appropriate model to fit the experimental curves.

2. Material and methods

2.1. *Description of the solar dryer*

In this study, the experimental approach that consists of determining the kinetic behavior of carob pulp, dried in a thin film solar dryer has been conducted. The experimental set-up, installed at the laboratory of the ''Ecole Normale Supérieure'' of Marrakesh (Figure 1), was described in detail in the reference [10]. This is mainly an indirect forced convection solar dryer. The components of the dryer are:

- A single solar collector with single glazing, 2.5 m² surface, inclined 30° from the horizontal plane and facing south. The dimensions of the collector are 2.5 m long and 1 m wide. The absorber is opaque in 0.5 mm thick black galvanized iron sheet with a non-selective surface. The rear thermal insulation is 0.05 m thick polyurethane foam sandwiched between two steel sheets. The absorber-insulator distance is 0.025 m and the absorber-to-glass distance is 0.02 m;
- An aspirating aeraulic channel constituted of tunnel whose section parallelepiped;
- A drying chamber consists of ten floors in trays;
- A centrifugal fan to circulate the air inside the dryer;
- A thermo-regulator with a precision range between 0-100 °C, connected to a PT100 platinum to measure drying temperature;
- Electrical resistances with 4 kilowatts of power acting as auxiliary source.

2.2. *Experimental protocol*

It is noted that drying is carried out under the controlled conditions of temperature and the flow rate of drying air. Indeed, the mass of the carob, comes from the region of Marrakesh (Morocco), used in drying experiments was (40 ± 1) g. They are spread in thin layers on trays in the drying chamber. Samples are uniformly and evenly distributed to ensure homogeneity of diffusion during the drying process.

The variation of the wet mass of the product $M_h(t)$ versus time was determined by a digital weighing apparatus (±0.1 g).

The duration of manipulations between the preparation of the product, loading the dryer and start-up of measures must be reduced to a minimum level; the evaporation phenomenon of pulp slices takes place very quickly due to the existence of the vapor concentration gradient between the product surface and the atmosphere.

During each drying experiment, the product mass on the tray was measured by removing it from the drying chamber for approximately thirty seconds. These measures were undertaken each 10 min at the beginning of the experiments until 30 min toward the end of the tests. Then, the temporal evolution of the wet mass M_h (t) of the product to dry has been tracked by successive weighings until it becomes stationary. This is followed by the dehydration of the product, at the end of each experimental test; in an oven whose temperature was fixed at 105 °C during 24 h in order to determine the dry mass M_d of product [11]. Throughout all the experiments, the temperature and the air flow drying rate are fixed. The water content in dry basis is defined by:

$$X(t) = \frac{M_h(t) - M_d}{M_d} \times 100 \quad (1)$$

Were:
X (t): Water content of the product during the experiment (% dry basis);
M_h (t): Wet mass of the product in a specific moment (g);
M_d: Dry mass at the end of the experiment (g);
t: Drying time (min).

Table. 1 Drying conditions during experiments in the solar dryer

Experiment	D_v (m^3/s)	Drying temperature Θ ± 0.1 (°C)	Air humidity R_h ± 2%
1	0.0833	50	15
2	0.0833	60	30.5
3	0.0833	70	45.5
4	0.0833	80	16.5

The different experimental experiments consisted in studying the independent influence of the drying air temperature on the drying process.

Figure. 1 Schematic diagram of the solar dryer: (1) Solar collector, (2) circulation fan, (3) fan, (4) air flow direction, (5) control box, (6) auxiliary heating system, (7) shelves, (8) drying chamber, (9) recycling air, (10) control foot, (11) exit of air, (12) humidity probes and (13) thermocouples

2.3. Modelling of drying curves

A literature review shows that several models have been developed to describe the drying process. They can be classified into two main categories corresponding to two different approaches, namely (i) phenomenological models arising from theoretical approach of the transport modes (diffusion, diffusion-sorption, and capillarity) and (ii) behavioral models reflecting a macroscopic approach to the problem.

In order to determine the most appropriate empirical equation to describe the shape of drying carob pulp from Marrakesh, four models were chosen: Newton, Page, Logarithmic and Middili & Kucuk. The most appropriate model was selected by the following statistical criteria [10]:

- Correlation coefficient (R);
- Mean bias error (MBE);
- Chi-square (χ^2).

These coefficients were defined by:

$$R = \sqrt{\frac{\sum_{i=1}^{N}(X^*_{pre,i} - \bar{X}^*_{exp,i})^2}{\sum_{i=1}^{N}(X^*_{exp,i} - \bar{X}^*_{exp,i})^2}} \quad (2)$$

$$MBE = \frac{1}{N}\sum_{i=1}^{N}(X^*_{pre,i} - \bar{X}^*_{exp,i}) \quad (3)$$

$$\chi^2 = \frac{\sum_{i=1}^{N}(X^*_{pre,i} - \bar{X}^*_{exp,i})^2}{N - n} \quad (4)$$

Where:

X^*_{exp} is the experimental value of reduced moisture content;

X^*_{pre} is the predicted value of reduced moisture content;

$\bar{X}^*_{exp}$ is the average of experimental reduced moisture values;

n is the number of model parameters;

N is the number of conducted experiments.

3. Results and discussion

Different experimental drying conditions are presented in the Table 1. Drying experiments have been conducted for four temperature values of drying air (50 ± 0.1 °C, 60 ± 0.1 °C , 70 ± 0.1 °C and 80 ± 0.1 °C) and for one value of drying air flow rate D_v (0.0833 ± 0.002 $m^3.s^{-1}$). The initial moisture content of the carob pulp ranged from 1.5508 % to 1.6065 % g water per g dry basis and was reduced to the final moisture content, which varies from 0.0887% to 0.1246% g water per g dry basis.

Table. 2 Drying time during experiments in the solar dryer

Experiment	t (min)
1	380
2	230
3	170
4	160

Figure 2 illustrates the changes in moisture content versus drying time. As shown in Figures 2 and 3, it was noted the presence of one drying phase in the drying curves of carob pulp. However, in this phase the acquired energy was used to evaporate water from the product and to increase its temperature. This effect is represented by the decrease in the drying rate [10]. As is showed in Table 2 we noted a small difference in drying time between 70 °C and 80 °C (decreasing of 10 min), so we can neglect the influence of air drying temperature for temperatures higher than 70 °C.

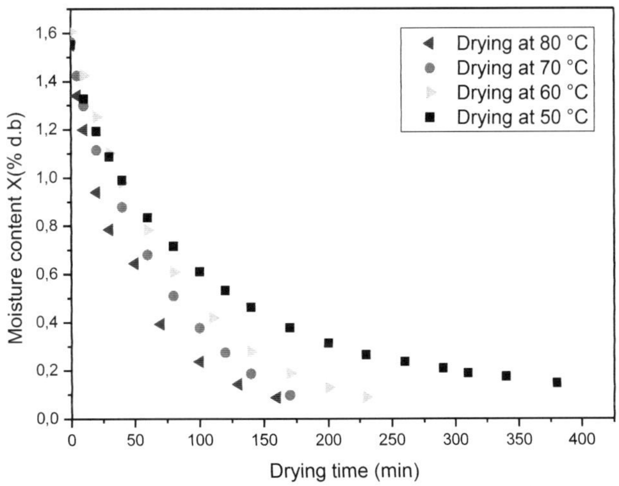

Figure. 2 Variation of moisture content versus drying time for different drying air conditions

As it can be seen from Figure 2, the drying air temperature is influencing the drying kinetics of pulp during the falling drying rate period, at the same flow rate of drying air.

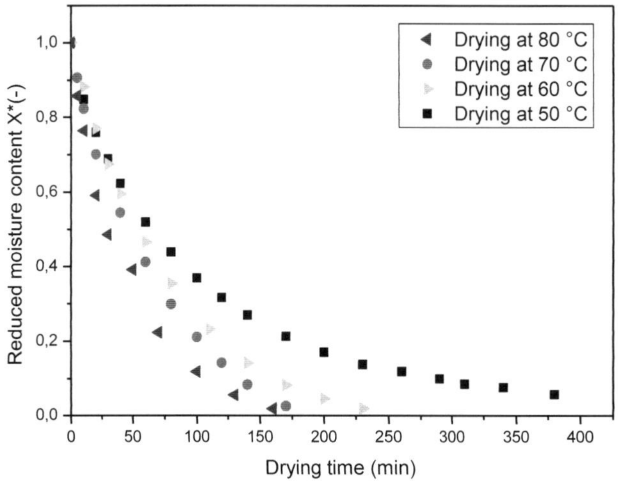

Figure. 3 Evolution of the reduced water content of the carob pulp as a function of the drying time for different drying air temperatures

An increase in drying air temperature resulted in a decrease in drying time. Indeed, as for the majority of porous hygroscopic products several investigators reported considerable increases in drying time when higher temperatures were used for drying various vegetables such as Citrus aurantium leaves [12], Prickly pear fruit [13], and Eucalyptus globules [14].

Indeed, an increase of drying temperature leads to an acceleration of the drying process. This is due to the increase in the gap between the partial pressures of water vapor on the surface of the product and that of the drying air, which favors the evaporation of the water contained in the product.

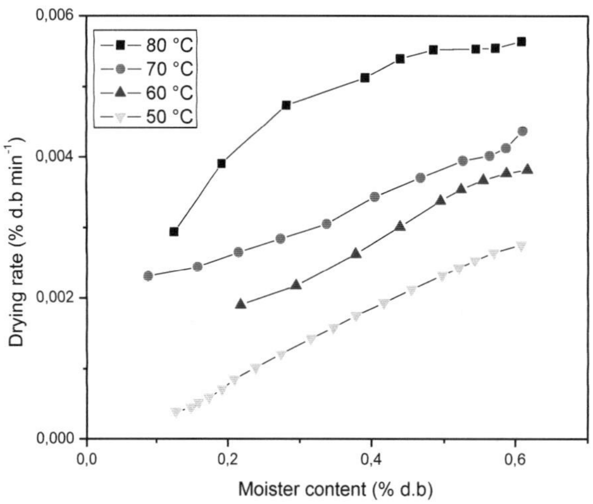

Figure. 4 Influence of drying air temperature on drying rate of Carob pulp at a fixed drying air flow rate 0.0833 m^3s^{-1}

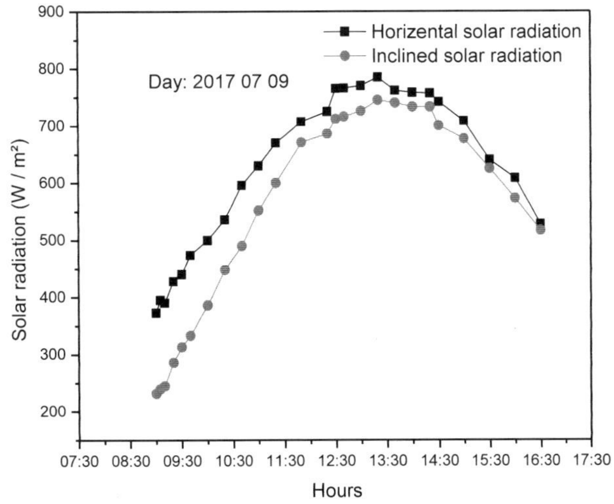

Figure. 5 Variations of horizontal and inclined solar radiation in Marrakesh

Thus, the drying rate evolution is presented in the Figure 4 for the four tests. This figure shows that for the same drying air flow rate, the drying rate increases with increase of the drying temperature values and hence the water content of the carob pulp decreases considerably. It is clear that the drying rate decreases continuously with moisture content and drying time.

There is no constant rate drying period in these curves, and all the drying process was seen to occur in the falling drying rate period. This shows that diffusion is the dominant physical mechanism governing moisture movement in the samples. Similar results were obtained by Kaymak-Ertekin [15] of green and red peppers, Madamba et al. [16] of garlic slices and Midilli et al. [17] of mushroom and pollen.

The solar collector of the experimental dryer is inclined with an angle of 30°. The variations of solar radiation in horizontal and surface of the collector is presented in Figure 5. The measurements are undertaken between 9:00 AM and 16:30 AM (Friday 09 July 2017) in Marrakesh city were the drying of carob pulp is processed. The horizontal solar radiation is higher than the inclined one as is presented in Figure 5.

Indeed, Table 3 shows that the Midilli and Kucuk model offers the best correlation with experimental curves of reduced moisture content (X*) with the highest value of R and the lowest values of χ^2 and MBE.

Table. 3 Modeling of drying curves and average statistical parameters

Model	θ (°C)	Parameters	R	X^2	MBE
Newton	50 °C	k=0.0098	0.9975	0.00085	0.02222
	60 °C	k=0.01335			
	70 °C	k=0.01595			
	80 °C	k=0.02247			
Page	50 °C	k=0.02457591; n=0.8036211	0.9986	0.00032	0.01181
	60 °C	k=0.01037443; n=1.05877689			
	70 °C	k=0.0169073; n=0.9863179			
	80 °C	k=0.0346247; n=0.8856654			

Logarithmic	50 °C	k=0.010863; c=0.0608973; a=1.0475	0.9962	0.00041	0.01187
	60 °C	k=0.01262186; c=-0.0495058; a=1.05323174			
	70 °C	k=0.0130; c=-0.0813; a=1.0532			
	80 °C	k=0.021; c=0.0021; a=0.9636			
Middili and Kucuk	50 °C	k=0.02318; n=0.8173; a=0.9965; b=0.0000196	0.9995	0.000153	0.00686
	60 °C	k=0.01264; n=0.9987; a=0.9997; b= -0.0001792			
	70 °C	k=0.02591; n=0.8376; a=0.9997; b= -0.00076614			
	80 °C	k=0.0433; n=0.8045; a=1.0017; b= -0.0004135			

The model Middili and Kucuk is the appropriate model to describe the drying kinetics of carob pulp. To take into consideration the effect of temperature on the parameters, k, a, b and n are expressed as function of temperature (ɵ).

$$k = -2.2188 + 0.06980\theta - 0.0005\theta^2 \quad (5)$$

$$n = -1.2044 + 0.0677\theta - 0.0005\theta^2 \quad (6)$$

$$a = 0.977 + 0.0005\theta - 3\times10^{-6}\theta^2 \quad (7)$$

$$b = 0.0065 - 0.0002\theta + 10^{-6}\theta^2 \quad (8)$$

4. Conclusions

In this study, the experiments of convective solar drying of carob pulp were conducted for four temperatures (50 °C, 60 °C, 70 °C and 80 °C) and for one drying air flow rate (0.00833 m^3/s). The experimental drying curve occurred in the falling rate period. From the obtained results, it can be concluded that the drying kinetics of Carob pulp were influenced by the drying air temperature.

Statistical analysis of experimental results in reduced coordinates X*(t), fitted by four empirical models (Newton, Page, Logarithmic and Middili & Kucuk) allows to conclude that Middili & Kucuk was the most appropriate model to describe the drying kinetics of carob pulp (*Ceratonia siliqua L.*)

Acknowledgments

The first author gratefully acknowledges the CNRST (Morocco) for providing financial support of this work in the program Research Excellence Scholarships.

References

[1] Inserra L, Luciano G, Bella M, Scerra M, Cilione C, Basile P, Lanza M and Priolo A 2015 Effect of including carob pulp in the diet of fattening pigs on the fatty acid composition and oxidative stability of pork *Meat Sci.* **100** 256–61

[2] Mulet A, Fernández-Salguero J, García-Pérez J V and Bon J 2016 Mechanistic modeling to address process analysis: Kibbles of carob (Ceratonia siliqua, L.) pod extraction *J. Food Eng.*

176 71–6
[3] Ayaz F A, Torun H, Glew R H, Bak Z D, Chuang L T, Presley J M and Andrews R 2009 Nutrient Content of Carob Pod (Ceratonia siliqua L.) Flour Prepared Commercially and Domestically *Plant Foods Hum. Nutr.* **64** 286–92
[4] Moses J A, Norton T, Alagusundaram K and Tiwari B K 2014 Novel Drying Techniques for the Food Industry *Food Eng. Rev.* **6** 43–55
[5] Zarein M, Samadi S H and Ghobadian B 2015 Investigation of microwave dryer effect on energy efficiency during drying of apple slices *J. Saudi Soc. Agric. Sci.* **14** 41–7
[6] Raghavan G S V, Rennie T J, Sunjka P S, Orsat V, Phaphuangwittayakul W and Terdtoon P 2005 Overview of new techniques for drying biological materials with emphasis on energy aspects *Brazilian J. Chem. Eng.* **22** 195–201
[7] Erbay Z and Icier F 2009 Optimization of Drying of Olive Leaves in a Pilot-Scale Heat Pump Dryer *Dry. Technol.* **27** 416–27
[8] Maache-Rezzoug Z, Rezzoug S A and Allaf K 2002 DEVELOPMENT OF A NEW DRYING PROCESS – DEHYDRATION BY CYCLICAL PRESSURE DROPS (D.D.S.): APPLICATION TO THE COLLAGEN GEL *Dry. Technol.* **20** 109–29
[9] Yaldýz O and Ertekýn C 2001 THIN LAYER SOLAR DRYING OF SOME VEGETABLES *Dry. Technol.* **19** 583–97
[10] Ali I, Abdelkader L, El Houssayne B, Mohamed K and El Khadir L 2016 Solar convective drying in thin layers and modeling of municipal waste at three temperatures *Appl. Therm. Eng.* **108** 41–7
[11] Koukouch A, Idlimam A, Asbik M, Sarh B, Izrar B, Bah A and Ansari O 2015 Thermophysical characterization and mathematical modeling of convective solar drying of raw olive pomace *Energy Convers. Manag.* **99** 221–30
[12] Mohamed L A, Kouhila M, Jamali A, Lahsasni S, Kechaou N and Mahrouz M 2005 Single layer solar drying behaviour of Citrus aurantium leaves under forced convection *Energy Convers. Manag.* **46** 1473–83
[13] Lahsasni S, Kouhila M, Mahrouz M and Jaouhari J T 2004 Drying kinetics of prickly pear fruit (Opuntia ficus indica) *J. Food Eng.* **61** 173–9
[14] Kouhila M, Kechaou N, Otmani M, Fliyou M and Lahsasni S 2002 EXPERIMENTAL STUDY OF SORPTION ISOTHERMS AND DRYING KINETICS OF MOROCCAN EUCALYPTUS GLOBULUS *Dry. Technol.* **20** 2027–39
[15] Kaymak-Ertekin F 2002 Drying and Rehydrating Kinetics of Green and Red Peppers *J. Food Sci.* **67** 168–75
[16] Madamba P S, Driscoll R H and Buckle K A 1996 The thin-layer drying characteristics of garlic slices *J. Food Eng.* **29** 75–97
[17] Midilli A, Olgun H and Ayhan T 1999 Experimental studies on mushroom and pollen drying *Int. J. Energy Res.* **23** 1143–52

Optimum design of a small wind turbine blade for maximum power production

L Tenghiri[1], Y Khalil[1], F Abdi[2] and A Bentamy[1]

[1]School of Science and Engineering, Al Akhawayn University, 53000, Ifrane, Morocco

[2]Electrical Engineering Department, Faculty of Science and Technologies, P.O. box 2202 Route d'Imouzzer, Fez, Morocco

E-mail: Lhoussaine.tenghiri@usmba.ac.ma

Abstract. The rotor blades of a wind turbine are critical components that consist of aerodynamic shapes called airfoils. The interaction of these airfoils with the wind allows converting the power in the wind to mechanical power. The structure of the rotor blades must be able to extract maximum energy from the wind while being strong enough to stand steady, periodic and randomly changing loads. For maximum power extraction, an optimum design of the rotor blades is necessary. This paper presents a typical design methodology of the rotor blades of a small wind turbine with a power generation of 11 kW (rotor radius of 3.5 m). First, the design parameters were presented. Then, optimum blade geometry (chord length and twist angle distribution) was determined using optimum rotor theory. Finally, the wind turbine blade performance (power coefficient and power production) was assessed using Q-blade software.

1. Introduction

The rotor blades are the most important components of a wind turbine system. The rotor assembly consists of several blades (usually three blades) that are joined to a common hub, a nose cone and fasteners. Generally, the blades are designed to capture maximum energy from wind and convert it into rotating mechanical energy. To increase the reliability of a wind turbine system, two fundamental parameters must be considered: the aerodynamic performance of the rotor and the structural response of the blades under normal/extreme wind conditions. For example, after the overall shape of the blade is aerodynamically optimized, it is common that the blade root must be redesigned to meet structural requirements. Aerodynamically advantageous features such as nonlinear distribution of the chord length and twist angle are often difficult to build. For this reason, some compromises must be considered so that the blade design can be manufactured with minimum investment cost (low cost).

The efficiency of the turbine system is strongly dependent on the aerodynamic performance of the rotor blades. The section view of these components is basically defined by an airfoil shape (one or more airfoils) that is designed to generate high lift forces and low drag forces. In designing a small wind turbine blade for maximum power extraction, several design methodologies have been used in the literature. Dias do Rio Vaz et al. [1] have presented a model to optimize the distribution of the chord and the twist angle of Horizontal Axis Wind Turbine (HAWT) blades, taking into account the influence of the wake, and using a Rankine vortex. Wang L. et al. [2] have performed a practical methodology for an optimum blade design of Fixed Pitch Fixed Speed (FPFS) small wind turbines.

The optimum blade design is based on the aerodynamic characteristics of the airfoil which are the lift and drag coefficients, and the annual mean wind speed. The design parameters for this blade optimisation include design wind speed, design tip speed ratio, and design attack angle. A detailed review of the current state-of-art for wind turbine blade design has been presented by P.J. Schubel et al. [3]. This review provides a complete description of the wind turbine blade design methodologies of modern HAWTs. M. S. Campobasso et al. [4] have conducted an optimization strategy for the aerodynamic design of horizontal axis wind turbine rotors by including the variability of the annual energy production. In this study, the ultimate objective is to develop a conceptual design of a small wind turbine blade with an electrical generation of 11 kW. This is a direct drive, Fixed Pitch Variable Speed (FPVS) wind turbine for remote communities (to cover the individual electricity needs of a typical isolated household).

2. Rotor design parameters

The chord and the twist angle distributions in each blade section are strongly dependent on both the blade airfoil and the operating design parameters. Some of these key parameters are the design wind speed, the operating range of Reynolds numbers (low Reynolds numbers for small wind turbines), the design angle of attack, and the blade tip speed ratio. Once these parameters have been defined, the optimum rotor theory can be used to determine the optimum blade shape.

2.1. Design wind speed

The design wind speed is defined as the wind speed at which the wind turbine rotor spines with a maximum power coefficient [5]. According to the IEC61400-2 standard, the design wind speed, V_{design}, should be 1.4 times the Annual Mean Wind Speed, V_{AMWS} [6]. In the present study, a wind turbine class III was selected for the wind turbine design. This corresponds to a V_{AMWS} of 7.5 m/s. Therefore the design wind speed can be easily calculated using the following equation [7]:

$$V_{design} = 1.4\, V_{AMWS} = 10.5\, m/s \quad (1)$$

Given that the wind turbine is designed to operate at its maximum power coefficient (C_p), from cut-in speed to rated wind speed, the design wind speed is selected to be the same as the rated wind speed.

2.2. Design Reynolds number

The Reynolds number is defined as the ratio of the inertial force to the viscous forces of the air. This can be determined using the following equation [8]:

$$R_e = \frac{U_{rel} . C}{\mu} \quad (2)$$

Where μ is the kinematic viscosity of the air, U_{rel} is the relative wind velocity to the airfoil, and C is the chord length of the airfoil. By using a chord length of 0.25 m (an average value from the literature) at the middle section of an 11 kW wind turbine blade, the design Reynolds number can be obtained from the relative wind velocity [9]. For sufficiently high tip speed ratios, the relative wind speed can be approximated using the following formula [7]:

$$U_{rel} = \lambda_{0.5} . V_{design} \quad (3)$$

The tip speed ratio of the middle blade section, $\lambda_{0.5}$, can be determined using equation (4).

$$\lambda_{0.5} = \lambda . \frac{r}{R} \quad (4)$$

Where r is the blade radius along the radial direction and R is the rotor radius which is taken as 3.5 m. As the majority of small wind turbines are usually designed to achieve maximum efficiency at tip speed ratios between 6 to 8 [10], the corresponding Reynolds numbers are shown in Table 1.

Table 1. Reynolds numbers for different tip speed ratios

Tip speed ratio	C (m)	U_{rel} (m/s)	R_e
6	0.25	31.5	525 000
7	0.25	36.75	612 500
8	0.25	42	700 000

2.3. Airfoil selection

The airfoil selection is a critical step of the design process of a small wind turbine blade. The aerodynamic characteristics associated with low Reynolds number regime can severely degrade the performance of the blade if the selected airfoil is not suitable for low Reynolds number applications [11]. Most of the unusual aerodynamic characteristics at low Reynolds numbers can be justified by the presence of a laminar separation bubble on the airfoil. Several airfoils have been used in the small wind energy industry. These include the SG604x airfoil family that has been developed specifically for small variable-speed wind turbines sized from 1 to 5 kW [12], the S series airfoils (especially the S823 and the S822) that have been designed for small stall- regulated wind turbines ranging from 2 to 20 kW [13], the SD7062 airfoil which started its life as an aerofoil for model gliders, and the DU series airfoils (Delft University of Technology) which are popular in middle and high Reynolds wind turbine blades [14].

For the selection process of an airfoil, there is no single consistent database or airfoil selection method in the literature. For the present study, the selection process of the airfoil was conducted based on following criteria: the airfoil should be thick enough to provide a good reserve of strength in bending while maintaining good lift and drag characteristics. The DU 93-W-210 airfoil was selected. The DU 93-W-210 airfoil (thickness of 21%) is a thick airfoil providing more options for strengthening the blade structure. It has good lift to drag coefficient, at typical design Reynolds numbers for small wind turbines.

2.4. Design tip speed ratio

Blade tip speed ratio is a critical parameter to be considered in a blade design procedure. This parameter is chosen between 6 and 8, for modern wind turbine systems. The optimum value remains uncertain for different airfoil shapes and blade numbers. For an initial selection of the design tip speed ratio, an empirical relation between the power coefficient and the tip speed ratio was used [15]. This gives the maximum power coefficient of a wind turbine rotor with a finite number of blades.

$$C_{pmax} = \left(\frac{16}{27}\right).\lambda.\left[\lambda + \frac{1.32 + \left(\frac{\lambda - 8}{20}\right)^2}{B^{\frac{2}{3}}}\right]^{-1} - \frac{0.57.\lambda^2}{\frac{C_l}{C_d}.\left(\lambda + \frac{1}{2B}\right)} \quad (5)$$

This multivariable relationship can be summarized to a high degree of accuracy of 0.5% for tip speed ratios from 4 to 20, lift to drag ratios (Cl/Cd) from 25 to infinity, and from one to three blades (parameter B). Based on XFOIL solver, the maximum lift to drag coefficients for the selected DU93-W-210 airfoil is 93.76. This maximum value occurs at an attack angle of 5°. The plot of the power coefficients against different tip speed ratios is given in Figure 1.

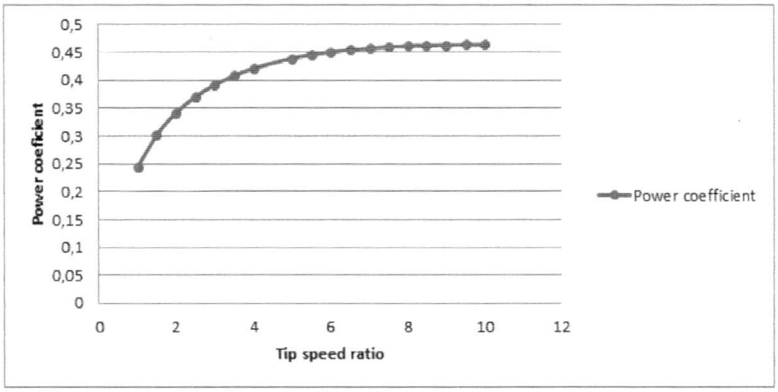

Figure 1. Variation of the power coefficient with the tip speed ratio

High tip speed ratios for small wind turbines presents many drawbacks such as noise generation in the audible and non-audible ranges, excessive rotations of the rotor, and problems of vibration. As the maximum efficiency does not vary sharply between 6 and 10, a value of 6 was selected as the design tip speed ratio in this study.

2.5. Design angle of attack

The design angle of attack is often selected at the critical angle of attack where the lift to drag ratio is maximum. However, the design angle of attack should be selected to be slightly lower than the optimum angle of attack (which gives the maximum lift to drag coefficient). This practice allows the blade to operate at angles of attack, near the optimum value while avoiding the airfoil to come rapidly into the stall regime when the wind speed increases. For the selected DU 93-W-210 airfoil, the plot of the lift and the drag coefficients against different angles of attack is shown in Figure 2.

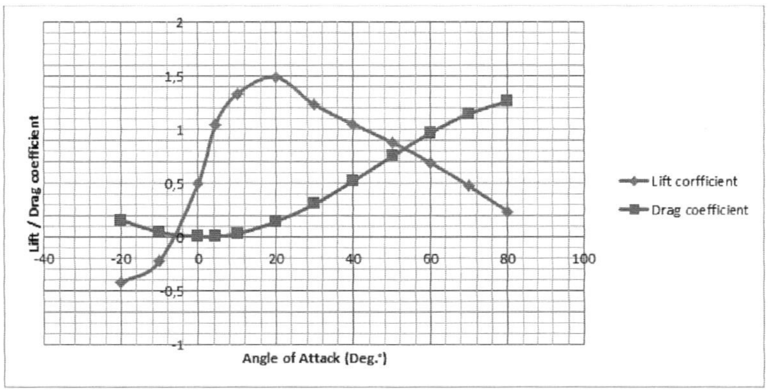

Figure 2. Lift and drag coefficients of DU 93-W-210 at Re=600,000

The maximum lift-to-drag ratio occurs at an angle of attack of 5°. However, the design angle of attack should be selected to be slightly lower than the optimum angle of attack (which gives the maximum lift to drag coefficient). Therefore, the design angle of attack, α_{design}, is chosen to be 4.5°.

3. Rotor blade design

For the aerodynamic design of the turbine blades, the optimum rotor theory (for blade shape design) and the BEM theory (for performance prediction) are widely used [16]. The BEM theory divides the blade into 30 sections from root to the tip and the total power coefficient is calculated by integrating the power coefficients at these sections, as described by J.F. Manwell et al. [17].

$$C_p = \left(\frac{8}{\lambda^2}\right).\int_{\lambda_h}^{\lambda} F_t.\sin^2\phi.(\cos\phi - \lambda_r \sin\phi).(\sin\phi + \lambda_r\cos\phi).\left[1 - \left(\frac{C_d}{C_l}\right).\cot\phi\right].\lambda_r^2 d_{\lambda r} \tag{6}$$

Where F_t is the blade tip losses factor. The most straightforward approach to calculate the blade tip losses is the one developed by Prandtl. This tip loss factor is an approved procedure for the correction of profile data to get a better agreement between measured and computed data. Based on Prandtl's method, the blade tip losses can be modelled using the following equation [18].

$$F_t = \left(\frac{2}{\pi}\right).\cos^{-1}\left[\exp\left(-\left\{\frac{\frac{B}{2}.\left(1-\frac{r}{R}\right)}{\frac{r}{R}.\sin\phi}\right\}\right)\right] \tag{7}$$

If the main part of the equation (6) is at its maximum, the total power coefficient is maximized.

$$\mathrm{Max}\left[Ft.\sin^2\phi.(\cos\phi - \lambda_r\sin\phi).(\sin\phi + \lambda_r\cos\phi).\left[1 - \left(\frac{C_d}{C_l}\right).\cot\phi\right].\lambda_r^2 d_{\lambda r}\right] \tag{8}$$

Ignoring the drag-to-lift coefficient ratio and setting the partial derivative of the main part to zero, the chord, C_i, and the twist angle, θ_i, can be obtained by the following equations [19]:

$$\lambda_{ri} = \lambda_{design}.\left(\frac{r_i}{R}\right) \tag{9}$$

$$\phi_i = \left(\frac{2}{3}\right).\tan^{-1}\left(\frac{1}{\lambda_{ri}}\right) \tag{10}$$

$$C_i = \frac{8.\pi.r_i}{B.C_{ldesign}}(1 - \cos\phi_i) \tag{11}$$

$$\theta_i = \phi_i - \alpha_{design} \tag{12}$$

Where i indicates the ith blade section, λ_{ri} is the speed ratio of the ith blade section, r_i is the distance from the ith blade section to the rotor center, ϕ_i is the angle of relative wind at the ith blade section. Cl_{design} and α_{design} are the design lift coefficient and the design angle of attack at the ith blade section, respectively. Figures 3 and 4 give the chord and twist angle distributions based on equations (9)-(12).

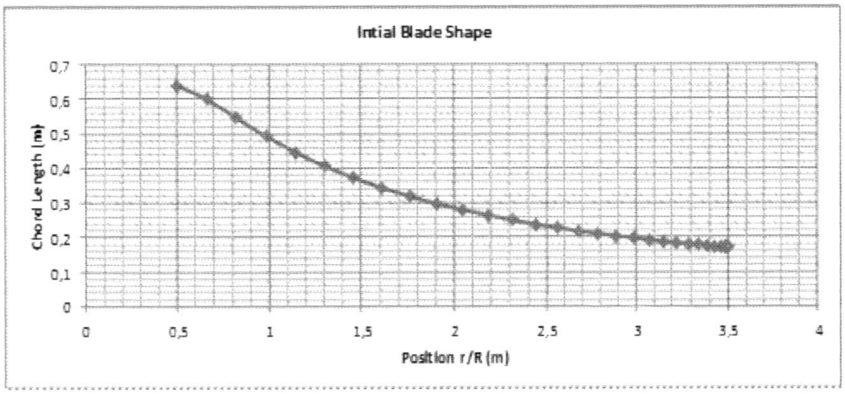

Figure 3. Blade chord length distribution

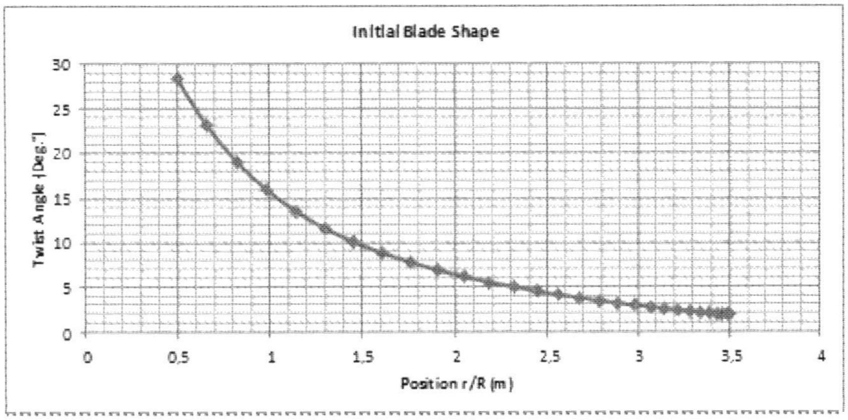

Figure 4. Blade twist angle distribution

The chord length is larger at the inner sections of the blade (closer to the root) and smaller at the outer sections of the blade. The twists of the sections close to the root are larger than those close to the tip. The chord at the blade root section (0.14 R) is around 0.62 m, and the twist angle varies from 28.43°, at the blade root to 1.8° at the blade tip. Figure 5 gives the blade geometry using the optimum rotor theory.

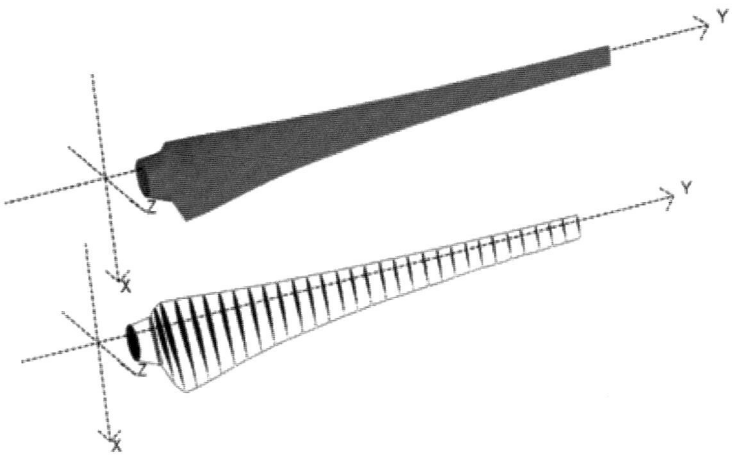

Figure 5. Optimum rotor blade shape

4. Wind turbine rotor performance analysis

The chord and twist distributions obtained from the above equations are just initial design values, and an iterative process should be followed normally. The most widely used procedure in this case is summarized in the following five steps [17,19]:

1) Estimate initial values of the axial induction factor, a, and angular induction factor, a'. The first values can be taken from the initial design blade:

$$\phi_{i,1} = \left(\frac{2}{3}\right)\tan^{-1}\left(\frac{1}{\lambda_{r,i}}\right) \quad (13)$$

$$a_{i,1} = \frac{1}{\left[1 + \frac{4\sin^2\phi_{i,1}}{\sigma'_{i,1}\, C_{L\,design}\cos\phi_{i,1}}\right]} \quad (14)$$

$$a'_{i,1} = \frac{1 - 3a_{i,1}}{4a_{i,1} - 1} \quad (15)$$

Where $\sigma'_{i,1}$ is the local solidity of the ith blade section expressed as:

$$\sigma'_{i,1} = \frac{BC_i}{2\pi r_i} \quad (16)$$

2) Start the iterative process for the jth. For the first iteration $j = 1$, the following equations are used:

$$\tan\phi_{i,j} = \frac{U(1 - a_{i,j})}{\Omega r(1 + a'_{i,j})} = \frac{1 - a_{i,j}}{(1 + a'_{i,j})\lambda_{ri}} \quad (17)$$

$$F_{ti,j} = \left(\frac{2}{\pi}\right).\cos^{-1}\left[\exp\left(-\left\{\frac{\frac{B}{2}\cdot\left(1 - \frac{r_i}{R}\right)}{\frac{r_i}{R}\cdot\sin\phi_{i,j}}\right\}\right)\right] \quad (18)$$

3) Calculate the new angle of attack, $\alpha_{i,j}$, and determine the corresponding lift coefficient, $C_{Li,j}$, and the drag coefficient, $C_{Di,j}$ from the polar data of the used airfoil.

$$\alpha_{i,j} = \phi_{i,j} - \theta_{i,j} \quad (19)$$

Where $\theta_{i,j}$, is the twist angle of the blade the ith blade section.

4) Calculate the local thrust coefficient by using the following equation:

$$C_{Thi,j} = \frac{\sigma'_i\left(1 - a_{i,1}\right)^2\left(C_{Li,j}\cos\phi_{i,j} + C_{Di,j}\sin\phi_{i,j}\right)}{\sin^2\phi_{i,j}} \quad (20)$$

5) Update the values of a and a' for the next iteration. If $C_{Thi,j} < 0.96$, then:

$$a_{i,j+1} = \frac{1}{\left[1 + \frac{4F_{ti,j}\sin^2\phi_{i,1}}{\sigma'_i C_{Li,j}\cos\phi_{i,j}}\right]} \tag{21}$$

If $C_{Thi,j} > 0.96$, then the axial induction factor is computed using the following equation:

$$a_{i,j+1} = \left(\frac{1}{F_{ti,j}}\right)\left[0.143 + \sqrt{0.0203 - 0.6427(0.889 - C_{Thi,j})}\right] \tag{22}$$

$$a'_{i,j+1} = \frac{1}{\frac{4F_{ti,j}\cos\phi_{i,j}}{\sigma'_i C_{Li,j}} - 1} \tag{23}$$

If the difference between the j^{i+1} and j^i iteration is within the acceptable convergence criteria, then the process converges, and the power performance can be computed. The convergence criterion, ε, defines when an iteration has converged. The maximum of the difference of axial and angular induction factor between the last and the current iteration has to be below ε. A typical value that can be used is given by [20]:

$$\max\left(|a_{i,j+1} - a_{i,j}|, |a'_{i,j+1} - a'_{i,j}|\right) < 10^{-5} \tag{24}$$

In this study, the iterative process in the blade design was conducted using Q-blade software. This design tool is realized being a part of the wind energy group at the Berlin Technical University Department of Experimental Fluid Mechanics. Q-blade is free-of-charge wind turbine calculation software that enables the user to rapidly design the turbine blades by determining the chord and the twist angle distribution. Q-blade software is based on BEM and Double-Multiple Streamtube (DMS) model. These are integrated into an XFOIL graphical user interface. Figures 6 and 7 give the chord length and the twist angle distribution along the blade span using both the optimum rotor theory and Q-blade software.

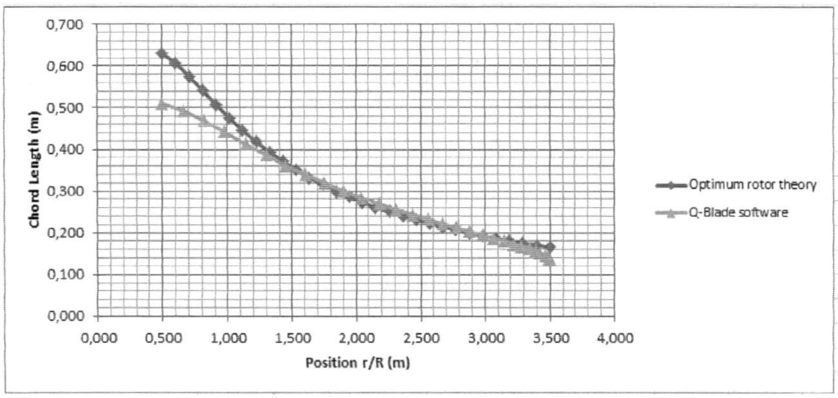

Figure 6. Optimized chord length distribution

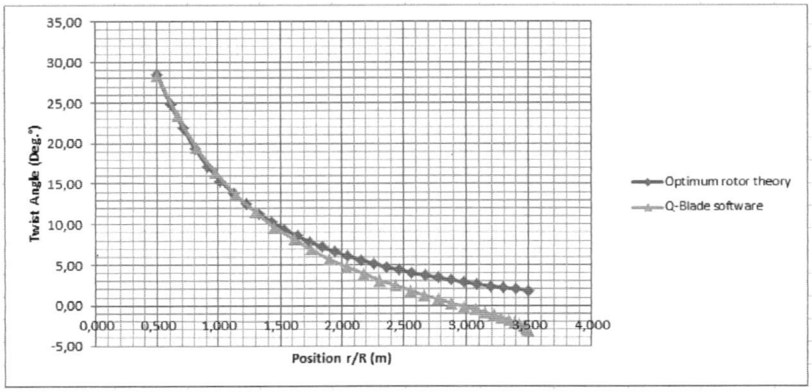

Figure 7. Optimized twist angle distribution

The chord distribution was optimized according to Schmitz method (equation 11). The twist angle distribution was selected to keep the lift to drag ratio at its maximum along the blade span. The evolution of the power coefficient as a function of the blade tip speed ratio is given in figure 8.

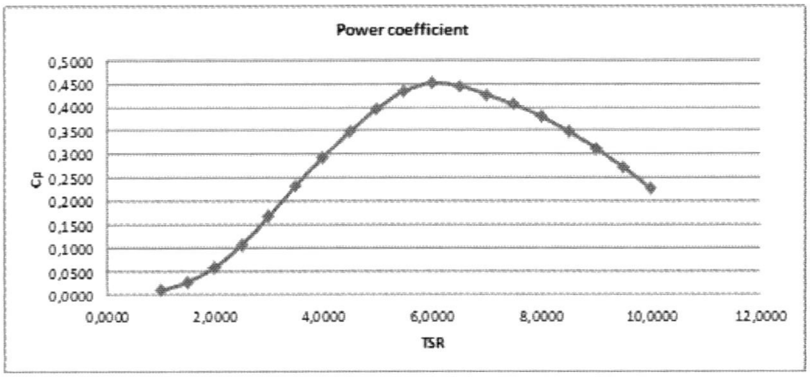

Figure 8. Power coefficient with the tip speed ratio from Q-blade

The maximum achievable power coefficient is 0.45, which corresponds to a tip speed ratio of 6. The value of C_p was computed for an ideal rotor by taking into account wake rotation, drag effects and tip losses (Prandtl correction factor). Figure 9 shows the power output of the wind turbine rotor using the optimized rotor blades.

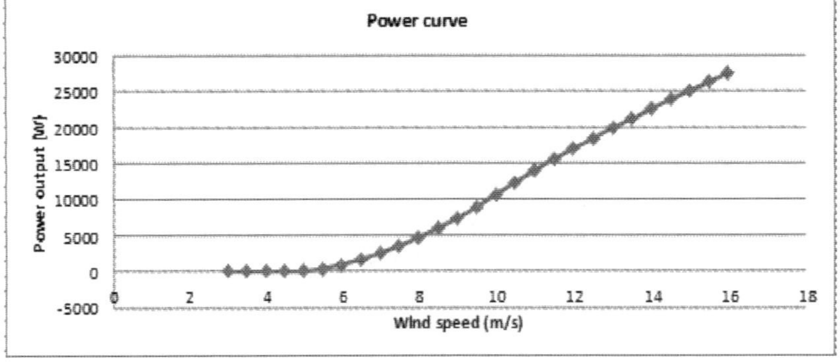

Figure 9. Power output from the small wind turbine rotor

From figure 9, three main conclusions can be drawn:
1) The power output from the turbine increases with the wind speed. At a design wind speed of 10.5 m/s, the rotor power generation is 12.3 kW. This high power production can be justified by the fact that the turbine simulation was performed for a rotor blades having a maximum achievable aerodynamic efficiency (C_p=0.45).
2) The starting time of the rotor is relatively high (5 m/s). This makes the wind turbine design not suitable for geographical areas where the mean wind speed is very low (the design cut-in wind speed is usually set to 3 m/s). For an optimum design of the rotor blades, starting time should be optimized too.
3) For relatively high wind speeds, the power production increases and exceeds the maximum overloading limit of the generator which is usually set around 120 %. This problem can be solved using an appropriate control strategy of the wind turbine: below the rated wind speed, the control system maximizes the energy production from the rotor, and above the rated wind speed, the control system limits the generated power using both passive stall regulation and a control algorithm (e.g. limitation of the rotational speed of the rotor using a damp load through a PWM modulation)).

5. Conclusion

The aim of this paper was to define an optimum blade shape, at a certain design tip speed ratio, for maximum power extraction. The proposed methodology does not take into account other constraints such as the manufacturing considerations, the blade mass and the blade starting performance. The nonlinear distribution of the chord length and the twist angle are often difficult to build. For this reasons, linearization of the blade geometry should be considered so that the blade can be manufactured with minimum cost. The starting capabilities of the blade must also be considered. Improving the starting performance would reduce the amount of time needed to reach the turbine rated power. This would increase the overall energy yield from the turbine. Additionally, the blade mass must be reduced as much as possible while maintaining the required structural requirements. This will significantly reduce the cost of the used blade material. Therefore, determination of an optimum blade shape is a nonlinear exercise that should be solved using multi-objective optimizations such as genetic algorithm.

Acknowledgments: This work was funded by the «Institut de Recherche en Énergie Solaire et Energies Nouvelles-IRESEN».

References

[1] Dias do Rio Vaz D, Pinheiro Vaz JR, Amarante Mesquita AL, Pinho JT, and Pinho Brasil Junior AC. 2013. Optimum aerodynamic design for wind turbine blade with a Rankine vortex wake. *Renewable Energy, vol.55:296-304*

[2] Wang L, Tang X, and Liu X. 2012. Blade Design Optimisation for Fixed-Pitch Fixed-Speed Wind Turbines. *International Scholarly Research Network; ISRN Renewable Energy, Volume 2012, Article ID 682859, 8 pages, doi:10.5402/2012/682859*

[3] Schubel PJ and Crossley RJ. 2012. Wind Turbine Blade Design. *Energies, vol.5: 3425-3449; doi: 10.3390/en5093425*

[4] Campobasso MS, Minisci E and Caboni M. 2014. Aerodynamic design optimization of wind turbine rotors under geometric uncertainty. *Wind Energy. Published online in Wiley Online Library (wileyonlinelibrary.com). DOI: 10.1002/we.1820*

[5] Thumthae C. 2015. Optimum Blade Profiles for a Variable-Speed Wind Turbine in Low Wind Area. *Energy Procedia, Vol. 75: 651 – 657*

[6] International Electrotechnical Commission, Wind turbines-Part 2: requirements for small wind turbines, *Ed.3.0 88/399/CD: 2011-05*

[7] Wood D. 2011. Small wind turbines. *Springer-Verlag*

[8] Mctavish S, Feszty D and Nitzsche F. 2013. Evaluating Reynolds number effects in small-scale wind turbine experiments. *Journal of Wind Engineering, vol. 120, pp. 81–90*

[9] Ashrafi ZN, Ghaderi M and Sedaghat A. 2015. Parametric study on off-design aerodynamic performance of a horizontal axis wind turbine blade and proposed pitch control. *Energy Conversion and Management, vol.93, 349–356*

[10] Tabesh A and Iravani R. 2006. Small signal dynamic model and analysis of a fixed-speed wind farm — a frequency response approach. *IEEE Transactions on Power Delivery (Volume: 21, Issue: 2)*

[11] Singh RK and Rafiuddin Ahmed M. 2013. Blade design and performance testing of a small wind turbine rotor for low wind speed applications. *Journal of Renewable Energy, Vol. 50, pp. 812-819*

[12] Giguere P and Selig MS. 1998. New airfoils for small horizontal axis wind turbines. *Department of Aeronautical and Astronautical Engineering, University of Illinois at Urbana-Champaign vol. 120, 1998.*

[13] Tangier JL and Somers DM. 1995. NREL airfoil families for HAWTs. *National Renewable Energy Laboratory, NREL/TP-442-7109, 1995.*

[14] Timmer WA and Van Rooij RPJOM. 2003. Summary of the Delft University Wind Turbine Dedicated Airfoils. *AIAA-2003-0352, Delft University Wind Energy Research Institute, Faculty of Civil Engineering and Geosciences*

[15] De Paor AM. 1982. Aerodynamic design of optimum wind turbine. *Applied Energy, vol. 12, pp.221–228.*

[16] Sun Z, Chen J, Shen WZ, Zhu WJ. 2016. Improved blade element momentum theory for wind turbine aerodynamic computations. *Renewable Energy, vol. 96, pp. 824–831, 2016.*

[17] Manwell JF, McGowan JG, Rogers AL. Wind energy explained: theory, design and application. *2nd edn. Wiley, Chichester*

[18] Wood DH, Okulov VL and Bhattacharjee D. 2016. Direct calculation of wind turbine tip loss. *Renewable Energy, vol. 95, pp. 269–276, 2016.*

[19] Liu X, Wang L and Tang X. 2013. Optimized linearization of chord and twist angle profiles for fixed-pitch fixed-speed wind turbine blades. *Renewable Energy, vol. 57, pp. 111–119, 2013.*

[20] Marten D, Wendler J. 2013. QBlade Guidelines, v0.6 – software *[available at: http://www.q-blade.org]*

IOP Conf. Series: Earth and Environmental Science **161** (2018) 012009 doi:10.1088/1755-1315/161/1/012009

Removal of textile dyes from aqueous solutions using low cost Moroccan clay

Z Benchegroun[1], Z Chaouki[1], M Hadri[1,2], M Nawdali[3], K Draoui[2], H Valdés[4] and H Zaitan[1*]

[1]Laboratory LCMC, Faculty of Science and Technology, Sidi Mohamed Ben Abdellah University, B.P. 2202, Fez, Morocco
[2]Laboratory MSI, Faculty of Sciences, Abdel Malek Essaadi University, B.P. 2121, M'hannech II, 93002, Tetouan, Morocco
[3]Laboratory LCMC, Polydisciplinary Faculty, Sidi Mohamed Ben Abdellah University, B.P 1223, Taza, Morocco.
[4]Laboratorio de Tecnologías Limpias, Universidad Católica de la Santísima Concepción, Alonso de Ribera 2850, Concepción, Chile

*Email: hicham.zaitan@usmba.ac.ma

Abstract. This study aims to evaluate the technical feasibility to apply natural clay from Fez area as a low cost alternative adsorbent to remove dyes from contaminated waters. Methylene blue (MB) and congo red (CR) are used here as representatives of cationic and anionic dyes normally present in wastewater from textile industry. This material was characterised by different physical-chemical methods, including BET, X-ray fluorescence and pH_{PZC}. Additionally, the influence of operating conditions such as contact time, adsorbent dosages and pH were evaluated. Experimental results show that the adsorption processes takes place very rapid, reaching the equilibrium at 30 and 45 min for MB and CR, respectively. Maximum adsorption capacities result to be pH-depends. Hence, MB adsorption is favoured under basic pH conditions, whereas CR does it at acid pH. A pseudo-second-order kinetic model provides the best fit to the experimental data of MB and CR adsorption. Adsorption data are better described by Langmuir and Toth isotherm models. Adsorption occurs with a maximum monolayer adsorption capacity of 195 and 185 mg g^{-1} for MB and CR, respectively. Experimental results indicate that the Moroccan clay could be used as a potential adsorbent for the elimination of dyes from contaminated waters, at a lower cost.

1. Introduction

Currently, a great number of dyes are used in paper-making, textile, food, and pharmaceutical industries [1]. The presence of synthetic dyes in the environment poses serious environmental and health problems, because most of them are toxic, mutagenic and carcinogenic [2]. Such pollutants must be removed from wastewater streams to acceptable levels by appropriate methods. Adsorption is widely used process in the removal of dyes from contaminated waters due to its low cost and operational simplicity [3–4].

Activated carbons have been a common choice as adsorbents in many industrial processes; since they exhibit great affinity toward organic compounds. However, the use of activated carbons as adsorbents is limited by their high operating costs, pore blocking and safety issues. Synthetic zeolites overcome such drawbacks but their high acquisition cost limits their application at industrial scale (10 folds more expensive than activated carbons) [5-6]. In this perspective, effective and low cost alternative adsorbents for the removal of dyes have to be found.

In this regard, natural clays could be used as low cost adsorbents to clean contaminated waters from toxic dyes. In this study, the feasibility to apply inexpensive materials like Moroccan clay as adsorbent to remove two types of dyes, methylene blue (MB) and congo red (CR), from aqueous solutions is investigated. Moroccan clay is physically and chemically characterised. Moreover, the effect of different adsorption parameters such as pH, initial dye concentration, contact time and adsorbent dosage are evaluated. Additionally, kinetic and equilibrium models are applied to experimental data and the best models are selected.

2. Materials and methods

2.1 Materials

Natural clay was obtained from an extraction site situated in the nearness to Fez city, Morocco (named here as CFS) and was used without any previous activation. It was sieved into a range of 60-100 μm and washed with deionised water. Then, it was dried at 383 K for 24 h and stored in a dessicator until further use.

Methylene blue (MB; $C_{16}H_{18}ClN_3S$; Mw = 319.85 g mol^{-1}; HiMedia Laboratories, Mumbai, India) and congo red (CR, $C_{32}H_{22}N_6Na_2O_6S_2$, Mw = 696.66 g mol^{-1}; Sigma-Aldrich) were chosen as model dyes for adsorption experiments. Dyes of commercial purity were used without any further purification; their chemical structures are given in Figure 1. The pH was adjusted to a given value by addition of HCl (1N) or NaOH (1N) and was measured using a JENCO – Electrocinics LTD pH-Meter.

(a)

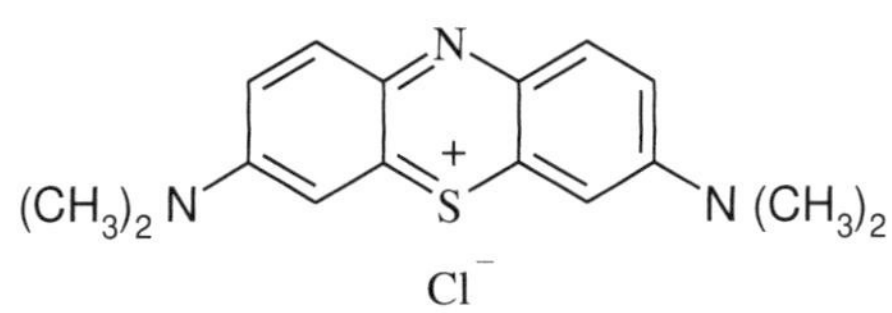

(b)

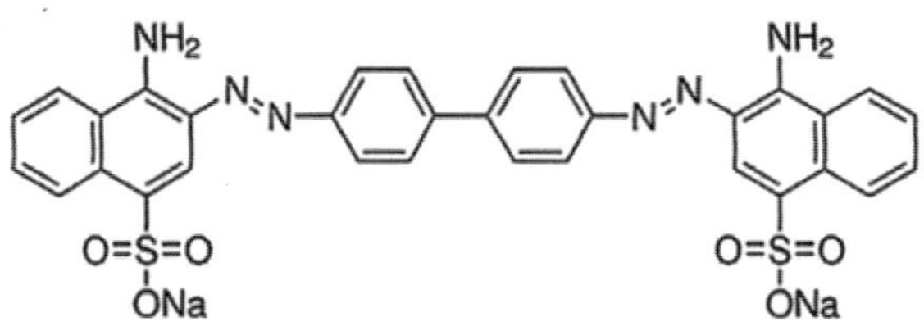

Figure 1. Chemical structures: (a) MB, and (b) CR.

2.2 Physical-chemical characterisation of the Moroccan clay

Specific surface area and pore volume of the CFS clay were obtained from N_2 adsorption isotherms at 77 K, using a Micromeritics Asap 2010 (Micromeritics Corporate Headquarters, U.S.A). Mineralogical composition was determined by X-ray diffraction analysis using X'Pert PRO Philips diffractometer (Philips Japan, Ltd, Japan), equipped with CuKα radiation (λ= 1.5406 Å). Bulk chemical composition was determined using a X-ray fluorescence (XRF) in a Bruker S4 Pioneer spectrometer (Bruker AXS GmbH Karlsruhe, Germany).

Basic and acid surface sites were determined by the titration method [7]. A mass of 1.5 g CFS sample was placed into vials and 40 mL of solutions of NaOH (0.1N) and HCl (0.1 N) were poured Vials were sealed and stirred during 72 h and then filtered. A volume of 10 mL of each filtrate was back-titrated using HCl (0.1 N) or NaOH (0.1 N), respectively. The number of acidic sites was determined under the assumption that NaOH neutralises acid sites. The number of basic sites was calculated from the amount of HCl that reacted with basic surface sites of present on CFS.

The pH of the point of zero charge (pH_{pzc}), was determined using the acidimetric–alkalimetric titration method [8]. Vials were filled with 20 mL of NaCl (0.1 N) and 0.1 g of CFS was placed into each vial. Different amount of NaOH (0.1N) and HCl (0.1 N) were added to each vial, varying the pH range and one was kept without any base or acid. Then, the suspensions were shaken and the equilibrium pH was measured by using a pH-meter (JENCO – Electrocinics LTD).

2.3 Adsorption experiments

Dye adsorption experiments were carried out using the batch method. A mass of 10 mg of CFS was dispersed in 20 mL solution with an initial dye concentration in the range from 0 to 1000 mg L^{-1} at pH= 7. The mixture was shaken (~200 rpm) until the equilibrium was reached using a water shaker bath. After that, the solid phase was separated from the liquid phase by centrifugation (5000 rpm during 10 min). Finally, the remaining concentration of MB and CR were determined by measuring the absorbance at 664 nm and at 500 nm, respectively by using a spectrophotometer (UV 1600PC)

The quantity of adsorbed dye at the equilibrium per gram of CFS (q_e, mg g^{-1}) and the amount of dye adsorbed per gram of CFS at specific contact time t (q_t, mg g^{-1}) were calculated using Eq. (1):

$$q_{t,e} = \frac{(C_0 - C_{t,e})V}{m} \quad (1)$$

where C_0 (mg L^{-1}) and $C_{t,e}$ (mg L^{-1}) are the dye concentration at the beginning and at equilibrium or specific time, respectively. m is the mass of adsorbent (g), and V is the total volume of liquid (L).

The influence of operating conditions on the adsorption of dye onto CFS clay was studied in a batch system by varying the CFS dosage from 1 to 35 g.L^{-1}, the initial dye concentration from 10 to 1000 mg L^{-1} and the pH of the dye solution in the range of 2-12. The pH was adjusted to a given value by the addition of HCl or NaOH.

3. Results and discussion

3.1. Characterisation of CFS clay

Figure 1 displays N_2 adsorption-desorption data of CFS clay. Results show a type IV isotherm according to the IUPAC classification. The initial part of the isotherm ($P/P_0<0.4$) could be attributed to physical adsorption. At a high relative pressure range, a clear hysteresis loop is observed that could be related to capillary condensation, taking place in mesopores structures formed between the elementary clay particles named tactoids.

The CFS clay used in this study has a specific surface area of 29 m^2 g^{-1}, with a highly developed microporous structure. Average pore diameter is around 2.1 nm. A summary of physical-chemical properties of CFS are listed in Table 1.

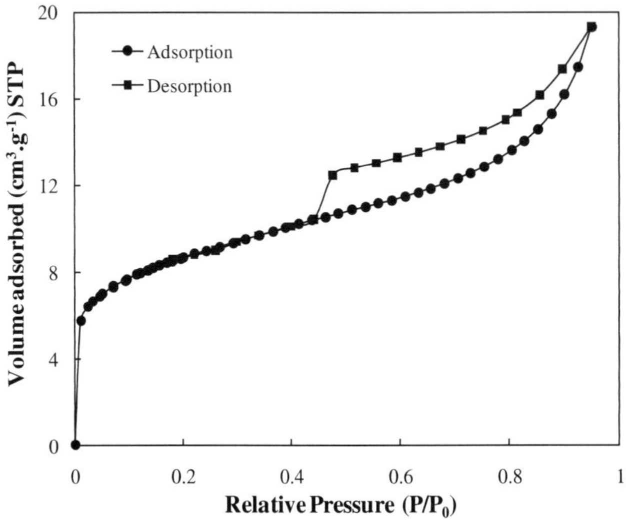

Figure 2. N_2 adsorption-desorption isotherm for CFS clay.

It can be observed that the amount of basic sites is significantly higher than the amount of acidic sites. These results suggest that CFS clay has a basic character. These results are in agreement with the measured value of pHpzc. When the pH of the solution is higher than the pHpzc, CFS surface will be negatively charged. In the pH region below pHpzc, positive charges will dominated on CFS surface.

Table 1: Physical–chemical properties of CFS clay.

Property	**CFS clay**
Origin	Fez area (Morocco)
Colour	grey
Apparent density (g dm^{-3})	700
Specific surface area S_{BET} (m^2 g^{-1})	29
Total pore volume V_T (cm^3 g^{-1})	0.03
Mesopore volume V_{meso} (cm^3 g^{-1})	0.025
Micropore volume V_{micro} (cm^3 g^{-1})	0.005
Average pore diameter D_p (nm)	2.1
SiO_2/Al_2O_3 (mol mol^{-1})	5.71
pH_{PZC}	9.4
Total acidic sites (mmol g^{-1})	0.32
Total basic sites (mmol g^{-1})	2.67

Figure 3 shows the mineralogical composition of CFS clay. On one hand, results reveal a complex heterogeneous structure, mainly composed of quartz (Q) combined with calcite (C). On the other hand, X-ray fluorescence (XRF) results reveal its chemical composition, being mainly composed by SiO_2 (41.28 %), Al_2O_3 (12.29 %), CaO (14.23%), Fe_2O_3 (4.13 %) and by traces of other oxides such as MgO, K_2O, Na_2O, P_2O_5, TiO_2. with a SiO_2/Al_2O_3 of 3.4.

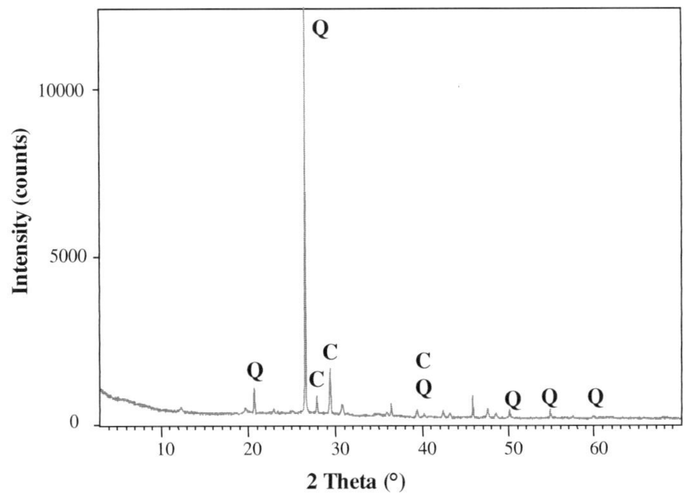

Figure 3. X-ray diffraction of CFS clay: (Q) quartz, (C) calcite.

3.2. Effect of adsorbent dosage

Figure 4 shows the results of the influence of different dosage of CFS clay (0-20 g.L^{-1}) on the removal of MB and CR dyes, using 20 mL of dye solution (500 mg L^{-1}) at 25 °C during 1.5 h.

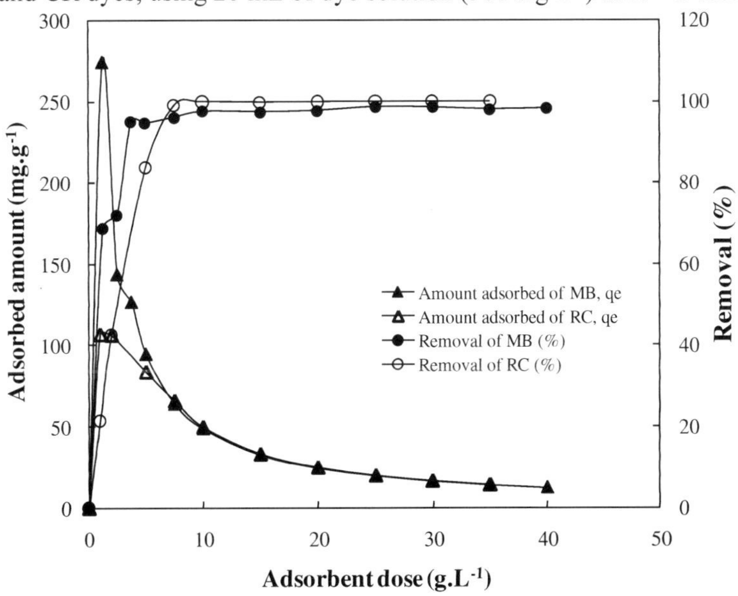

Figure 4. Effect of adsorbent dosage on MB and CR removal. Experimental conditions: 500 mg L^{-1} of initial concentration of MB or CR, 0–40 g.L^{-1} of CFS, 24 h of contact time, 200 rpm at 25 °C.

As it can be seen in Figure 4, dye removal efficiencies increase as the adsorbent dosage increases, which is related to the increase on the availability of adsorption sites on CFS surface. The optimum adsorbent dosages toward MB and CR result to be 7.5 g.L^{-1} and 5 g.L^{-1}, respectively. Removal efficiencies for MB and CR reach values up to 99.9% and 98.6%, respectively.

IOP Conf. Series: Earth and Environmental Science **161** (2018) 012009 doi:10.1088/1755-1315/161/1/012009

3.3 Effect of pH

Results of the effect of pH values (2-12) on the adsorption of MB and CR are illustrated in Figure 5. The initial solution pH was adjusted by the addition of HCl or NaOH solutions. Results show that MB adsorption onto CFS clay is clearly pH-dependant. MB adsorption increases at pH conditions higher than the pHPZC. Maximum adsorption capacity toward MB, qe, increases from 88 to 119.9 mg g-1, when the solution pH increases from 2 to 12. Such results may be due to the fact that at pH > pHPZC CFS surface becomes negative charged, enhancing electrostatic attractions with positively charged MB cations [9].

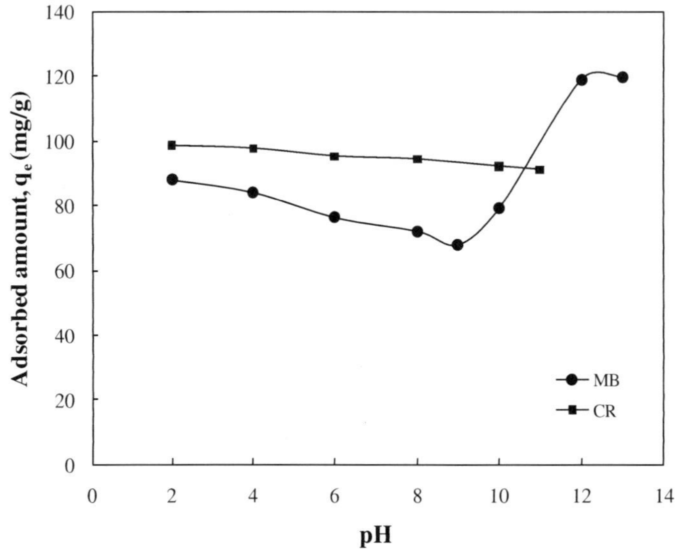

Figure 5. Effect of pH on the adsorption of MB and CR onto CFS: Initial conc.: 500 mg.L^{-1}, contact time= 6 h, R= 5 g.L^{-1} for MB and CR, agitation speed= 200 rpm and T = 25°C.

However, as pH decreases, CFS surface increases its number of sites positively charged; leading to a decrease on the adsorption capacity toward MB. Repulsion forces among a positively charged surface and MB cationic molecules could be responsible of such results. In the case of the adsorption of CR, adsorption capacities shows a maximum in the pH range between 2 and 4 that could be related to an increase on electrostatic attraction forces among positively charged sites of CFS surface and CR anionic molecules.

3.4 Adsorption kinetics

Adsorption kinetics provides important information for designing batch adsorption systems. Information on the kinetics of solute uptake is required for selecting optimum operating conditions for full-scale batch process. The effect of contact time on the adsorption of dyes using CFS clay is shown in Fig. 6. The rate of MB and CR adsorption onto CFS clay was studied by measuring the adsorbed amount as a function of time, keeping constant the volume (V= 200 mL), the initial concentration (C_0= 200 mg L^{-1}) and the mass of adsorbent per liter of aqueous solution (R= 7.5 g L^{-1} for MB and R= 5 g L^{-1} for CR).

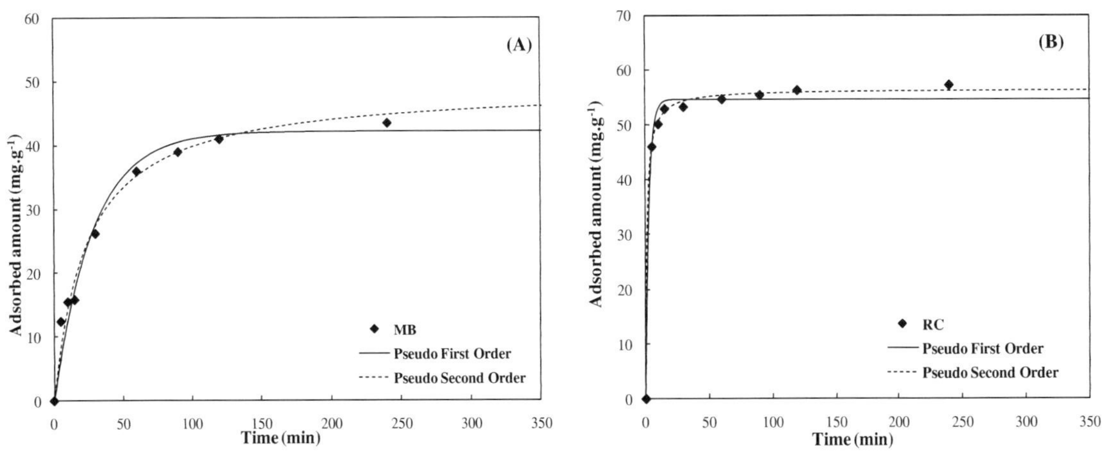

Figure 6. Effect of contact time on the adsorption of dyes onto CFS clay: (A) MB and (B) CR. Experimental conditions: initial dye concentration of 200 mg L^{-1}, R= 7.5 g L^{-1} for MB and 5 g L^{-1} for CR, 200 rpm, at 25 $^{\circ}$C.

Experimental data show that the adsorption capacity of CFS toward MB and CR increases sharply during the first minutes of the process. Then, it increases slower over time, until equilibrium is reached at 120 min. These results suggest that the adsorption CR and MB onto CFS clay is a fast process. The fast adsorption rates imply strong interactions among the active adsorption sites of CFS surface and the dye molecules. Adsorption data fit very well pseudo-first-order (Eq. 2) and pseudo-second order (Eq. 3) kinetic models, representing the kinetic and mechanism of solid-liquid adsorption interactions [10-11].

$$q_t = q_e(1- e^{-k_1 t}) \quad (2)$$

$$q_t = \frac{k_2 q_e^2 t}{1+k_2 q_e t} \quad (3)$$

where q_t and q_e (mg g^{-1}) are the adsorption capacity at any time t and at the equilibrium, respectively. k_1 (min^{-1}), k_2 (g. $mg^{-1}.min^{-1}$) are the adsorption rate constants for the pseudo-first order and pseudo-second order kinetic models, respectively.

Kinetic parameters are summarised in Table 2. Pseudo-second-order kinetic model gives a better fit to the adsorption data of MB and CR onto CFS clay, with higher values of the correlation coefficient (R^2>0.98). Additionally, q_e values determined using the pseudo-second-order kinetic model are more in agreement with the experimentally values (q_{exp}) than those obtained with the pseudo-first-order model. Such results suggest that chemical interactions are responsible for the adsorption of MB and CR onto CFS clay [12].

Table 2: Parameters of the kinetic models for the adsorption of MB and CR onto CFS clay.

Dyes	$q_{e,\ exp}$ (mg g^{-1})	Pseudo-first-order			Pseudo-second-order		
		$q_{e,cal}$ (mg g^{-1})	k_1 (min^{-1})	R^2	$q_{e,cal}$ (mg g^{-1})	k_2 (g mg^{-1} min^{-1})	R^2
MB	49	42.24	0.036	0.96	49.21	0.0008	0.98
CR	57	54.73	0.432	0.99	56.4	0.015	0.99

3.5 Adsorption isotherm

Adsorption isotherms indicate how adsorbate molecules are distributed between the liquid and solid phases when the equilibrium is reached [13]. Experimental data fit to different adsorption isotherm models is an important step to determine the most suitable model.

Figure 7 represents the total amount of adsorbed dyes onto CFS clay at the equilibrium (q_e). Three adsorption isotherms are fitted to the experimental data, including Langmuir (Eq. 4), Freundlich (Eq. 5) and Toth (Eq. 6) isotherms [14-16], as follows:

$$q_e = \frac{q_m K_L C_e}{1+K_L C_e} \tag{4}$$

$$q_e = K_F\, C_e^{1/n} \tag{5}$$

$$q_e = \frac{q_{mT} K_T C_e}{(1+(K_T C_e)^t)^{\frac{1}{t}}} \tag{6}$$

where K_L (L.mg^{-1}) is the Langmuir equilibrium constant related to the energy of adsorption. q_m and q_e (mg.g^{-1}) are the maximum and the equilibrium adsorption capacities of the Langmuir model, respectively. C_e (mg. L^{-1}) is the equilibrium concentration of the adsorbate. K_F (L.mg^{-1}) is the Freundlich constant and $1/n$ is the heterogeneity factor. q_{mT} (mg.g^{-1}) is the Toth maximum adsorption capacity. K_T is the Toth equilibrium constant, and t is the Toth model exponent.

The dimensionless constant of Langmuir isotherm model (equilibrium parameter, R_L) for the adsorption process of MB and CR dyes is calculated from Eq (7). R_L values is useful to understand the nature of the adsorption process. All the calculated R_L values are higher than zero and lower than the unity ($0<R_L< 1$) which reflects favourable adsorption of MB and CR onto CFS clay.

$$R_L = \frac{1}{(1+K_L C_0)} \tag{7}$$

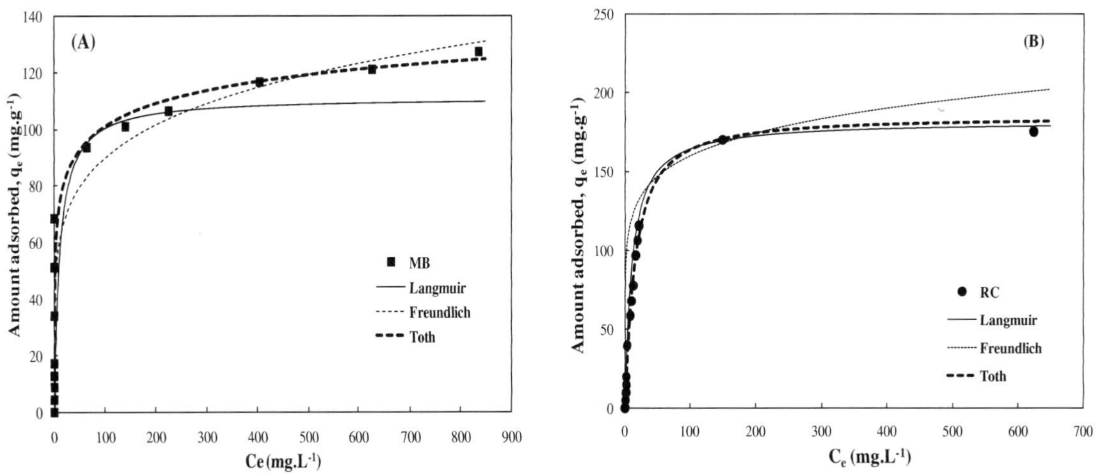

Figure 7. Equilibrium data and non-linear fitted curves isotherms: (A) MB and (B) CR onto CFS adsorbent at 25°C.

Parameters of Langmuir, Freundlich and Toth isotherm models and their respective correlation coefficients are listed in Table 3. As it can be seen, the obtained values of R^2 of the Toth and Langmuir isotherm models (R^2~0.98) indicate a better fit to the experimental data. Toth isotherm model has been developed from the potential theory in order to enhance the outcomes of Langmuir isotherm and be able to be used in heterogeneous adsorption systems. The adsorption of CR dye is well represented by the Langmuir isotherm model, followed by Toth isotherm model. Such results could indicate that the adsorption of CR onto CFS clay could occur forming a monolayer. On the other hand, the adsorption of MB is very well fitted to Toth isotherm model, indicating that different kind of interactions take places.

Table 3: Langmuir, Freundlich and Toth parameters for MB and RC adsorption onto CFS clay.

Models	Isotherm parameters	Dyes	
		MB	CR
Langmuir	q_m (mg g^{-1})	117.65	181.8
	K_L (L mg^{-1})	0.093	0.093
	R^2	0.90	0.99
	R_L	0.007-0.26	0.01-0.3
Freundlich	K_F (mg g^{-1}) (L mg^{-1})$^{1/n}$	53.11	90.01
	n	7.73	8
	R^2	0.89	0.96
Toth	q_{mT} (mg g^{-1})	195	185
	K_T (L mg^{-1})	414	0.056
	t	0.19	1.12
	R^2	0.98	0.99

Maximum adsorption capacities predicted by Toth isotherm for MB and CR onto CFS clay are 195 and 185 mg.g^{-1}, respectively and are compared here to maximum monolayer adsorption capacities of different types of adsorbents used in the removal of such dyes (see data presented in Table 4). It can be observed that the obtained values of q_m in the present study are higher than those reported in the literature using different kind of adsorbents.

Table 4: Comparison of the values of maximum adsorption capacity of different kinds of adsorbents toward MB and CR.

Adsorbent	Adsorption capacity (mg g^{-1})		References
	MB	CR	
CFS clay	195	185	Present work
Banana peel	20.80	--	[17]
Rice husk	40.60	--	[17]
D. harra	185.59	--	[18]
G. coronaria L.	258.76	--	[18]
Natural ilittic clay	24.87	--	[19]
Kaolin	--	5.44	[20]
Bagasse fly ash	--	11.89	[21]
Montmorillonite	--	12.70	[22]
Ca-bentonite	--	107.41	[23]
Mesoporous activated carbon	--	189	[24]

4. Conclusions

Results obtained in this study show that natural clay from Fez area (CFS) can be used as a low cost adsorbent to remove cationic and anionic dyes from polluted waters. Kinetic studies reveal that an equilibrium time of 120 min is achieved in the adsorption of methylene blue and Congo red dyes. The adsorption processes are very well represented by a pseudo-second-order kinetic model. Congo red adsorption equilibrium data are better described by Langmuir model, suggesting the formation of a monolayer onto the surface of CFS clay; whereas methylene blue adsorption data are depicted by Toth model, indicating that different kinds of interactions are involved. Results also show that the adsorption capacity of CFS clay is highly controlled by the pH values. Maximum removal efficiency toward MB is obtained at alkaline conditions; while better performances for CR uptake are attained at acidic conditions. Natural clays seem to be promising candidates for fast removal of ionic pollutants.

References

[1] Pokhrel D and Viraraghavan T 2004 *Sci. Total Environ.* **333** 37–58.
[2] O'Neill C, Hawkes F, Hawkes D, Lourenco N, Pinheiro H and Delee W 1999 *J. Chem. Tech. Biotech.* **74** 1009-1018.
[3] Jain A K, Gupta V K, Bhatnagar A and Suhas 2003 *J. Hazard. Mater.* **101** 31–42.
[4] Ho Y S and McKay G 2003 *Process Biochem.* **38(7)** 1047–1061.
[5] Elmoubarki R, Mahjoubi FZ, Tounsadi H, Moustadraf J, Abdennouri M, Zouhri A, El Albani A and Barka N 2015 *Wat. Resour. Ind.* **9** 16–29.
[6] Hadri M, Chaouki Z, Draoui K, Nawdali M, Barhoun A, Valdés H, Drouiche N and Zaitan H 2017 *Desal. Water Treat.* **75** 213–224
[7] Boehm HP 1994 *Carbon* **32** 759-764.
[8] Noh JS and Schwarz JA 1989 *J. Colloid Interface Sci.* **130** 157–164.
[9] Ozcan A, Omeroğlu C, Erdoğan Y, Ozcan AS 2007 *J. Hazard. Mater.* **140** 173–179.
[10] S. Lagergren 1898 *Kungliga Svenska Vetenskapsakademiens. Handlingar* **24(4)** 1–39.
[11] Ho YS, McKay G, Wase D A J and Forster CF 2000 *Adsorp. Sci. Technol.* **18** 639–650.
[12] Malash GF and E K Mie 2010 *J Colloid Interf Sci.* **348(2)** 537–545.
[13] Fu J, Chen Z, Wang M, Liu S, Zhang J, Zhang J, Han R and Xu Q 2015 *Chem Eng J.* **259** 53–61.
[14] Langmuir I 1918 *J. Am. Chem. Soc.* **40(9)** 1361-1403.
[15] Freundlich H M F 1906 *Z. Phys. Chem.* **57** 385–470.
[16] Toth J 1971 *Acta Chem. Acad. Hung.* **69** 311–317.
[17] Annadurai G, Juang R and Lee D 2002 *J. Hazard. Mater.* **92** 263–274.
[18] Tounsadi H, Khalidi A, Abdennouri M and Barka N 2016 *Desal. Water Treat.* **57** 16633–16642.
[19] Ozdes D, Duran C, Senturk H. B, Avan H and Bicer B 2014 *Desal. Water Treat.* **52 (1-3)** 208-218.
[20] Vimonses V, Lei S, Jin B, Chow C, and Saint C 2009 *Chem. Eng. J* **148** 354–364.
[21] Mall I, Srivastava V, Agarwal N and Mishra I 2005 *Chemosphere* **61** 492–501.
[22] Wang L and Wang A 2007 *J. Hazard. Mater.* **147** 979–985.
[23] Lian L, Guo L and Guo C 2009 *J. Hazard. Mater* **161** 126–131
[24] Grabowska E and Gryglewicz G *2007 Dyes Pigments **74** 34–40.*

Acknowledgment
Authors are gratefully for the financial support provided from the Ministry of Environment Morocco (Project DE-LIX).

IOP Conf. Series: Earth and Environmental Science **161** (2018) 012010 doi:10.1088/1755-1315/161/1/012010

Assessment of a quality control procedure of hourly solar irradiations at Fez city, Morocco

Z BOUNOUA[1], M MARZOUQ[2] and A MECHAQRANE[1]

[1] Laboratory of Renewable Energies and Intelligent System Electrical Engineering department, Faculty of Sciences and Technologies, Sidi Mohamed Ben Abdellah University of Fez PO. Box 2202 Fez Morocco

[2] Laboratory of Computer Sciences and Interdisciplinary Physics, National School of Applied Sciences, Sidi Mohamed Ben Abdellah University –Fez

E-mail: **zineb.bounoua@usmba.ac.ma**

Abstract. In order to properly dimension solar power plants, it is necessary to have long series of relevant measurements of the available solar components at the site chosen for the installation. It is, however, well known that the quality of the measurements of solar irradiations can fail from time to time. Therefore, a quality control of the recorded data must be carried out before their use. This paper describes an assessment of a quality control procedure which is implemented for the hourly horizontal global and diffuse solar irradiations. A set of four quality control tests are applied to remove the erroneous data.

1. Introduction

The knowledge of the solar irradiation received by a surface of the earth is important for the use and good management of this energy in several fields of research. The measurement of the global solar irradiation on a horizontal plane is carried out by using a pyranometer for a defined time step. Obviously, pyranometer measurements are often sensitive to errors which may be due to technical or operation and maintenance problems of the equipment [1, 2]. It is for these reasons that the quality control of solar measurements has become a concern of most researcher's scholars who have proposed different methods to ensure good reliability of the data. In the case of data representing a time step of one hour (our case), errors due to the cosine response are more relevant [2, 3]. The ENDORSE project [4] proposed a quality control procedure (QCP) based on extreme and rare observations, considering the three components of solar irradiance: Global Horizontal Irradiance (GHI), Diffuse Horizontal Irradiance (DHI) and Direct Normal Irradiance (DNI). This QCP will only be applied for solar altitude data below 7° to process the intrinsic cosine error of the pyranometers. Younes et al. [2] developed a ten-minute and hourly irradiations quality control procedure that went through four different levels of testing. They proposed physical and statistical indices to develop a semi-automated procedure based on the clearness and diffuse indexes (K_t and K_d respectively).

In the present work, we applied the quality control procedure of Younes et al. [2] to our solar irradiation measurements at Fez city in Morocco.

2. DATABASE

Our work consists of a quality control of the global and diffuse hourly horizontal solar irradiation measurements. These measurements are made using two identical Kip & Zonen pyranometers (model CM11) whose one is equipped with a shading ring to measure the DHI. These equipments are part of a meteorological station placed on the roof of the Faculty of Sciences and Technology of Fez (Latitude: 33°59'58"N, longitude: 4°59'22"W and altitude: 450 m). The measurement period is from 01/01/2009 00:00:00 GMT to 11/06/2015 at 12: 00:00 GMT. Figure1 shows the hourly variations of the measured GHI and DHI for the whole considered period. Given that the diffuse solar irradiance is measured by a pyranometer equipped with a shadow band, we applied the correction procedure proposed by Drummond [5] to take into account the part of diffuse component blocked by the shadow band. To calculate solar position at each instant, we use the Solar Position Algorithm (SPA) [6]. SPA is an algorithm based on precise astronomical equations allowing to calculate the solar position with a very low uncertainty, it was developed at the National Renewable Energy Laboratory (NREL) of the U.S. Department of Energy to calculate the sun's zenith and azimuth angles and other related parameters with unmatched low uncertainty of +/- 0.0003 degrees, in the period of years from -2000 to 6000 [7].

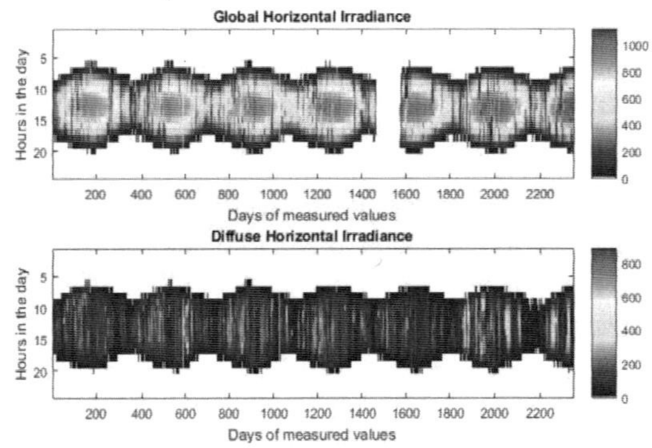

Figure 1. Hourly variations of the measured GHI and DHI for the whole considered period

Figure 1 shows that:

- There is a significant data hole in the GHI measurements due to a problem in the connection cable between the pyranometer and the acquisition chain.
- There are significant values (red color) of DHI.

The second problem is due to the fact that the solar shadow band suffers from misalignments from time to time. This can be seen in figure 2 illustrating the scatter plot of the hourly diffuse fraction K_d against the clearness index K_t, defined as:

$$K_t = \frac{GHI}{G_{TOA}} \qquad (1)$$

$$K_d = \frac{DHI}{GHI} \quad (2)$$

Where G_{TOA} is the horizontal solar irradiation available at the top of atmosphere.

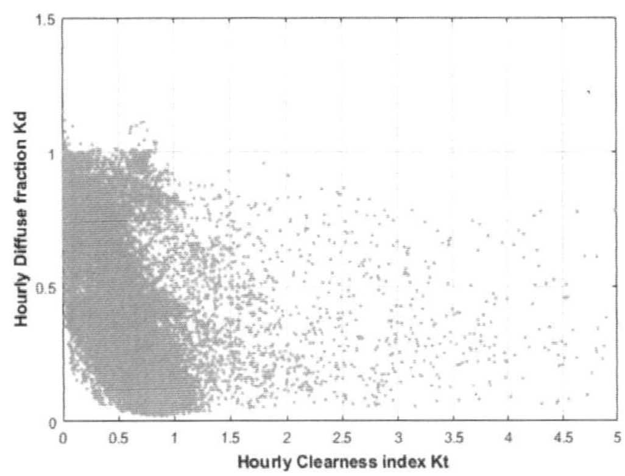

Figure 2. Scatter plot of the diffuse solar fraction according to the clearness index before quality control for the hourly solar data.

This figure shows several anomalies, mainly several values of K_d and K_t are greater than 1. This phenomenon can be explained by the fact that the clouds albedo could increase the global radiation to a value greater than that of the extraterrestrial radiation, which gives a value of the clearness index greater than one [3]. To construct an accurate and reliable dataset, we must apply a check of data quality control.

3. Quality Control Procedure

The erroneous solar data are clearly visible in Figure 2. To solve this problem, we applied a set of four tests of quality control.

3.1. The first test

Authors of references [2, 3, 8] recommended to exclude the recorded data at low sun altitudes (h_{solar}< 7°). This is in order to process the intrinsic cosine error caused by the pyranometer. The cosine response is the response of the sensor to the angle at which the solar radiation touch the detection zone, this angle is generally important during sunrise and sunset, which introduces a higher cosine error [2]. However, by plotting the clearness index K_t according to the solar altitude (Figure 3), we can see that a choice of a limit of 5° is very plausible.

According to this illustration, it has been observed that certain values of the clearness index have exceeded 2 for solar altitudes less than 5°. Figure 4 shows the (K_t,K_d) scatter plot after this test application.

3.2. The second test

This test is based on physical limits of clearness index K_t and diffuse fraction K_d:

$$0 < K_t < 1 \quad (3)$$

$$0 < K_d < 1 \quad (4)$$

Results of this test are illustrated in figure 5.

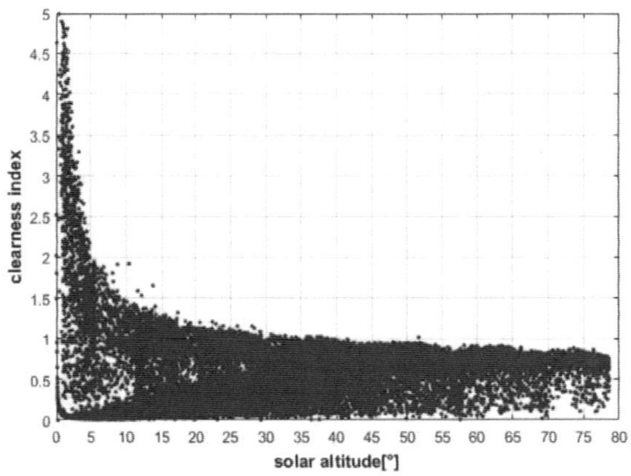

Figure 3. Clearness index according to the solar altitude for the measured data.

3.3. The third test

In this test, GHI and DHI are limited by their values which correspond respectively to a clear sky (figure 6) and a cloudy sky.

$$GHI < GHI_{clear\ sky} \quad (5)$$

$$DHI_{clear\ sky} < DHI < DHI_{cover\ sky} \quad (6)$$

To calculate these limits, the Page model [9] is used to determine the direct and diffuse hourly irradiations under clear and overcast sky conditions. This model was also validated by Younes et al [2] in 11 locations in the northern hemisphere from England to Japan. Figure 7 illustrates the (K_t, K_d) scatter plot after the application of the third test.

3.4. The fourth test

This test is based on a statistical analysis which consists in:

- Dividing the [0, 1] interval of the clearness index K_t in twenty equally spaced intervals.
- Calculating for each interval the average of the diffuse fraction K_d (K_{dm}) and its standard deviation (σ_d) .
- Retaining only values which respect the following range:

$$K_{dm} - 2 *\sigma_d \leq \ K_d \leq K_{dm} + 2 *\sigma_d \qquad (7)$$

Figure 8 represents the scatter plot after the application of this test.

4. Results

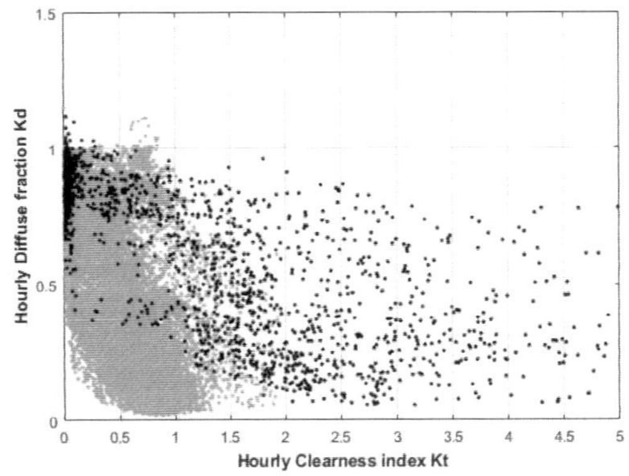

Figure 4. The scatter plot of the diffuse solar fraction according to the clarity index after the first test of quality control for the hourly solar data between 2009 and 2015. The grey dots are the values that passed the test, the black dots are the values that didn't pass the test.

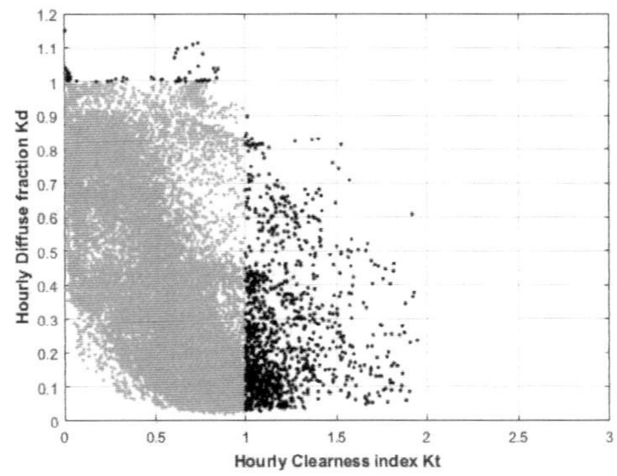

Figure 5. The scatter plot of the diffuse solar fraction according to the clarity index after the second test of quality control for the hourly solar data between 2009 and 2015. The grey dots are the values that passed the test, the black dots are the values that didn't pass the test.

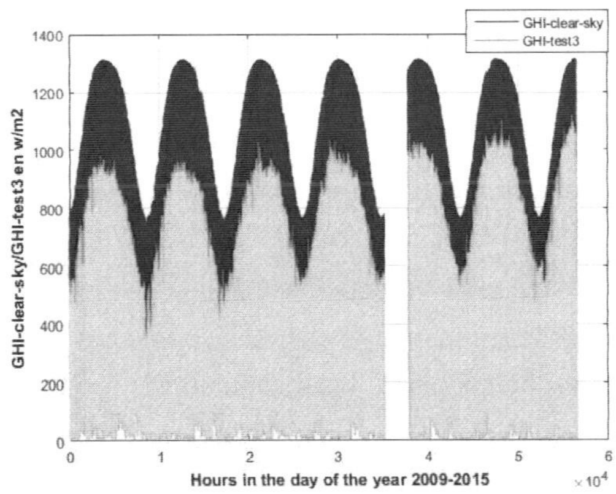

Figure 6. Comparison between GHI after the third test of quality control and GHI under clear sky conditions for the hourly solar data between 2009 and 2015.

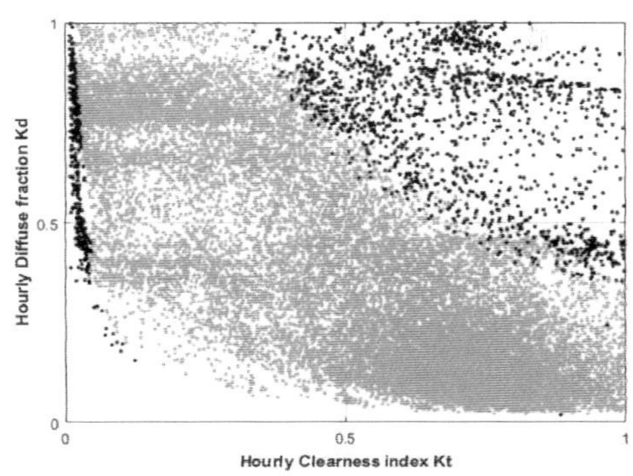

Figure 7. The scatter plot of the diffuse solar fraction according to the clarity index after the third test of quality control for the hourly solar data between 2009 and 2015. The grey dots are the values that passed the test, the black dots are the values that didn't pass the test.

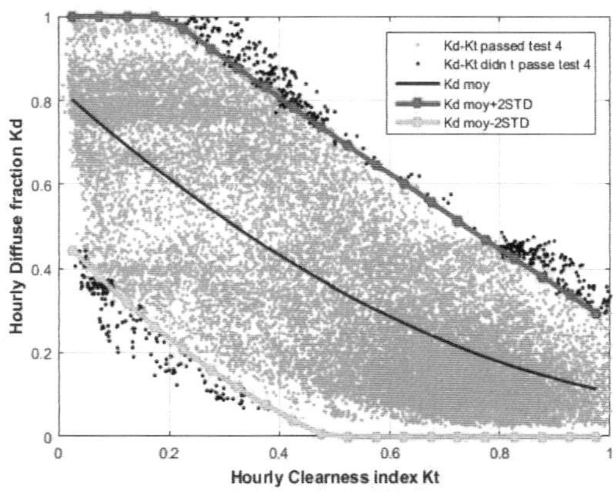

Figure 8. The scatter plot of the diffuse solar fraction according to the clarity index after the fourth test of quality control and average, standard deviation of the diffusion index for the hourly data between 2009 and 2015. The grey dots are the values that passed the test, the black dots are the values that didn't pass the test.

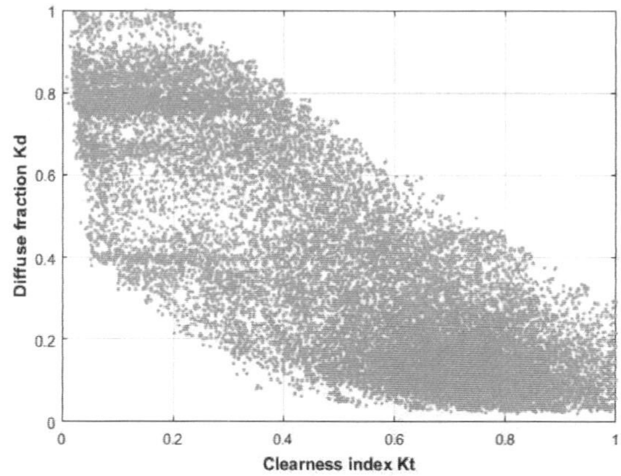

Figure 9. The scatter plot of the diffuse solar fraction according to the clarity index after quality control for the hourly solar data between 2009 and 2015

5. Conclusion

The hourly measurements of solar irradiations are often subject to different measurement errors. The aim of this work is to monitor the quality of these data in order to serve better scientific research. A quality control procedure was applied to the hourly solar data measured at Fez city for a period defined between the year 2009 and 2015. A set of four tests were applied and we noted that 1938 measures were eliminated by the first test, 2582 by the second, 1814 by the third and 562 by the fourth test. At the end, we have 21458 hourly measures that passed all tests successfully.

References:

[1] Moradi I 2009 Quality control of global solar radiation using sunshine duration hours *Energy* **34.1**: 1-6.

[2] Younes S, Claywell R and Muneer T 2005 Quality control of solar radiation data: present status and proposed new approaches *Energy* **30**:1533–49

[3] Ruiz-Arias J.A., Alsamamra H., Tovar-Pescador J and Pozo-Vazquez D 2010 Proposal of a regressive model for the hourly diffuse solar radiation under all sky conditions *Energy Convers Manag*, **51**:881–893

[4] Espinar B, Wald L, Blanc P, Hoyer-Klick C, Schroedter-Homscheidt M and Wanderer T 2011 Report on the Harmonization and Qualification of Meteorological Data: Project ENDORSE—*Energy Downstream Service*: Providing Energy Components for GMES—Grant Agreement No. 262892. *Paris, France: Armines. Accessed November, 11*, 2014.

[5] Drummond A.J. 1964 Comments on "sky radiation measurement and a.corrections", *J Appl Meteorol* **3**:810–811

[6] Reda I. and Andreas A. 2003 Solar position algorithm for solar radiation applications: Technical Report, *National Renewable Energy Laboratory*, Golden, Co, USA. Revised version: January 2008.

[7] Blanc P and Wald L 2012 The SG2 algorithm for a fast and accurate computation of the position of the Sun for multi-decadal time period. *Solar Energy*, ***86*(10)** 3072-3083.

[8] Ihya B , Mechaqrane A , Tadili R and Bargach M N 2014 Prediction of hourly and daily diffuse solar fraction in the city of Fez (Morocco), *Theoretical and Applied Climatology*, 1-13

[9] Greif, J, Scharmer, K, Dogniaux, R. and Page J 2000 ESRA, European Solar Radiation Atlas. *Including CD-ROM (fourth ed.), Presses de L'ecole des Mines de Paris, Paris, France*.

IOP Conf. Series: Earth and Environmental Science **161** (2018) 012011 doi:10.1088/1755-1315/161/1/012011

Dimensioning of a battery system to store energy from a hybrid PV/wind/diesel system at 3 kVA

K EL Hammoumi[1], R EL Bachtiri[1], M Boussetta[2], and M Khanfara[1]

[1]CED STI, FST, PERE Laboratory EST-University of Sidi Mohamed Ben Abdellah (USMBA) Fez, Morocco,

[2]FSDM, LESSI Laboratory, REPEER Group EST-University of Sidi Mohamed Ben Abdellah (USMBA) Fez, Morocco,

karima.elhammoumi@usmba.ac.ma,

Abstract. Battery storage presents a considerable challenge in the development of hybrid PV/wind systems. In this work, we present a battery storage dimensioning for a hybrid PV/wind/diesel 3 kVA system. An economic study was carried out by fixing as a criterion the price minimization of one stored kWh. For this study we have taken into account the main battery technologies on the market: GEL batteries, AGM batteries, AGM deep cycle batteries, lithium-ion batteries and lithium iron phosphate batteries. In order to consider a realistic price, we have heeded losses in the battery. We have developed a suitable Matlab program allowing the price calculation of one stored kWh taking into account the battery losses. Two solutions have been identified.

1. Introduction

The current energy situation, characterized by dwindling fossil fuel reserves, global warming due to the emission of greenhouse gases and environmental damage, is encouraging the development of renewable energy sources such as solar and wind. But this energy has a major Achilles heel: its character that can be described as random and intermittent because its production depends entirely on climatic conditions that have become increasingly random. Hybrid systems combining several energy sources with a storage system (combining solar, wind, diesel and batteries), present a promising solution to overcome this random nature. Indeed, such hybrid systems are increasingly used to meet the electricity needs of isolated small and medium cities ensuring total energy autonomy. The combination of solar and wind energies enables us to exploit complementarities between these two sources which are best shared, whether at the daily or annual scale [1]. The most used storage medium for small and medium power hybrid systems are chemical accumulators or batteries. Indeed, chemical storage systems have a good efficiency (close to 100%) and can provide a flexible energy management that can improve the power quality.

The sizing of battery storage in hybrid systems (whether or not connected to the grid) has been the subject of several research projects in recent years. Much of this work concerns the evaluation of optimal storage capacity based on different minimization criteria. The first methods proposed are based on the statistical study of the production deposit data (wind speed, sunshine and ambient temperature) and consumption profile [2] - [3]. Recent methods [4] - [6] are rather based on dynamic simulations or use sizing software tools such as HOMER developed by the National Renewable Energy Laboratory [7]-[9]. For example, Wang X et al. propose a methodology to determine the capacity of the battery storage system (BESS) required to obtain the optimal level of distributed power of a wind farm and to mitigate the effects of unsteady input power from the farms [10]. Ru Y et al. examine the problem of determining the size of the battery for PV systems connected to the grid for power arbitration and advanced peak shaving. The goal is to minimize the investment in storage of the battery while minimizing the purchase of net power from the grid by storing excess electricity generated by photovoltaic systems when the price is low to sell it to the grid when the price is high [11]. However, few studies detail the practical dimensioning of battery based on their prices and comparing the technologies and capacities available on the market

The case study in this paper aims to size the battery storage for a 3 kVA hybrid PV/wind/diesel system. In order to allow a judicious choice, we carried out an economic study based on the calculation of the price of one stored kWh according to the cost price of the battery and its lifetime. For this price to be as realistic as possible, we have taken into account the losses in the battery.

2. Presentation of the hybrid system

The studied system is a 3 kVA hybrid system developed within the framework of the CNRST-funded GISER project consisting of a 2.5 kW peak PV generator, a 1.5 kVA wind turbine, a 1 kW battery pack in order to ensure the electrical energy needs during the night and a 1.5 kVA diesel generator as a backup power supply in the absence of PV-wind generation and if the batteries are discharged (e.g. figure1).

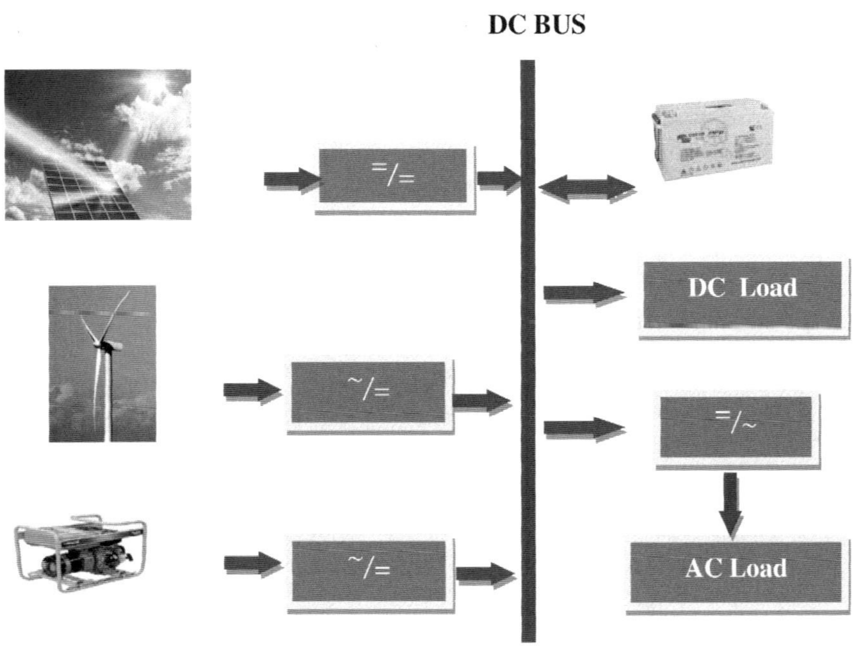

Figure 1. *Structure of the studied hybrid system*

3. Dimensioning batteries

Once the overall storage power has been set, the sizing of the batteries for a given use consists in determining the correct unitary battery capacity (referring to the offers on the market) and the number of batteries needed to meet the energy requirements requested, in respecting the imposed constraints [12].

3.1. Battery features
Before proceeding to the sizing of batteries, it is advisable to present the main characteristics and the battery model to be considered.

3.1.1. Battery model considered. For sizing, it is especially the storage capacity that is important; so we take the simple model of battery shown in figure 2.

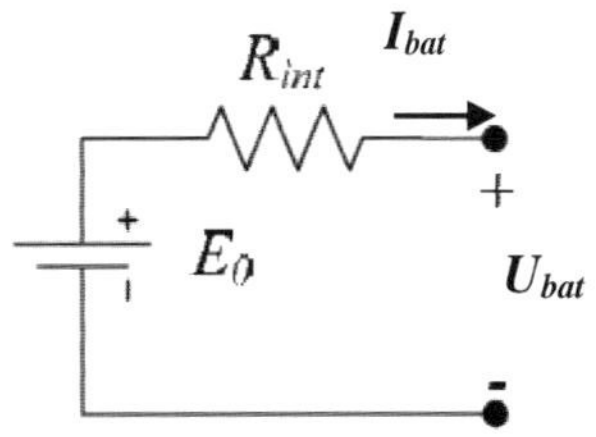

With:
E_o : Open circuit battery voltage,
R_{int} : Equivalent internal battery resistance,
U_{bat} : Voltage across the battery
I_{bat} : Battery current

Figure 2. *Battery simple model*

The power provided by the battery is expressed by:

$$P_{bat} = U_{bat} \times I_{bat} \tag{1}$$

3.1.2. Battery capacity. The nominal capacity (expressed most often in Ah) of a battery is the amount of energy that can be withdrawn from it at a particular constant current, starting from a fully charged state. It is given by the manufacturers under specified conditions of use (temperature $\boldsymbol{T}$°, $\boldsymbol{I_{bat}}$ et $\boldsymbol{E_o}$) under two terms $\mathbf{C_{nom}}$ and $\mathbf{C_n}$. $\boldsymbol{C_{nom}}$ is the capacity in Ah and $\boldsymbol{C_n}$ (with "n" hours) indicates how long the battery is discharged.

*3.1.3. Depth of Discharge (**DoD**).* The depth of discharge is the percentage of the total capacity of the battery that is used during a charge/discharge cycle.

3.2. Calculation of current and total energy to be supplied by batteries
To ensure a useful power $\mathbf{P_u}$= 1 kW during a period **T** = 12 h (the night between 6 PM and 6 AM), the battery pack must provide a current $\mathbf{I_{bat}}$ and a global useful energy $\mathbf{W_u}$ given by:

$$I_{bat} = \frac{P_u}{U_{bat}} = 83.33\ \text{A} \tag{2}$$

$$W_u = P_u.T = 12\ \text{kWh} \tag{3}$$

With: $\mathbf{U_{bat}}$ is the nominal voltage of the batteries set at 12V for our case.

By dividing this useful energy by the nominal voltage of the batteries U_{bat} , we obtain the useful storage capacity of the batteries C_u in Ah

$$C_u = \frac{W_u}{U_{bat}} = 1000\ \text{Ah} \tag{4}$$

3.3. Sizing
The sizing of the batteries is made from the maximum storage capacity C_{bat} whose expression according to the useful storage capacity is given by the following formula [13]:

$$C_{bat} = \frac{C_u}{DoD\,.\,f_t} \tag{5}$$

Where C_{bat} is the batteries storage capacity in Ah, DoD : Maximum depth of discharge allowable by the batteries and f_t : Temperature reduction factor given by:

$$f_t = \frac{C}{C_0} = 0.01035 \times T_a + 0.724 \tag{6}$$

C : Battery capacity at temperature T_a (in °C) and C_0: Battery capacity at temperature 26.7 °C.

Thus, the storage capacity of the batteries to be installed depends both on the maximum discharge depth allowed by the batteries considered and on the temperature [13].

We consider an operating temperature of 30 °C and a discharge depth of 80 %, so the battery storage capacity in our case will be:

$$C_{bat} = 1208.3\text{Ah} \tag{7}$$

If C_{nom} is the nominal capacity of a battery, the battery pack will consist of N_{bat} batteries associated in parallel and calculated by:

$$N_{bat} = \frac{C_{bat}}{C_{nom}} \tag{8}$$

4. Technical-economic aspects

To choose the technology and the nominal capacity of the batteries to use, we conducted an economic study. As a first step in this study we have established a database for the sizing of batteries by referring to the data of the different websites specialized in the sale of batteries [14] - [20]. For this study we have taken into account the main battery technologies on the market: GEL batteries, AGM batteries, AGM deep cycle batteries, lithium-ion batteries and lithium iron phosphate batteries. For these two last the price proposals are very limited. For the nominal capacities, we took capacities between 80Ah and 220Ah. The main characteristics of the batteries considered are given in Table 1.

Table1. Main characteristics of the batteries considered.

Battery technology		Main Features
Lead Acid Availability on the market Low price	***AGM*** (Absorbent Glass Mat) ***AGM deep cycle***	Low power density Low life (200 to 900 cycles)
	GEL	Low power density Lifetime between 300 and 1300 cycles
Lithium Limited proposals on the market High price	***Lithium-ion***	High energy density Large number of cycles (1500 to 2500). Risk of explosion
	Lithium iron phosphate (LiFePO4)	High energy density Large number of cycles (2000 to 5000). Fully stable

Then we realized a Matlab program for calculating:

- The number of batteries N_{bat} needed to provide the useful energy W_u according to equation (8).
- The price of one stored kWh according to the price of the battery ***Prbat,*** its lifetime in cycle number ***Ncy***, the depth of discharge ***DoD*** and the battery nominal capacity C_{nom} for equation (9) and the total storage capacity C_{bat} for equation (10).

$$Pr = \frac{Prbat}{C_{nom} \times U_{bat} \times DoD \times 2 \times Ncy \times 10^{-3}} \quad (9)$$

$$Pr_t = \frac{Prbat \times N_{bat}}{C_{bat} \times U_{bat} \times DOD \times 2 \times Ncy \times 10^{-3}} \quad (10)$$

The factor 2 is used to take into account the charging and discharging of the accumulator [1].

The results obtained are represented in the figures from 3 to 6 below.

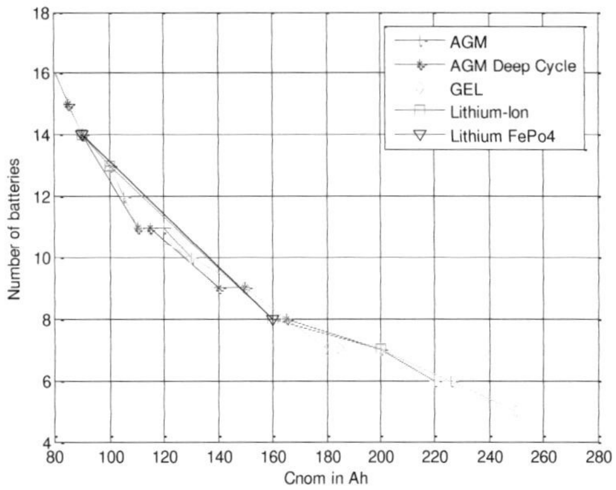

Figure 3. Number of batteries needed to provide storage capacity $C_{bat} = 1208.3Ah$, for the 5 technologies considered.

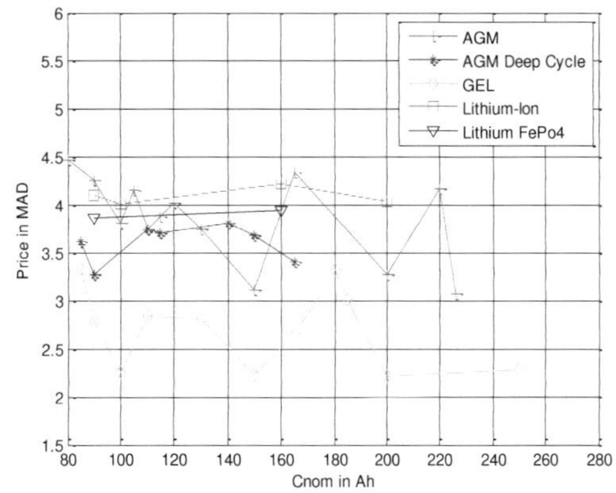

Figure 4. The price of one stored kWh according to the nominal capacity of the batteries and for the 5 technologies considered (using equation 9).

Considering figure 3, we find that the number of batteries decreases as the rated capacity increases. For reasons of congestion, we will limit the number of batteries to 10 batteries maximum so we should choose a nominal capacity greater than 130 Ah.

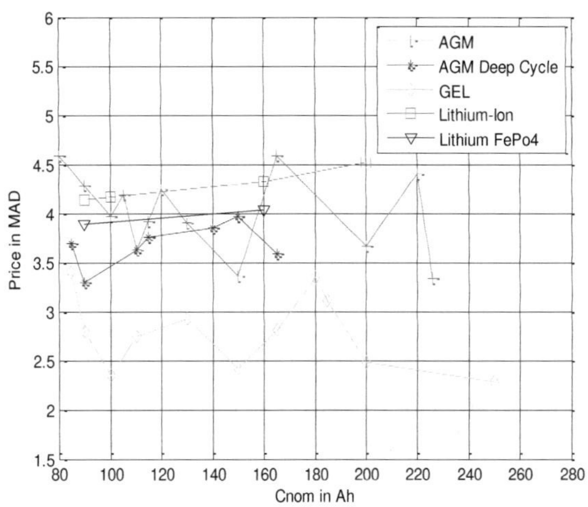

Figure 5. Price of one stored kWh according to the nominal capacity of the batteries and for 5 technologies (using equation 10).

Based on the curves in figures 4 and 5, we find that GEL technology batteries have the lowest kWh price. Especially those rated capacity 100Ah, 150Ah and 200Ah.The slight difference in price between the two curves is explained by the fact that the number of calculated batteries is rounded up to the nearest whole number.

Referring to the battery model of figure 2 we can calculate the losses in the battery by:

$$Ls_{bat} = R_{int} \times {I_{bat}}^2 \quad (11)$$

However, the internal resistances of the different batteries considered are less than 6 mΩ [14]-[20]. So for the calculation of the battery losses we will use for all the batteries $R_{int} = 6$ mΩ.

According to (2) and (11) we have at most:

$$Ls_{bat} = 41.6\,W \quad (12)$$

The dissipated energy W_{dp} in 12h is $W_{dp} = 500Wh$ (13)

To consider losses in the price calculation of one stored kWh, equation (10) becomes:

$$Pr_{md} = \frac{Prbat \times N_{bat}}{(C_{bat} \times U_{bat} - W_{dp}) \times DoD \times 2 \times Ncy \times 10^{-3}} \quad (14)$$

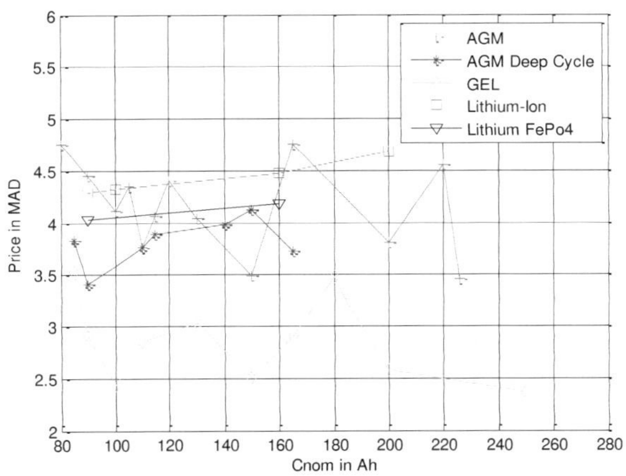

Figure 6. Price of one stored kWh according to the nominal capacity of the batteries and for 5 technologies, calculated taking into account the losses in the batteries.

By introducing the battery losses in the calculation of the price of the one stored kWh we note a very slight increase in the price of the order of 3.5% (The curves in figures 5 and 6 are almost identical) because the losses represent only 4. 2%.

According to the results obtained, two solutions appear clearly, either 7 GEL batteries of 200 Ah nominal capacity, or 9 GEL batteries of 150 Ah nominal capacity.

5. conclusion

Battery storage accounts for a large part of the cost of photovoltaic/wind/diesel hybrid installations, which is why optimal dimensioning is required. This work presents an economic study allowing the sizing of battery storage for a 3 kVA hybrid system using as criterion the minimization of the price of the stored kWh. For this study we were interested in the main battery technologies on the market: GEL batteries, AGM batteries, AGM deep cycle batteries, lithium-ion batteries and lithium iron phosphate batteries. A Matlab program calculating the price of one kWh stored taking into account the battery losses has been realized. We have found that GEL technology batteries have the lowest kWh price.

Especially two solutions have been identified: 7 GEL batteries of 200 Ah nominal capacity in parallel or 9 GEL batteries of 150 Ah nominal capacity in parallel.

References

[1] Dhaker A 2012 Contribution au dimensionnement et à l'optimisation des systèmes hybrides éoliens-photovoltaïques avec batteries pour l'habitat résidentiel autonome, *Thèse de Doctorat en Génie électrique. Poitiers : Université de Poitiers*

[2] Borowy B and Salameh Z 1996 Methodology for optimally sizing the combination of a battery bank and pv array in wind/pv hybrid system *IEEE Transactions on Energy Conversion* **vol 12** n° 11 pp. 367-375

[3] Shrestha G and Goel L Dec. 1998 A study on optimal sizing of stand-alone photovoltaic stations, *IEEE Transactions on Energy Conversion.* **vol 13**, no. 4, pp. 373–378

[4] Bernal-Agustín J. L and Dufo-López R 2009 Simulation and optimization of stand-alone hybrid renewable energy systems, *Renewable and Sustainable Energy Reviews*, **vol 13**, n° 18, pp.2111-2118

[5] Badawy M O, Cingoz F and Sozer Y 1-7 2016 Battery storage sizing for a grid tied PV system based on operating cost minimization, *in proc.IEEE Energy Conversion Congress and Exposition (ECCE), IEEE*

[6] Ghenai C, Salameh T, Merabet A and Hamid A-K. 4-6 April, 2017 Modeling and Optimization of Hybrid Solar-Diesel-Battery Power System, *Seventh International Conference on Modeling, Simulation, and Applied Optimization - (ICMSAO'17), Sharjah, American University of Sharjah, UAE*

[7] Boussetta M, El Bachtiri R., El Hammoumi K and Khanfara M 2016 Optimal Sizing Of Grid-Connected Pv-Wind System Case Study: Agricultural Farm In Morocco," *Journal of Theoretical and Applied Information Technology*, **vol. 86**, no. 2, p. 196

[8] MaT, Yang H and LuL 2014 A feasibility study of a stand-alone hybrid solar–wind–battery system for a remote island *Applied Energy* **vol 121** pp. 149–158

[9] Boussetta M, El Bachtiri R., Khanfara M and El Hammoumi K 2017 Assessing the potential of hybrid PV–Wind systems to cover public facilities loads under different Moroccan climate conditions *Sustainable Energy Technologies and Assessments* , **vol. 22**, 74–82

[10] Wang X, Vilathgamuwa D.M and Choi S Sep. 2008 Determination of battery storage capacity in energy buffer for wind farm, *IEEE Trans. Energy Convers*, **vol. 23**, no. 3, pp. 868–878

[11] Ru Y, Kleissl J and Martinez S January 2013 Storage Size Determination for Grid-Connected Photovoltaic Systems, *IEEE Transactions on Sustainable Energy*, **vol. 4**, no. 1

[12] Devillers N 2012 caractérisation et modélisation de composants de stockage électrochimique et électrostatique, *Thèse de Doctorat en Génie électrique. : Université de Franche-Comté*

[13] Labouret A and Villoz M 2005 *Energie solaire photovoltaïque*, 2nd ed., Dunod

[14] Wattuneed, [online].Available: https//www.wattuneed.com/fr/11-batteries-solaires#/ (july 2017)

[15] atbatt, [online]. Available: http://www.atbatt.com/. (july 2017)

[16] solaris-store, [online]. Available: https://www.solaris-store.com/ (july 2017)

[17] batterie-solaire, [online]. Available: http://www.batterie-solaire.com/ (july 2017)

[18] solar-kit, [online]. Available: http://www.solar-kit.com/batterie-solaire(july 2017)

[19] newenergyeco, [online]. Available: https://www.newenergyeco.com/(july 2017)

[20] masolise, [online]. Available: http://www.masolise.com/(july 2017)

IOP Conf. Series: Earth and Environmental Science **161** (2018) 012012 doi:10.1088/1755-1315/161/1/012012

A comparison between Series Dynamic Resistors and CROWBAR circuit protection for LVRT capability of Doubly-Fed Induction Generator

Yassir EL KARKRI[1,*], Hassane EL MARKHI[1], Tijani LAMHAMDI[1], Hassan EL MOUSSAOUI[1]

[1]The Signals, Systems and Components Laboratory, Sidi Mohamed Ben Abdellah University, FST Fez, Morocco

*yassir.elkarkri@usmba.ac.ma

Abstract. In this paper, we will focus on a comparison between two strategies (the Series Dynamic Resistors (SDR) and CROWBAR circuit for Low Voltage Ride Through (LVRT) capability of Doubly-Fed Induction Generator (DFIG), in order to determine different responses of reactive power and active power during the grid faults in line with the requirements determined by the grid codes (GCs). The two strategies are simulated by using MATLAB/Simulink environment

1. Introduction

The electricity produced from nuclear and fossil energy has grown significantly across these last years. In fact, those classical energy sources have a serious environment impact. Therefore, the emerging ecological knowledge has greatly enhanced the interest in renewable energies integration [1]. Wind energy is one of these several clean energies sources. This technology has been having a fast development and has been attained a level of maturity during these last years. So its integration in the grid has a big interest by researchers [2].

Actually, DFIG is the most used generator in the wind power area, the wind turbine is connected to the DFIG rotor through a gearbox that adapts the wind speed generated from wind blades to the DFIG. The grid is directly connected to the DFIG stator windings and to the DFIG rotor windings via a back-to-back power converter [3] as shown in Figure 1. In addition, a filter is coupled to the grid to attenuate harmonic effects created by these converters (Figure 1). The Rotor Side Converter (RSC) is used in order to regulate the active power transported between DFIG and the grid [4]. However, the Grid Side Converter (GSC) is utilized to regulate the reactive power transferred to the grid. The aim of the Maximum Power Point Tracking (MPPT) block is to optimize the energy conversion of the Wind Turbine Generator (WTG) and then estimate torque reference [5]. It is used with reactive power reference and dq-axis rotor currents as inputs to the RSC control block. Therefore, the Pulse Width Modulation (PWM) signals are generated from the RSC control block output after been transformed to abc-axis (Figure 1). The GSC control block proposes to regulate the power factor and to control DC-link voltage. Thus, the block inputs are: the DC-link measured, dq-axis filtered grid currents, DC-link voltage reference and active power reference. Therefore, PWM signals are generated from the GSC control block outputs after been transformed to abc-axis [6].

The command used for the converter is the vector control (Figure 1), also called field-oriented control, is a method of controlling power inverters in which the stator and rotor three-phase alternating currents are transformed into two orthogonal components which can be considered as vectors. The first vector allows adjustment of the magnetic flux of the motor, while the second regulates the torque. Thus, the vector control comes to solve this problem of decoupling the flux inside the machine from that of the torque.

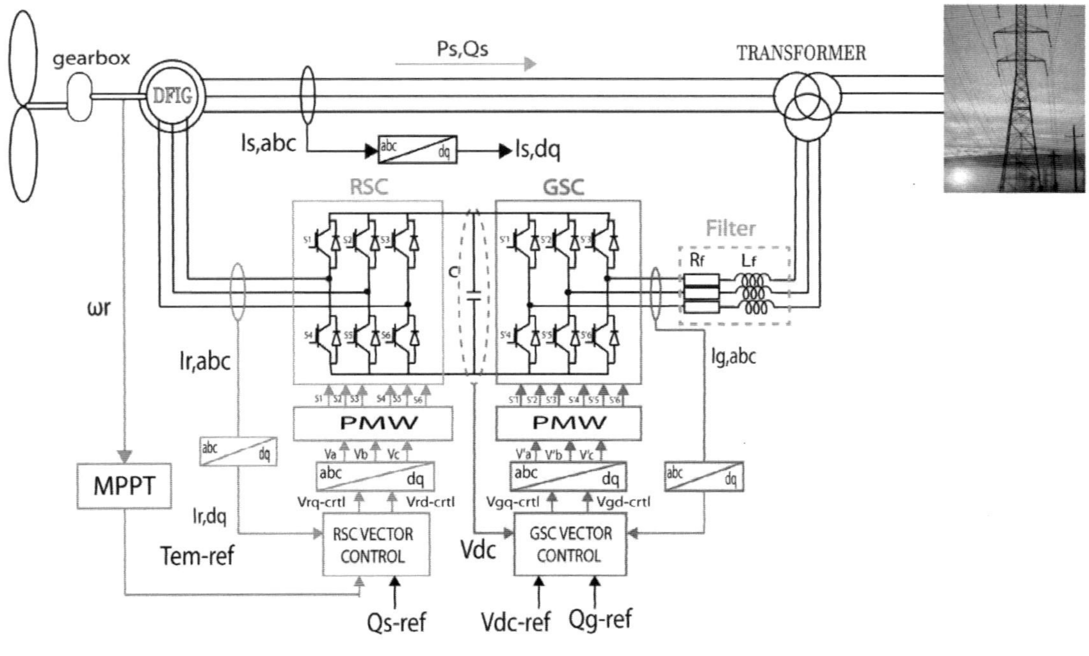

Figure 1. DFIG based wind turbine connected to the grid.

In the past, during a grid fault, WTGs has disconnected from the GRID. Actually, to remain these WTs connected to GRID even if a fault occurs, new requirements exist. In fact, during these last years, many countries have proved their GCs in transmission and distribution. These codes specify voltage conditions ranges for which WTGs have to keep connected to the power electric system [7]. As WTGs become higher and level of penetration becomes larger in the grid, so the grid operators had to change the GC [8].

LVRT is a section of the GC fixed for the grid voltage dips, such as WTGs are required to stay connected to the grid for a considered time before being allowed to disconnect. This considered time can be changed from one GC to another depends on the severity of the fault. Figure 2 shows differences between these codes during voltage dips in different countries [9, 10]. This paper presents a comparison between two LVRT strategies. (i): the SDR with bypass switching devices coupled with stator windings. (ii): the CROWBAR circuit which consists of a controllable switch and a dissipation resistor coupled between the RSC and the rotor windings.

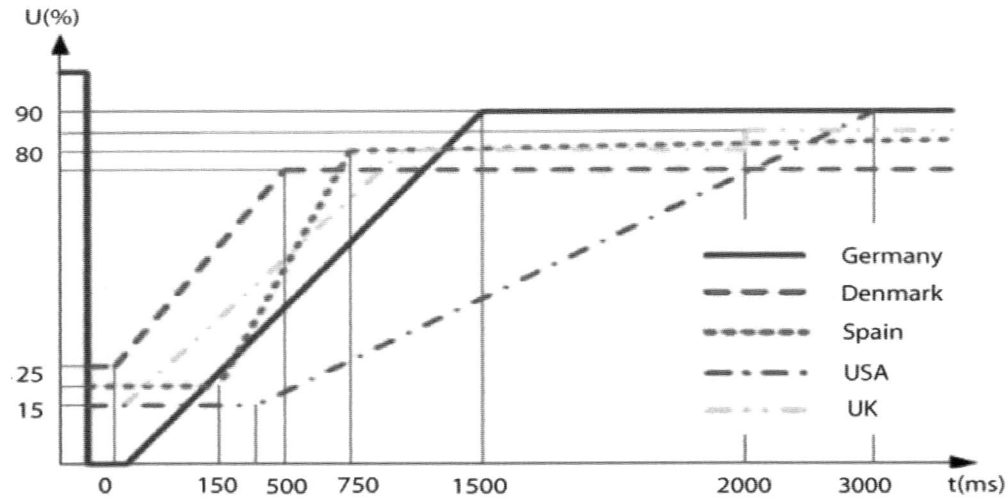

Figure 2. LVRT Requirements of the wind farms in different countries [8].

This paper is structured as: In the second section, the scheme of the CROWBAR circuit is presented. In the third section, the principle of the SDR strategy is presented. The simulation results of the SDR compared with the CROWBAR strategy is given in the fourth section. Finally, the conclusion is summarized in the fifth section.

2. Crowbar circuit protection

A voltage limiting circuit "CROWBAR" is usually implemented between the rotor windings and its electronic power converter to provide an electrical bypass circuit for the transient current induced by the voltage drops.

Initially, the solution implemented was simply to short-circuit the rotor with a CROWBAR circuit. The CROWBAR circuit is connected to the rotor windings and the RSC. A triac can be used as a controllable switch for CROWBAR circuit as shown in Figure 3. The semiconductors are supposed ideal, and a global connection function Sc is fixed for this three-phase electronic switch. It takes values 1, when the switch is closed and 0 when it is open. Thus, the voltage limiting circuit can be modelled by a simple equation:

$$v_{c-bar} = S_c R_{crowbar} i_{crowbar} \tag{1}$$

The behavior of this type of system is very impressed by the value of the dissipation resistances. It has been demonstrated by simulation that a low-value resistor drives to an increase in electromagnetic torque, high over-currents, and low rotor voltage [11]. In opposite, a high value for the resistance will be translated by a decrease in the electromagnetic torque and electric currents, but higher voltages crossed the rotor. Therefore, the resistors must be low enough to avoid over-voltages on the power converters. Also, they must be high enough to limit rotor current. In our study, this resistance value is taken to be equal to 30 Rr (the resistance of the rotor windings)

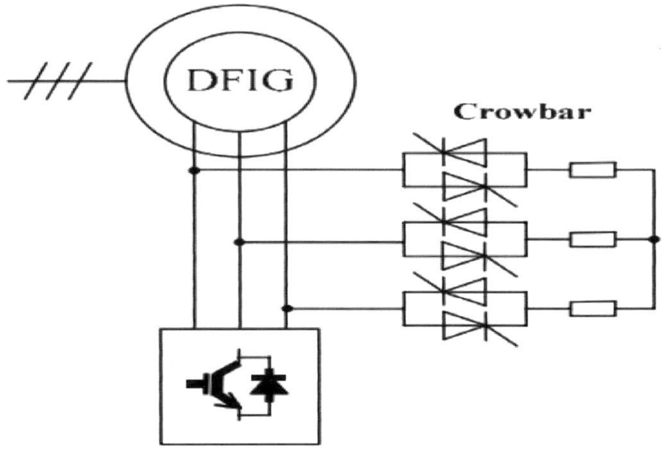

Figure 3. DFIG with CROWBAR coupled with ROTOR windings.

3. Series Dynamic Resistors

This method consists to coupled resistors in series with the stator (Figure 4), in order to raise the stator voltage and to decrease the DC component at the flux during the voltage sags, this would limit over-voltages at the rotor and therefore avoid peaks in rotor current [12].

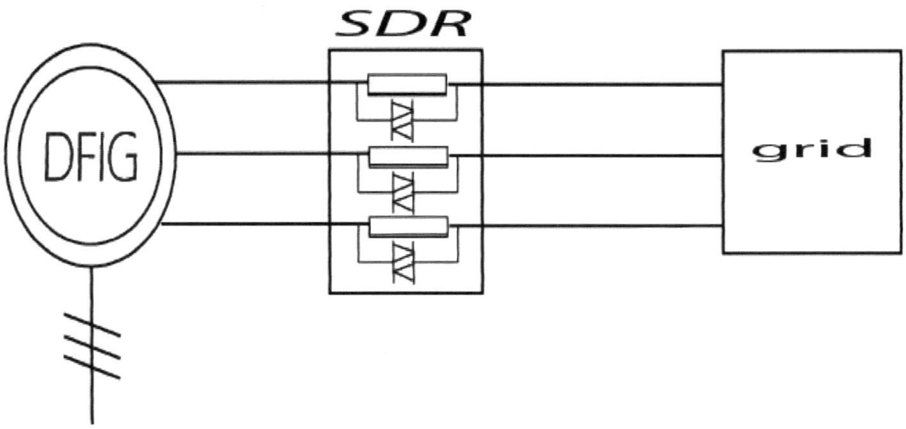

Figure 4. DFIG with SDR coupled with stator windings.

The determination of the value of the dynamic resistance is necessary, a high-value resistor will produce a large dissipation of energy and can pass the maximum value of the stator windings voltages. On the opposite, a low-value resistor cannot limit the fault current. So the dimensioning of the resistance is defined according to two conditions:

The resistors must be high enough to limit the rotor voltage to the maximum input voltage allowed by the converter, in order to avoid the damage of RSC control. Equation (2) expresses the case where the value of rotor voltage have to not exceed the maximum RSC voltage.

$$v_{r,max} < V_{RSC,max} \quad (2)$$

Since the RSC can only produce a voltage lower than the DC voltage, its maximum output value $V_{RSC,max}$ is calculated as follows:

$$V_{RSC,max} = \frac{V_{dc}}{2\sqrt{\frac{2}{3}}} \tag{3}$$

The resistors must be small enough to limit the stator voltage from exceeding his maximum value. Equation (3) indicates the condition which the SDR maximum voltage value adding to the grid voltage has to not pass the maximum stator voltage.

$$V_{SDR,max} + v_{grid} < V_{S,max} \tag{4}$$

Replacing V_{SBR} by $V_{SBR} = R_{SBR}\, I_S$, and we considered the worst case for grid fault when $v_{grid} = 0V$.

$$\frac{1}{I_s}\left\{(1-g)V_1 - \frac{M}{L_s}\frac{V_{dc}}{2\sqrt{\frac{2}{3}}}\right\} \le R_{SDBR} \le \frac{V_{Smax}}{I_s} \tag{5}$$

We found: $0.05pu \le R_{SDBR} \le 1.25\, pu$

4. Simulation and results

The system under study, presented in Figure 1, consists of a 1.5MW wind turbine. In this comparison, the tests were conducted using MATLAB/SIMULINK software, coherent model of the generator is used. The DFIG ride through capability is simulated for a three-phase voltage dip, in which the grid voltage in three-phase drops to 10% of its nominal value (90% voltage dip) at t=0.7s and remains for 0.9s (Figure 5).

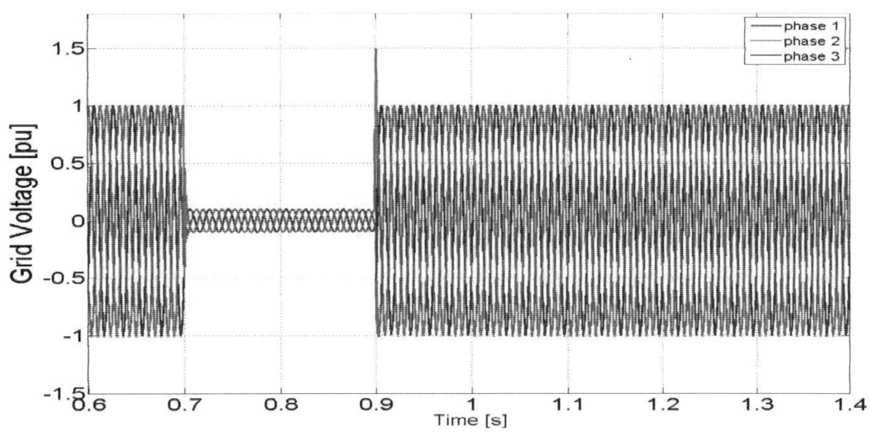

Figure 5. Simulation of voltage dips applied.

In this paper, two distinct strategies are investigated and compared in figure 6 and figure 7. In the first case, the SDR is applied (the blue curve), and the second case demonstrates the DFIG with the CROWBAR circuit protection (the red curve).

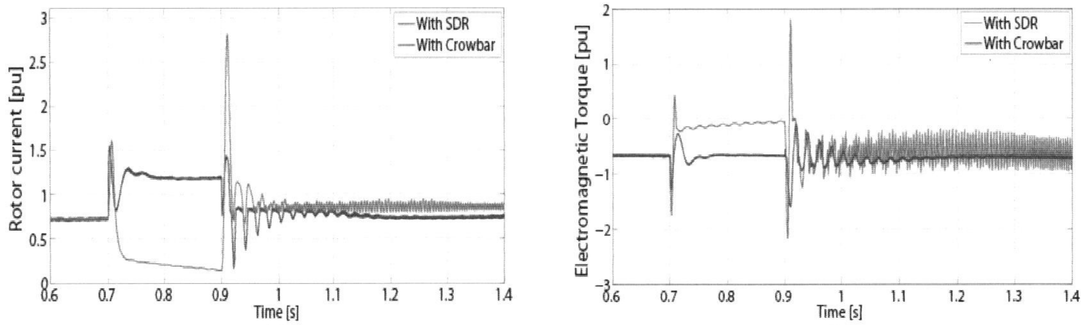

(a) (b)

Figure 6. Simulation results of (a) rotor current and (b) Electromagnetic torque with SDR (blue) and with CROWBAR (red).

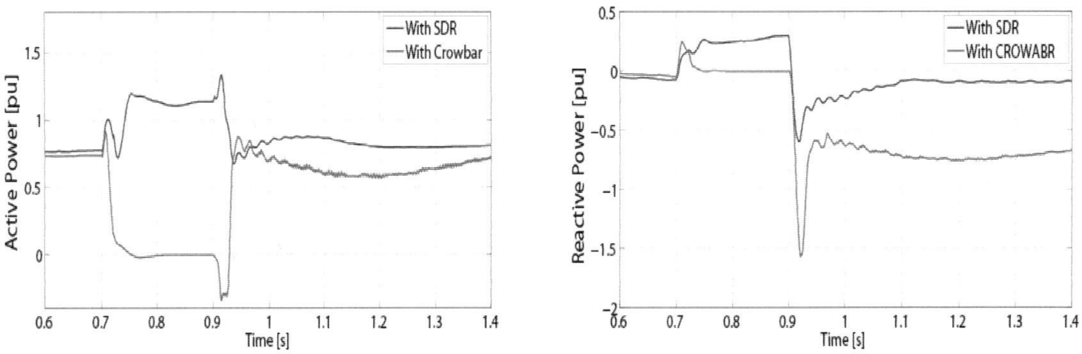

(a) (b)

Figure 7. Simulation results of (a) Active power and (b) Reactive power with SDR (blue) and with CROWBAR (red).

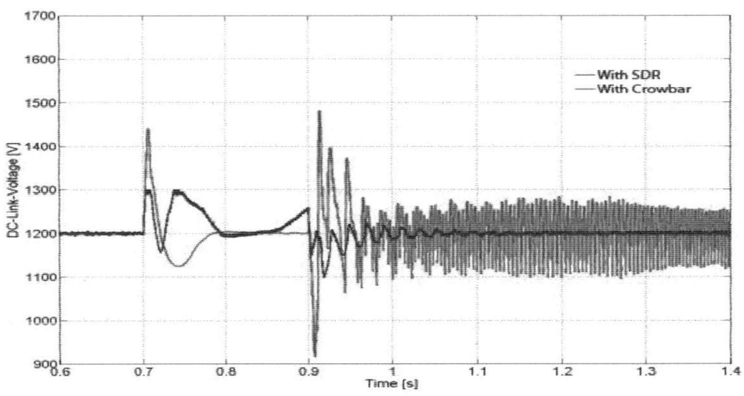

Figure 8: Simulation results of DC-voltage with SDR (blue) and with CROWBAR (red).

Figure 6 illustrates (a) Rotor current, (b) Electromagnetic torque and Figure 7 represent (a) Active Power, (b) Reactive power, with the SDR (the blue curve) and with the CROWBAR circuit protection (the red curve). By utilizing the SDR method, the simulations show positive results, thus the electrical values of active power, rotor current, electromagnetic torque and reactive power have passed to reasonable values and much better than those obtained using the CROWBAR circuit protection, it is noted that the improvement of the electrical values above-mentioned will allow: The wind turbine to still connected to the grid for a long times without much stress and without making any damage to the devices of the wind turbine. Also, to avoid mechanical stress due to the important oscillations during a grid fault, so with SDR solution the oscillations and the influence on the functioning of the DFIG including power converters and capacity are significantly degraded. In addition, the the system with can support grid power quality through the fault. From Figure 8 we remark that the DC-link voltage is almost substantial, by using SDR which remains the safety and the stability of the DC link.

As result, the wind turbine equipment will be well protected and the important cost of replacement of such devices will be avoided, in addition, the wind turbine working time is maximized. Also, with the SDR, the reactive power can restore quickly its normal state in order to ensure the electrical system stability. On the other hand, with the use of CROWBAR circuit protection it is observed that during the grid fault the rotor current and the active power almost zero due to separating of the converters from the rotor which misses the control of DFIG, this impact will result frequent disruptions of the wind turbine, current and accidental disconnections of the wind turbine from the grid, poor quality of service, GC not respected, very high maintenance cost and the cost of the wind turbines disconnection is very penalizing for both the producers and the grid operators.

5. Conclusion

This paper presents a comparison between the CROWBAR circuit protection and the SDR. The second strategy which based on the use of SDR coupled with stator windings is significant decreases the height values of the rotor currents, reduces the oscillations of the electromagnetic torque at the instants of occurring and clearing the grid faults. Also, the active power stays in the admitted DFIG operation range (0.7pu and 1.3pu) and the reactive power takes its normal value quickly after the fault disappear compared to CROWBAR circuit protection, which improves the LVRT capability of the DFIG. Therefore, the SDR solution is more efficient than the CROWBAR solution.

References

[1] Ostolaza, Joseba Xabier, Aitor Etxeberria, and Itziar Zubia. 2015 Wind farm node connected DFIG/back-to back converter coupling transient model for grid integration studies. *Energy Conversion and Management*. **106** 428-439.

[2] Cheng M, and Zhu Y 2014 The state of the art of wind energy conversion systems and technologies: A review. *Energy Conversion and Management*. **88** 332-347.

[3] Dinesh S, Meenakshi R, Suhanya M S, Kumaran M S and Muthu R 2014 Modeling and direct power control of DFIG for wind energy conversion system with a back to back converter. *Green Computing Communication and Electrical Engineering ICGCCEE International Conference on* IEEE 1-6.

[4] Abad G, Lopez J, Rodriguez M A, Marroyo L and Iwanski G 2011 Vector Control Strategies for Grid-Connected DFIM Wind Turbines. *Doubly Fed Induction Machine: Modeling and Control for Wind Energy Generation* 303-361.

[5] Mehdipour C, Hajizadeh A, and Mehdipour I 2016 Dynamic modeling and control of DFIG-based wind turbines under balanced network conditions. *International Journal of Electrical Power and Energ Systems* **83** 560-569.

[6] Xiang D, Ran L, Tavner P. J, and Yang S 2006 Control of a doubly fed induction generator in a wind turbine during grid fault ride-through. *IEEE Transactions on Energy Conversion* ***21*(3)** 652-662.

[7] Mullane A, Lightbody G, & Yacamini R 2005 Wind-turbine fault ride-through enhancement. *IEEE Transactions on Power Systems* **20(4)** 1929-1937.

[8] Boukhris Y., El Makrini A., El Moussaoui H and El Markhi H. 2015. Low Voltage Ride-through Capability Enhancement of Doubly Fed Induction Generator Based Wind Turbines under Voltage Dips. *International Journal of Power Electronics and Drive Systems* **6(4).**

[9] Wilch M, Pappala V S, Singh S N and Erlich I 2007 Reactive power generation by DFIG based wind farms with AC grid connection. *Power Tech, IEEE Lausanne* 626-632.

[10] Lima F K, Luna A, Rodriguez P, Watanabe E H and Blaabjerg F 2010 Rotor voltage dynamics in the doubly fed induction generator during grid faults. *IEEE Transactions on power electronics* **25(1)** 118-130.

[11] Lohde R, Jensen S and Knop A 2007 Analysis of three phase grid failure on doubly fed induction generator ride-through using Crowbars. *Power Electronics and Applications European Conference on IEEE* 1-8.

[12] El Makrini A, El Karkri Y, Boukhriss Y, El Markhi H and El Moussaoui H 2017 LVRT Control Strategy of DFIG Based Wind Turbines Combined the Active and Passive Protections. *International Journal of Renewable Energy Research IJRER*, **7(3)** 1258-1269.

IOP Conf. Series: Earth and Environmental Science **161** (2018) 012013 doi:10.1088/1755-1315/161/1/012013

Experimental validation of a photovoltaic panel model

Y Boukili[1], R El Bachtiri[1], H Chafouk[2], I Idrissi[1, 2] and M Khanfara[1]
[1] CED STI, FST, PERE Laboratory EST-USMBA, Fez, Morocco
[2] IRSEEM, ESIGELEC, Normandie University / UNIRouen, Rouen, France

Email: yassine.boukili@usmba.ac.ma ; rachid.elbachtiri@usmba.ac.ma

Abstract. For operation optimizing, it's important to have internal parameters of a photovoltaic (PV) panel. Experimental measurements were carried out, in order to determine a PV panel internal parameters. From simple measurements, two resistances values: series R_s and shunt R_{sh} have been determined. The two experimental characteristics I-V and P-V have been plotted. And afterward, the system operation has been simulated in Matlab Simulink, to validate the model. Otherwise, PV current equation resulting from the PV cell model has been used to determine 5 parameters. A Matlab program has been developed with the function "fsolve" for solving the system of five equations with five unknowns. Theoretical values were compared with experimental results.

1. Introduction

Electricity needs in the world are constantly increasing at the rate of industrialization and the improvement of people's level of living. In face of this, the climate issue implies an increasingly clean electricity production.

However, the solution can no longer go only through the centralized production of electricity. It must also concern the decentralized production of renewable energy. Photovoltaic has a great potential, among the sources of renewable energy. In fact, the demand for solar energy has increased by 20% to 25% over the past 20 years [1].

In order to improve the use of photovoltaic systems, there are research activities to reduce their costs and improve their efficiency and reliability.

A PV panel in our laboratory will be used to determine by measurements the internal parameters. After that, theoretical characteristics will be validated.

Starting with experimental measurements to determine two resistances values: R_s and R_{sh} of the equivalent electrical circuit. Then, two theoretical characteristics P-V and I-V will be plotted in Matlab Simulink environment using values found in the experimentation. Afterward, the experimental characteristics P-V and I-V will be plotted to validate simulation results.

To be sure of experimental results, a program on Matlab with "fsolve" function will be created, to solve PV current equation for determining the PV panel parameters [11]. The equation has been obtained from the equivalent electrical circuit of a PV cell model, it's a nonlinear equation. A system will be created by repeating this equation five times according to voltage and current measurements. With the "fsolve" function, we will determine five parameters (R_s, R_{sh}, A, T, and I_{sat}) from the system. Finally, a comparison between theoretical and experimental results will be presented.

2. Modeling

PV cells convert sunlight directly into electricity. When the photons of the sun reach the cell, it releases electrons from the negative layer. This electron moves to the positive layer, which gives an electrical current.

Theoretically, the voltage generated by a PV cell is between 0.5 and 0.8 volts, depending on the semiconductor technology. This voltage is very low; this is why several PV cells are made in series or in parallel. The voltage is increased if cells are connected in series and current if they are connected in parallel.

The parameters of a monocrystalline PV panel have been used [2], with 36 cells connected in series and Table 1 shows the characteristics of this panel. The model used in the simulation is PM0115.

Table1: Photovoltaic panel feature

Characteristic	Value
Maximum power	115 Watt
Voltage at maximum power	17,7 V
Current at maximum power	6,5 A
Open Circuit Voltage	21,6 V
Short circuit current	6,96 A
Tolerance	±5%

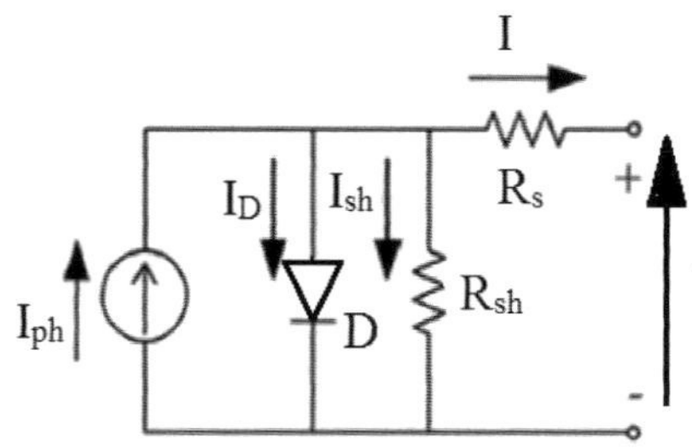

Figure 1. Equivalent electrical circuit of a PV cell

It is possible to represent a cell PV with an electrical equivalent circuit in Figure 1. It consists of a shunt resistance, a series resistance, a diode and a current source [3], [4] and [5].

To implement this PV cell model in simulation software, the following values should be known: Series Resistance R_s, shunt resistance R_{sh}, saturation current I_0, ideality factor of junction A and temperature T. The manufacturer's values have been taken for I_0 and A. Then experimental measurements were done to determine R_s and R_{sh}.

3. Experimental measurements

The series resistance of the PV cell is very low [6], and its value can influence the variation on the output of PV cells [7], it causes the deviation of the maximum power point.

The shunt resistance of any PV cell should be large enough for higher output power. In fact, if the shunt resistance is low, it will be a higher power loss.

Experimental measurements have been made to determine the two resistances R_s and R_{sh}, the measurements were carried out so rapidly to consider temperature as constant.

To determine the series resistance R_s, the Figure 2 scheme has been used, which consists of a DC generator, a variable resistance, an ammeter and a voltmeter. The PV panel was fully covered with paper to have: $I_{ph} = 0$.

IOP Conf. Series: Earth and Environmental Science **161** (2018) 012013 doi:10.1088/1755-1315/161/1/012013

3.1 Measurement of R_S

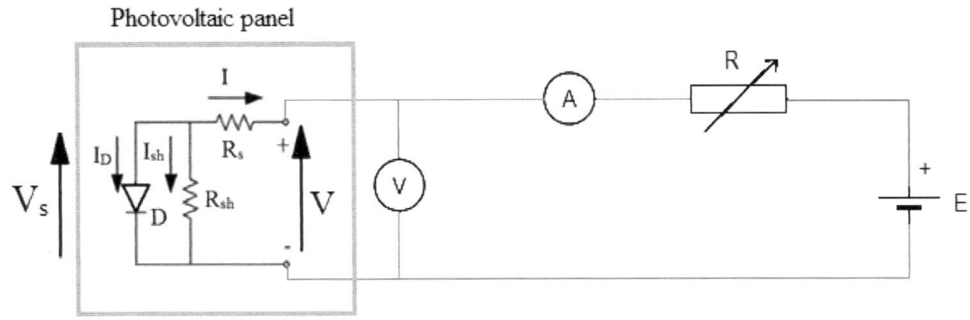

Figure 2. Experimental scheme for determining R_S

By decreasing the variable resistance, the values have been taken of voltage and current and the values are presented in Table 2.

Table 2: Experimental measurements

V (V)	13.6	15.3	16.1	17	18	19	19.2	20	20.95	21.4	22	22.6
I (A)	-0.01	-0.02	-0.03	-0.05	-0.1	-0.2	-0.3	-0.5	-1	-1.5	-2	-3

According to the electrical wiring in Figure 2:

$$R_S = \frac{V - V_S}{I} \tag{1}$$

To determine V_s, the curve I-V characteristic has been plotted using TABLE 2 values and it is presented in Figure 3. This curve is likewise the diode characteristic. The extrapolation of the linear part of the curve gives the V_s value.

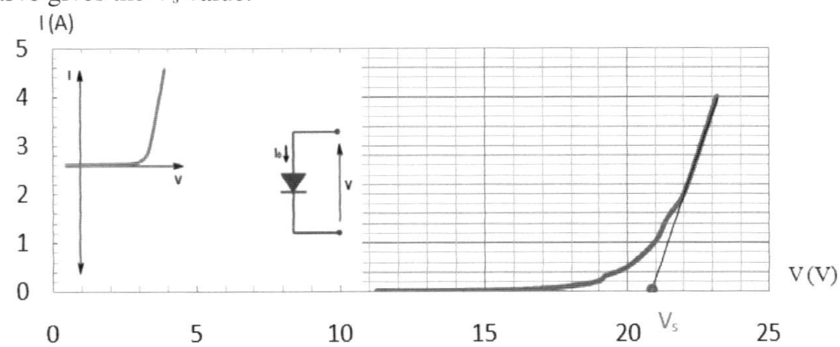

Figure 3. Curve I-V

For 36 cells, Vs = 20.8V.
R_{S_36} for 36 cells is calculated by:

$$R_{S\_36} = \frac{\Delta V}{\Delta I} = \frac{23{,}15 - 22{,}6}{4 - 3} = 0{,}55\Omega \tag{2}$$

And to calculate R_S for one cell, it is enough to divide by 36:

$$R_S = 0{,}015\Omega$$

3.2 Measurement of R_{sh}

The same previous electrical scheme will be used but the terminals of the DC generator were inverted, then this wiring presented in Figure 4 will be used. The diode is blocked in this case.

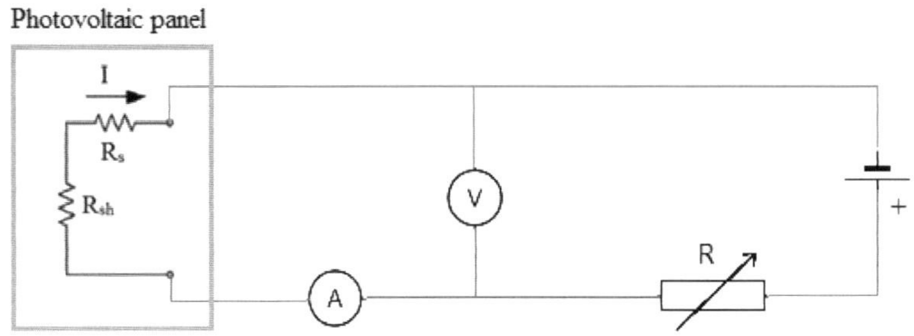

Figure 4. Experimental scheme to determine R_{sh}

According to the Law of Mail:

$$R_{sh\_36} + R_{s\_36} = \frac{V}{I} \tag{3}$$

Then an electrical measurement is taken: V = 50 V and I = 250 μA. So:

$$R_{sh\_36} + R_{s\_36} = 200\ k\Omega$$

For one cell:

$$R_{sh} \simeq 5550\ \Omega$$

4. Simulation

From the equivalent circuit of a PV panel represented in Figure 1, the following equations are deduced [8], [9] and [10]:

$$I = n_p I_{ph} - n_p I_{sat}\left(\exp\left(\left(\frac{q}{AKT}\right)\left(\frac{V}{n_s} + R_s I\right)\right) - 1\right) \tag{4}$$

$$I_{ph} = (I_{sso} + k_i(T - T_r))\frac{S}{1000} \tag{5}$$

$$I_{sat} = I_{rr}\left(\frac{T}{T_r}\right)^3 \exp\left(\left(\frac{qE_{gap}}{kA}\right) * \left(\frac{1}{T_r} - \frac{1}{T}\right)\right) \tag{6}$$

Where the Table 3 presents the values of measured parameters and Table 4 presents the values of the characters in the equation 4.
Then, these equations are used to do simulation on Simulink with the internal parameters found in the experimental measurements and the manufacturer's values.

Table 3: measured parameters

Symbol	Description	Value
T	Surface temperature of the panel	303° K
R_s	Series resistance of the PV cell	0,015 Ω
R_{sh}	Parallel resistance of the PV cell	5500 Ω

Table 4: parameters values

Symbol	Description	Value
q	Electron charge	1.602×10^{-19} C
A	Ideality factor	1.74
k	Boltzman constant	1.38×10^{-23} j/K
I_{sso}	Short-circuit current	6.96 A
k_i	SC current temperature coefficient	1.7×10^{-3}
T_r	Reference temperature	298.13 °K
I_{rr}	Reverse saturation current at T_r	2.0793×10^{-6} A
E_{gap}	Energy of the band gap for Silicon	1.1 eV
n_p	Number of cells in parallel	0
n_s	Number of cells in series	36
S	Solar radiation level	0 ~ 1000 W/m²
I_{ph}	Photocurrent	
I_{sat}	Module reserve saturation current	
S	Solar radiation level	1000 W/m²

5. Programming

To validate in a second way the values found in the experimental part and to give the study more accuracy, a program under Matlab software have been created with "fsolve" function to determine the values of the internal panel: R_s, R_{sh}, T, A and I_{sat}.

Matlab's "fsolve" subroutine is very effective in solving the nonlinear equations. Thus, the proposed method is user-friendly, with the use of initial values for to start the simulation.

Equation (4) has been used for a single cell, and also under the same conditions in the experimental measurements by covering the PV panel, so I_{ph}= 0 and the equation becomes:

$$-\mathbf{I_{sat}}\left(\exp\left(\frac{q(V+\mathbf{R_s}I)}{\mathbf{AKT}}\right)-1\right)-\frac{V+\mathbf{R_s}I}{\mathbf{R_{sh}}}-I=0 \tag{7}$$

The values of experimental measurements of voltage and current have been used, and five measurements were taken to create a system (8) of five equations:

$$\begin{cases} -I_{sat}\left(\exp\left(\frac{q(V_1+R_sI_1)}{AKT}\right)-1\right)-\frac{V_1+R_sI_1}{R_{sh}}-I_1=0 \\ -I_{sat}\left(\exp\left(\frac{q(V_2+R_sI_2)}{AKT}\right)-1\right)-\frac{V_2+R_sI_2}{R_{sh}}-I_2=0 \\ -I_{sat}\left(\exp\left(\frac{q(V_3+R_sI_3)}{AKT}\right)-1\right)-\frac{V_3+R_sI_3}{R_{sh}}-I_3=0 \\ -I_{sat}\left(\exp\left(\frac{q(V_4+R_sI_4)}{AKT}\right)-1\right)-\frac{V_4+R_sI_4}{R_{sh}}-I_4=0 \\ -I_{sat}\left(\exp\left(\frac{q(V_5+R_sI_5)}{AKT}\right)-1\right)-\frac{V_5+R_sI_5}{R_{sh}}-I_5=0 \end{cases} \tag{8}$$

Table 5 presents the five measurements of voltage and current used in the program:

Table 5: measurements used in the programme

	1	2	3	4	5
V (V)	13.6	16.1	19	21.4	23.15
I (A)	- 0.01	- 0.03	- 0.2	- 1.5	- 4

6. Results and discussions

To determine the two experimental characteristics of the PV panel I-V and P-V, the Figure 5 scheme have been used, which consists of a variable resistance, an ammeter and a voltmeter. The PV panel was illuminated by a projector then the power produced is very low.

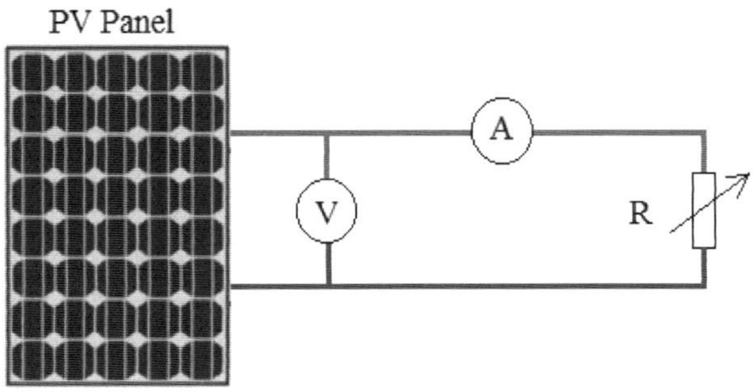

Figure 5. Experimental scheme to determine experimental I-V and P-V

By decreasing the variable resistance we have plotted two experimental characteristics I-V and P-V presented in a. and b. successively in Figure 6.

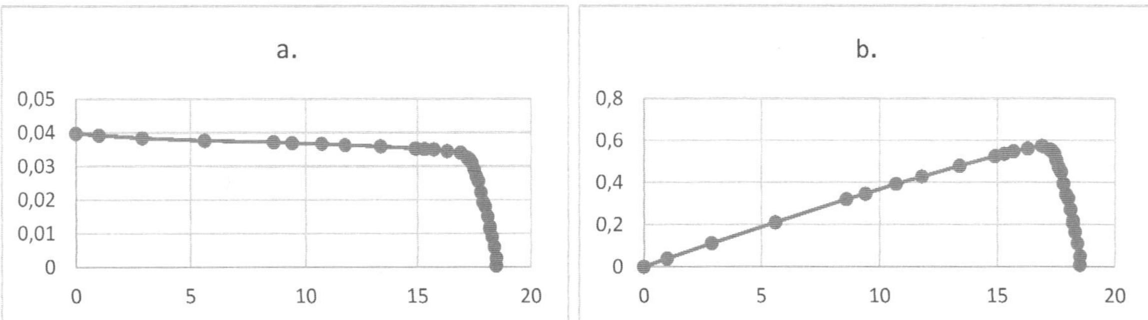

Figure 6. Experimental characteristics a. I-V and b. P-V

The simulation of the PV panel model on Simulink software was done and figure 7 presents two characteristics I-V and P-V of this PV panel. The results found are exactly what we have in the nameplate presented in table 1.

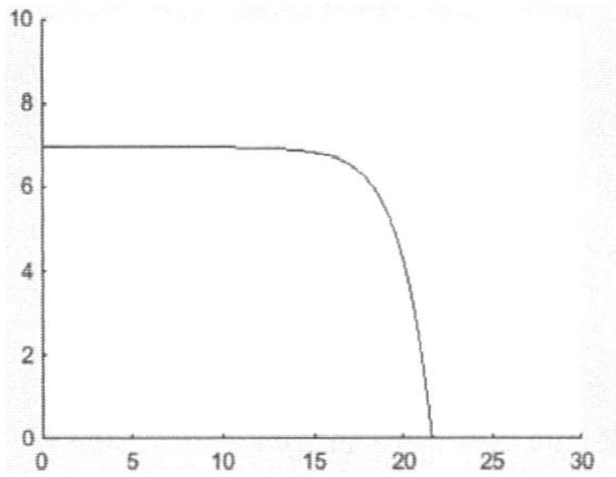

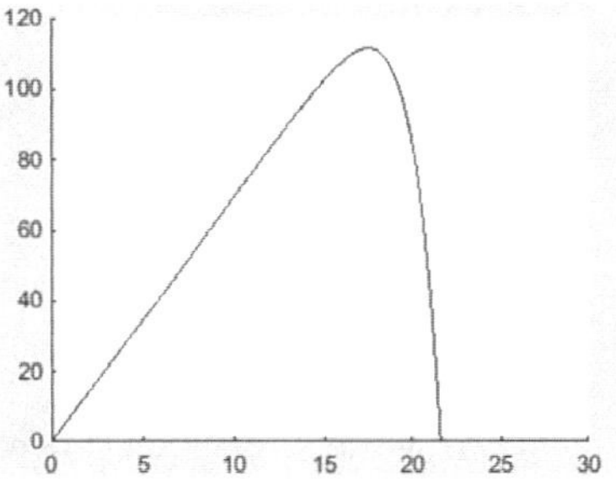

Figure 7. Characteristics of the PV panel

The results of the programming are presented in table 6:

Table 6: Program simulation results

Theoretic data	Measurements data
R_S = 0.016 Ω	R_S = 0.015 Ω
A = 1.733	A= 1.74
T = 304.143°K = 29,008 °C	T = 27 °C
R_{sh} = 5548.210 Ω	R_{sh} ≃ 5550 Ω
I_0 = 2.44953e-06 A	I_0 = 4.842*10×10^{-6} A

The results of the simulation and measurement are very close, which gives our study more accuracy.

7. Conclusion

In this paper, experimentally internal resistances of a PV panel have been determined: series and shunt resistances.

The PV panel was simulated using Simulink to determine theoretical I-V and P-V characteristics.

Also Experimental characteristics of the PV panel: I-V and P-V were plotted.

In order to determine all internal characteristics, a program on Matlab with "fsolve" function have been created to solve the nonlinear equation. With "fsolve" function, parameters (R_S, R_{sh}, A, T, and I_{sat}) from a system of five equations have been determined. The system is obtained from experimental measurements, by using the values of the voltage and current.

Finally, theoretical results are very close to the experimental values.

Acknowledgment

This research work is within the framework of the project "intelligent management of renewable energy sources" GISER, funded by the CNRST (for the MESRSFC, Morocco).

References

[1] J. Selvaraj and A. R. Nasrudin, "Multilevel Inverter For Grid-Connected PV System Employing Digital PI Controller", IEEE Transactions On Industrial Electronics, vol. 56, No. 1 2009, pp. 149-158.

[2] S. Nema, R.K. Nema and G. Agnihotri, "MATLAB/Simulink based study of photovoltaic cells / modules / array and their experimental verification", International journal of Energy and Environment, vol.1, No.3, pp.487-500, 2010.

[3] H. L. Tsai, C. S. Tu and Y. J. Su, "Development of Generalized Photovoltaic Model Using MATLAB/SIMULINK", Proceedings of the World Congress on Engineering and Computer Science WCECS, San Francisco, USA, 2008.

[4] M. Francisco and G. Longatt, "Model of Photovoltaic Module in Matlab™", 2do congresoı beroamerı cano de estudıantes de ıngenıeríaeléctrı ca, electrónıca y computacıón pp.1-5, 2005.
[5] M. Salhi, R. El-Bachtiri and E. Matagne, "The development of a new maximum power point tracker for a PV panel". International Scientific Journal for Alternative Energy and Ecology (ISJAEE), 2001.
[6] S. Nema, R. K. Nema and G. Agnihotri, "MATLAB/Simulink based study of photovoltaic cells / modules / array and their experimental verification", International journal of Energy and Environment, vol.1, No.3, pp.487-500, 2010.
[7] T. Salmi, M. Bouzguenda, A. Gastli and A. Masmoudi. « Matlab/simulink based modeling of photovoltaic cell". International Journal of Renewable Energy Research (IJRER), 2(2), 213-218. 2012
[8] M. E. Ropp and S. Gonzalez, "Development of a MATLAB/Simulink model of a single-phase grid-connected photovoltaic system," IEEE Trans. Energy Conv., vol. 24, no. 1, pp. 195–202, Mar. 2009.
[9] K. H. Chao, C. J. Li and S. H. Ho, "Modeling and fault simulation of photovoltaic generation systems using circuit-based model," in Proc.
[10] M. Salhi, and R. El-Bachtiri, "Determination of equivalent circuit parameters of a photovoltaic panel," Seventh International Symposium on Environnement Catalysis and Processes Engineering (ECGP'7), ENIM-Rabat (Morocco), April 2008, pp. 149.
[11] M. ARJUN, K. U. RAO and A. B. RAJU, "A novel simplified approach for evaluation of performance characteristics of SEIG". International Conference In: Advances in Energy Conversion Technologies (ICAECT), IEEE. p. 31-36 2014.

Structural, morphological and optical properties of copper-tin-sulfide thin film synthesized by spin coating technique

F Welatta[1,2], A El kissani[1], M Aggour[2] *and* A Outzourhit[1]

[1]Nanomaterials for Energy and Environment Laboratory, Faculty of Sciences Semlalia, Cadi Ayyad University.
[2]Renewable Energies and Environment Laboratory, Faculty of Sciences, Ibn Tofail University.

E-mail: Fatimetou_83@yahoo.fr

Abstract. Copper-tin-sulfide, thin films were fabricated by the sol-gel method on soda lime glass. The structural, morphological and optical properties of these thin films were studied by x-ray diffraction, scanning electron microscopy, and ultraviolet-visible spectroscopy. X-ray diffraction studies revealed the formation of the Cu_3SnS_4 phase for annealed thin films at 350°C and 400°C in argon atmosphere for one hour. The scanning electron microscope study revealed a good surface for solar cells. The optical transmission measurements showed that the band gap of these films decrease from 0.97 eV to 0.82 eV after annealing.

1. Introduction

Copper indium gallium selenide (CIGS), CdTe and copper indium selenide Cu_2InSe_2 (CIS) thin films absorbers are used today in commercial thin film photovoltaic devices. High conversion efficiency of approximately 21.7 % was achieved in CIGS-based cells [1]. The high cost and scarcity of indium (In), gallium (Ga) and tellurium (Te) besides the toxicity of cadmium (Cd) and its environment impact limit the large scale production of CIS, CIGS and CdTe photovoltaic panels that can compete with the silicon technology. Copper tin sulfide (CTS) is potential candidate for low-cost thin film solar cells, because of its suitable direct band gap between 0.82 and 1.7 eV [10,11], large absorption coefficient over 10^4 cm^{-1} and the abundance and non-toxicity of its elements [12]. Several techniques were used or the deposition CTS thin films, namly sputtering [2], chemical bath deposition (CBD) [3], spray pyrolysis [4], electro-deposition [5], pulsed laser [6], co-evaporation [7], sol-gel [8], sulfurization [9], etc. However, most of these techniques require a sulfurization step. In the present article, we propose the synthesis of Cu_3SnS_4 thin films by the sol-gel process without any further sulfurization. The sol-gel is a versatile, potentially low-cost technique that can be used to deposit CTS thin films.

2. Experimantal

2.1. Sample preparation
The copper tin sulfide thin films were deposited by the spin coating technique on soda lime glass substrates. The substrate was first cleaned by successive ultrasonic bath of water, acetone and water respectively during 15 min and dried in an oven. The precursor solution consisted of 2 M copper (II) chloride $CuCl_2.H_2O$; (>99% from Sigma Aldrich), 1.3 M tin (II) chloride dehydrate $SnCl_2.2H_2O$; (98% from Sigma Aldrich) and 8 M thiourea H_2NCSNH_2; (>99% from Sigma Aldrich) added in a solvent mixture composed of dionized water/ethanol (v/v 30:70). The mixture was then stirred at 40°C during 40 min to yield a clear and homogenous solution. The resultant solution was then spin-coated using an SPI the spin coater Model P6700 onto glass substrate at 4000 rpm for 30 s followed by solvent-drying at 100°C for 5 min. The deposition step was repeated three times in order to achieve a suitable thickness. Subsequently, the films were annealed at two temperatures of 350°C and 400°C in an Argon atmosphere for one hour using a tubular furnace.

2.2. Characterization
The structural, morphological and optical properties were studied. The crystalline structure was analysed by X-ray diffraction using copper a CuKα radiation (λ=1.5418 A°) and an X'Pert difractometer (PANalytical, Almelo, The Netherlands). Sample morphology and uniformity were determined by a scanning slectron microscope (SEM) (TESCANVGA 3) at the center of analysis and characterization, Cadi Ayyad University. The elemental composition of the films was examined by an energy dispersive spectroscopy (EDS). The optical properties were measured on a UV-Vis- NIR double beam spectrophotometer (Shimadzu, Tokyo, Japan).

3. Resultats and Discussions

3.1. Structural charcterization
The X-ray diffraction (XRD) patterns of thin films deposited on soda lime glass substrate are shown in figure 1. These patterns contain three diffraction peaks indicating that these films are polycrystalline in nature. Only the peaks characteristic of the Kuramite phase of CTS can be observed in this case. The major XRD peaks are located at 2θ = 28.5 °, 47.4° and 56.5 °. Figure 1 shows that the films exhibit a high intensity peak at 2θ = 28.5 °, suggesting a preferred growth in the (112). The other peaks of this phase are also present with relatively lower intensity. We also note that annealing the samples at 400°C improved the intensity of the peaks and reduces their width. This suggests that annealing has improved the crystallinity of our films. Table 1 shows that the observed d-spacing values of our sample are in good agreement with the values the tetragonal (*I -42m)* Cu_3SnS_4 phase (JCPDS 33-0501 standard). The lattice parameters corresponding to tetragonal crystal structure were calculated using the relation [13] :

$$d_{hkl} = \frac{1}{\sqrt{\frac{h^2+k^2}{a^2}+\frac{l^2}{c^2}}} \quad (1)$$

Where (*hkl*) are the Miller indices and *d* is the interplanar distance. The lattice parameter values are a= b = 5.42 A°, c = 10.86 A°. The d_{hkl} is the interplanar distance which is determined by Bragg's Law [14]:

$$d_{hkl} = \frac{n\lambda}{2sin\theta} \quad (2)$$

Where n is a positive integer, λ =1.541 Å is the radiation wavelength and θ is Bragg's angle of the (hkl) planes associated to the structure.
The evaluated lattice parameters are in close agreement with the JCPDS 33-0501. The ratio of c/2a, which is almost equal to one, shows that the phase is tetragonal.
The full width at half maximum (FWHM) of the (112) diffraction peak for the thin films annealed at 350 °C and 400°C, allows to estimate the grain size D by using Scherrer's formula [15] :

$$D = \frac{0.9\lambda}{\beta\cos\theta} \tag{3}$$

Where λ is the X-ray wavelength and β is the FWHM of the (112) peak and θ is the corresponding Bragg diffraction angle. The estimated crystallite sizes are 7.5 nm and 9 nm for the films annealed at 350°C and 400 °C respectively.
Furthermore, the lattice strain effect can be calculated using the following equation (Williamson- Hall method) [16]:

$$\varepsilon = \frac{\beta}{4\tan\theta} \tag{4}$$

The dislocation density Ψ was calculated using the Williamson and Smallman formula [17]:

$$\Psi = \frac{1}{D^2} \tag{5}$$

The number of crystallites N per unit surface is determined using the following relation [18]:

$$N = \frac{t}{D^3} \tag{6}$$

Where t is the film thickness, which was estimated to 1400 nm. It is clear from Table 2 that with the increase of the annealing temperature, the crystallite size of CTS films increases while strain, dislocation density and number of crystallites decrease.

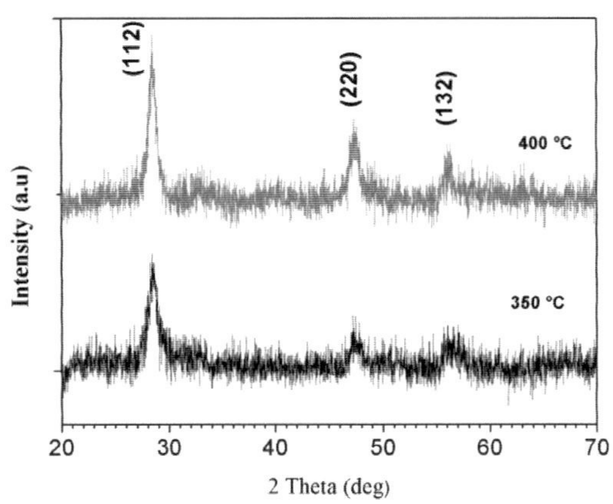

Figure 1. XRD patterns of the Cu_3SnS_4 (CTS) thin films annealed at 350°C and 400°C in argon gas for one hour.

Table 1. Comparison between observed d spacing values of Cu_3SnS_4 and JCPDS References.

2θ (°)	*hkl*	d-spacing (Å)			Compounds
		Observed values		JCPDS Ref	
		350°C	400°C		
28.5	112	3.13070	3.13361	3.13000	Cu_3SnS_4
47.4	220	1.91774	1.9286	1.91400	Cu_3SnS_4
56.5	132	1.62689	1.64030	1.64000	Cu_3SnS_4

Table2. Summary of calculations from Debye Scherer formula and results of Williamsone Hall analysis for lattice strains.

Samples	FWHM(°)	D(nm)	$\Psi(nm^{-2})$	$N(nm^{-2})$	$\varepsilon(10^{-3})$
350°C	0.0191	7.5	0.017	3.317	0.018
400°C	0.0161	9	0.012	1.920	0.015

3.2. Morphological characterization

Figure 2 shows SEM surface micrographs for samples annealed at 350°C and 400°C in argon atmosphere for one hour. Annealing at 350°C (figure 2a) resulted in grains having almost the same size. For 400°C (figure 2b), there is appearance of small grains and beginning of formation of crystallization. As it can be seen in the SEM images of Figure 2 (c,d,e and f), the surface of the annealed thin films is homogenous with no cracks or pinholes, which is suitable for photovoltaic applications. The compositional analysis of the films is investigated by the energy dispersive spectroscopy. The quantitative elemental analysis is carried out for Cu, Sn and S (Table 3). The atomic ratio S/(Cu+Sn) is close to that of the Cu_3SnS_4 stoichiometric. The weak Cu/Sn ratio indicates that there are phases that are not yet crystallized. In fact the EDS detects all the phases while the XRD detects only the crystalline phases.

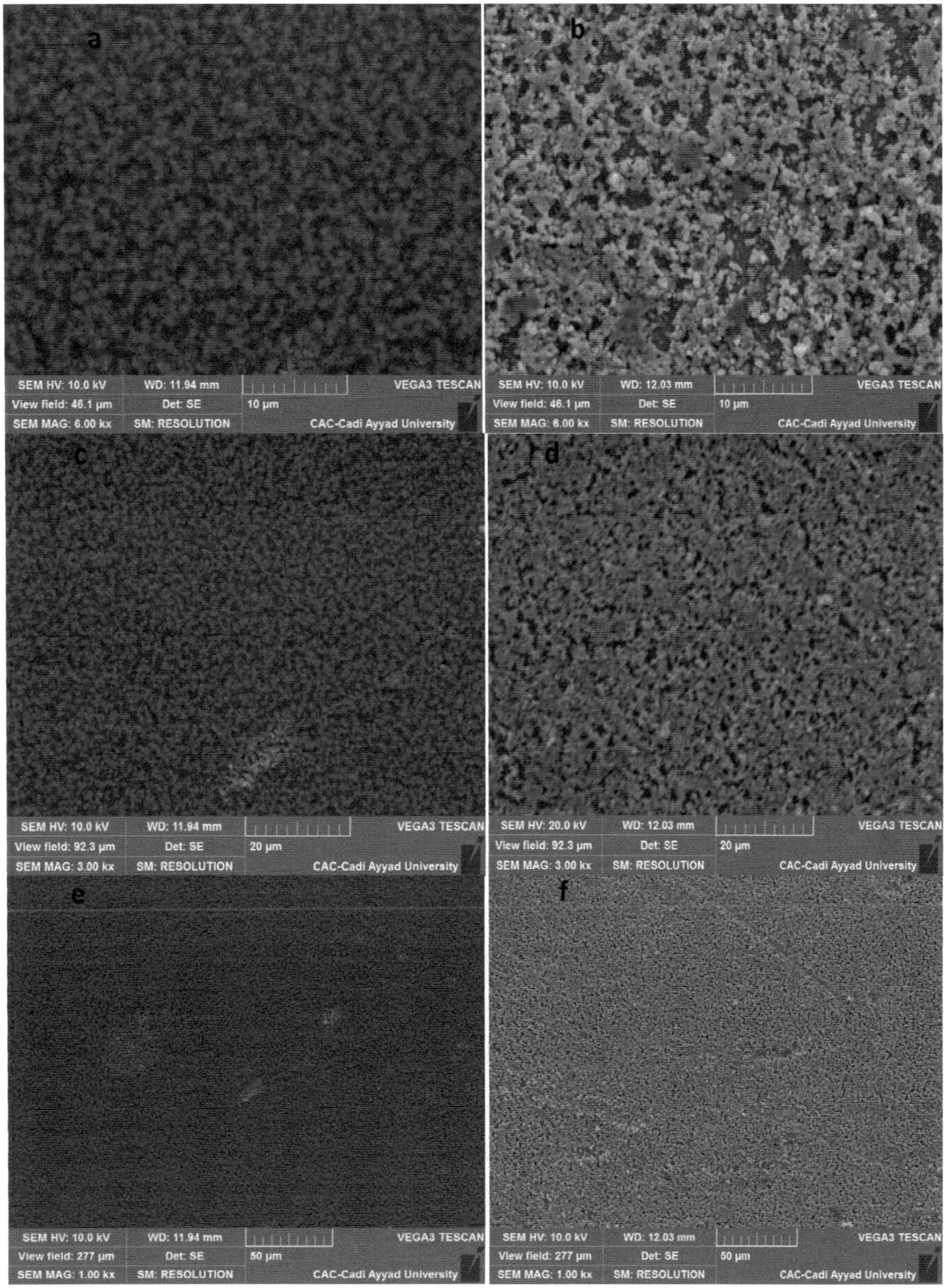

Figure 2. SEM micrographs of Cu_3SnS_4 absorbers annealed for (a), (c) and (e) at 350°C; and for (b), (d) and (f) at 400°C respctivily.

Table 3. Composition ratios of the annealed films from EDS.

Temperature (°C)	Atomic percent			Atomic ratios	
	Cu	Sn	S	Cu/Sn	S/(Cu+Sn)
350	24.97	26.73	48.30	0.73	0.93
400	21.26	24.15	54.58	0.38	1.20

3.3. Optical characterization

The absorption coefficient α was calculated from the transmittance T using the approximation $T = e^{-\alpha d}$, which is valid in the high-absorption region of the spectrum. Where, d is the film thickness which is estimated to be 1.4 μm. The optical gap (Eg) was evaluated from the plot of $(\alpha h\upsilon)^2$ against photon energy (hυ). This plot should be linear near the absorption edge in the case of a direct band gap semiconductor, which is the case for our films (Figure 3). The optical gap is obtained by extrapolating the linear region of this plot to hυ = 0. Hence, by using this method, we found that the band gap is 0.97 eV at 350°C and 0.82 eV at 400°C [10]. These results are in good agreement with the reported values.

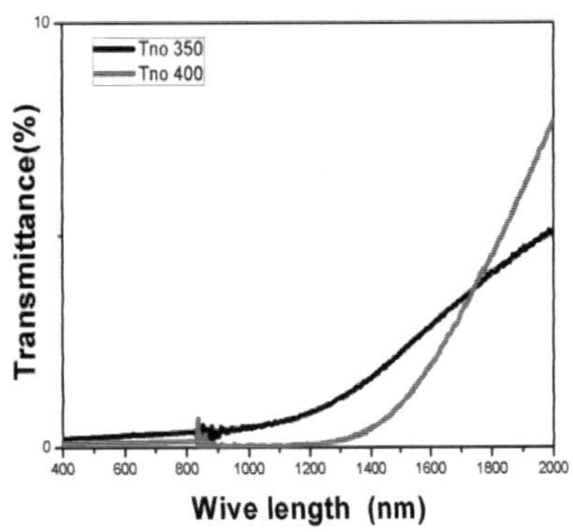

Figure 3. Optical transmittance as a function of photon energy of the CTS film annealed at 350 °C and 400 °C.

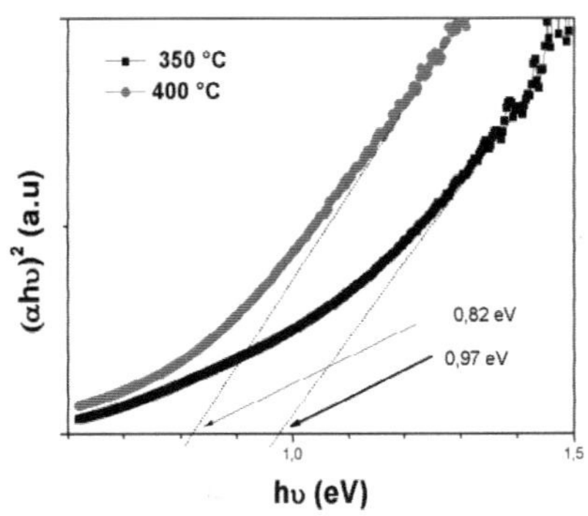

Figure 4. Square of the product of the absorption coefficients and photon energies of the annealed thin films as a function of photon energy.

4. Conclusion

The copper tin sulfide thin films were successfully synthesized by spin coating without sulfurization. The effects of annealing at 350°C and 400°C temperature for 1 hr on the structural, morphological and optical properties of the CTS thin films were investigated. The X-ray diffraction result shows that the obtained layer was composed by Cu_3SnS_4 Kuramite phase and has a (I-42m) tetragonal structure with (112) preferential orientation. The prepared films showed a good homogeneity for photovoltaic applications. The deposited films showed high optical absorption coefficient with a band gap between 0.82 eV and 0.97 eV.

References

[1] Oon CH and Ng KW. *Energy Procedia* 2017; 143: 733–738.
[2] Young I, Yeon J, Ghorpade U V, Suryawanshi M P, Seon D and Hyeok J. *J Alloys Compd* 2016; 688: 12–17.
[3] Anuar Kassim, Saravanan Nagalingam, Zulkefly Kuang, Atan Sharif, Tan Wee Tee and Ho Min S. *Chem Soc Ethiop* 2010; 24: 259–266.
[4] Chalapathi U, Jayasree Y, Uthanna S, and Sundara V. Phys. Status Solidi 2013; 2390: 2384–2390.
[5] Kassim A, Min HS, Tee TW, Sharif A, Kuang Z, Haron MJ and Nagalingam S. CMU.J.Nat.Sci 2008; 7: 317–326.
[6] Vanalakar SA, Agawane GL, Kamble AS, Hong CW, Patil PS and Kim JH. Solar Energy Materials & Solar Cells 2015; 138: 1–8.
[7] Vani VPG, Reddy MV and Reddy KTR. Hindawi 2013; 2013:6.
[8] Dias S, Murali B and Krupanidhi SB. *Mater Chem Phys* 2015; 167: 309–314.
[9] Wang CJ, Shei SC, Chang SC and Chang SJ. Applied Surface Science 2016; 388: 71–76.
[10] Chen Q, Dou X, Ni Y, Cheng S and Zhuang S. Journal of Colloid and Interface Science 2012; 376: 327–330.
[11] Aihara N, Kanai A, Kimura K, Yamada M, Toyonaga K, Araki H, Takeuchi A and Katagiri H. IOP SCience 2015; 13: 1–6.
[12] Tiwari D, Chaudhuri TK, Shripathi T, Deshpande U and Rawat R. *Sol Energy Mater Sol Cells* 2013; 113: 165–170.
[13] Belaqziz M, Medjnoun K, Djessas K, Chehouani H and Grillo SE. *Mater Res Bull* 2018; 99: 182–188.
[14] Tripathy SK. *African Rev Phys* 2011; 6: 111–119.
[15] Shelke HD, Lokhande AC, Patil AM, Kim JH and Lokhande CD. *Surfaces and Interfaces* 2017; 9: 238–244.
[16] Shelke HD, Lokhande AC, Kim JH and Lokhande CD. *J Colloid Interface Sci* 2017; 506: 144–153.
[17] Girija K, Thirumalairajan S, Mohan SM and Chandrasekaran J. *Chalcogenide Lett* 2009; 6: 351–357.
[18] Kariper IA. *J Mater Res Technol* 2016; 5: 77–83.

Estimating the Renewable Energy Potential in Morocco: solar energy as a case study

G Tazi[1], O Jbaihi[1], A Ghennioui[3], A Alami Merrouni[3], M Bakkali[2]

[1] School of energy engineering, International University Of Rabat, Sala El Jadida, Morocco.

[2] School of architecture, International University Of Rabat, Sala El Jadida, Morocco.

[3] Research Institute for solar energy and new energies (IRESEN), Green Energy Park, Benguerir, Morocco.

ouafae.jbaihi@uir.ac.ma

ghita.tazi.eng@uir.ac.ma

Abstract. The issue of climate change is expanding over the world. Thus, the interest in renewable energies is growing to minimize the effect of this phenomenon. Environmental policies are changing in different countries (including Morocco); they focus more on the integration and the development of renewable energy projects. The aim of this paper is to assess the potential of Morocco to host solar power plants from both technologies: Concentrated Solar Power (CSP) and Photovoltaics (PV). To have an accurate study, we need to choose the right criteria pursuing a specific methodology. For this reason, four selection criteria were chosen: climate, orography, location and water resource. To calculate the weight/importance of each one of the criteria the Analytical Hierarchy Process (AHP) has been used. Once obtained, the weights are applied on the map of each criteria to generate a final ranking map which classifies the regions according to their potential. Results show that Morocco possesses high potential in both technologies and especially in the south. Finally, this work is the first in its field to include all Morocco in the study area.

1. Introduction

The Conference of Parties, held in Morocco November 2016, sorted out the Marrakech Action Proclamation which underlines its task of moving forward to reduce greenhouse gas emissions and to encourage adaptation measures [1].

Morocco is aware of the importance of enhancing renewable energy investments and encouraging sustainable development. Thus, the Kingdom shows its willingness to increase the share of renewable energies from 42% of the installed capacity for 2020 to 52% for 2030 [2]. In solar, for instance, the capacity production will increase to reach 4560 MW.

The present study aims to estimate the solar potential in Morocco by elaborating a map highlighting the most suitable sites to host solar power plants. As the combination of the Multi-Criteria Decision Making (MCDM) and Geographic Information System (GIS) tools became an effective method [3] to do so, our results will represent a tool for the decision support in the Moroccan renewable energy strategy.

2. Criteria selection

2.1. Methodology

To realize a map of the correct potential of our study area, it is imperative to follow certain steps. According to the literature, the most used methodology is a combination of one of the Multi-Criteria decision making (MCDM) and Geographic Information System (GIS) software. The used steps are: the definition of the weight of each criterion, the development of exclusion maps, the combination of the restriction maps with the resource maps and the development of the technical resource potentials.

In the process, the criteria must be chosen wisely and their weight should be calculated. In this study to calculate the weights we used pair comparison matrix based on the analytical hierarchy process (AHP) method.

The following figure (Figure 1) resume the methodology used.

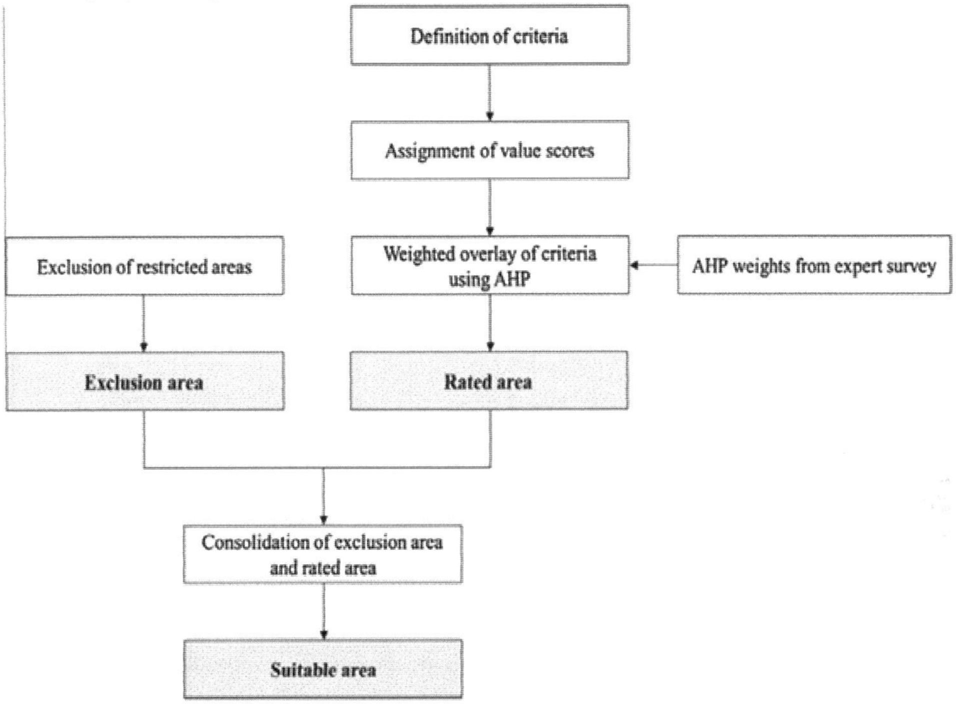

Figure 1. The architecture of the methodology. [4]

To make a decision over the criteria, we assembled all the values and choices of each case in the table below:

Table 1. Values of the criteria for PV.

Criteria	Sub-criteria	Limited suitability	Suitable	highly suitable	Excellent
Location	Distance to urban areas (km)	>40	20	15	10
	Distance from road and railway network (km)	>20	10-20	05-10	5
	Distance for electricity grid (km)	>20	15	10	5
Orography	Slope (°)	10°-45°	5°-10°	3°-5°	1°-3°
Climate	Temperature (°C)	>45	30-45	15-30	0-15
	GHI (kWh/m²)	1650-1950	1950-2100	2100-2150	2150-2500
Water source	Distance from water ways (km)	>15	15	10	5
	Distance from dams (km)	>25	15-25	10	5

Table 2. Values of the criteria for CSP.

Criteria	Sub-criteria	Limited suitability	Suitable	Highly suitable	Excellent
Climate	Solar radiance values (DNI) (kWh/ m2/year)	1800 - 2000	2000 - 2300	2300 - 2500	2500 - 2800
Orography	Suggested values of slopes (°)	>2.1	1.5-2.1	1 - 1.5	0.5-1
Location	Distance from cities and urban areas (km)	30.0 - 200.0	20.0- 30.0	15.0 - 20.0	10.0 - 15.0
	Distance from road and railway network (km)	> 40	10.0 - 40.0	5.0 - 10.0	5.0
	Distance from airports (km)	2.0 - 3.0	3.0 - 4.0	4.0 - 5.0	5.0 - 6.0
	Distance from Electricity grid (km)	>200	100 - 150	50 - 100	1.0 - 50.0
Water resources	Distance from water ways (km)	>15	10 -15	5--10	< 5
	Distance from dams (km)	> 25	15 - 25	10 - 15	<10

2.2. *Methodology*

Before conducting any Solar plant project, a pre-feasibility study needs to be done taking into account some different criteria. A criterion is a standard we take to make a judgment and decision. For both PV and CSP technologies the same criteria have been selected in this paper. Those criteria are climate, orography, water resources and location [5]. However, the difference between CSP and PV resides in the sub-criteria (like temperature, GHI and DNI [6]) and their weights. In what follow we will discuss the selected criterions:

- Climate: For this study, this criterion takes a big part in estimating the potential of PV and CSP in Morocco. For PV, two sub-criteria are evaluated: temperature and global horizontal irradiation. As for CSP, one criterion is needed which is the Normal Direct Irradiation (DNI). The efficiency of the panels and therefore the yield of solar plant is mostly affected by the irradiation.
- Orography: It is the study of the reliefs and elevation or in other words the terrain's slope analysis. For economic reasons, the lower the slope is the lower investment cost will be. Each of the technologies has its sensibility to the land slope, but the CSP is the most sensitive one to the angle of the slope and it requires a maximum slope value of 2.1% to work efficiently.
- Water resource: This resource is important for both technologies especially in arid regions. Indeed, soiling affects considerably the productivity of a solar plants and thus it need to be cleaned. Additionally, CSP power plants require water for cooling the steam turbines. In our study two sub-criteria were used: the distance from water ways and the distance from dams.
- Location: This criterion regroups three sub-criteria: the distance to urban areas, the distance from road and railway network and the distance for electricity grid. During the installation of a solar plant, the transportation of the materials as well as the labors are needed.

Consequently, the plant should be as close as possible to the roads to minimize the construction expenses. Also the best scenario is to have plant close to the electricity grid, to inject the produced electricity.

3. The analytical hierarchy process

3.1. Multi-Criteria Decision Making (MCDM)

Evaluating the potential of Morocco to host solar power plants, and classifying regions according to the suitability level needs to take into account many criteria as well as applying decision-making methods.

MCDM methods are often divided to multi-object decision making (MODM) and multi-attribute decision making (MADM). The difference between them is the domain of alternatives, while MODM concerns a continuous, MADM works with a discrete domain of alternatives. [4]

The AHP approach, which is adopted in this study, is a MADM method. It is one of the most commonly used MCDM methods despite the debate around it. This method has demonstrated its capacity at resolving many decision-making problems in several fields and renewable energy analysis is among them.

The AHP method, developed by Saaty in 1977[5], is a process of comparing the importance of a group of criteria and sub-criteria so as to reach a certain aim. Ours is to determine the most suitable sites for CSP in Morocco.

A matrix is generated after setting a goal and the number n refers to the number of criteria, which are set for comparison. The procedures are as following [7]:

- The matrix (P) is a (n x n) one where pij describes the importance of the criterion i to the criterion j. The relative importance is set according a numerical scale from 1 to 9.
    - pii=1
    - We assume that pji=1/pij
- For weight calculation, we need to normalize the matrix(P) by following these steps:
    - Sum the values of each column,
    - Divide each entry by its column sum [8],
    - Average across rows to get the relative weights. [5]
- To control the consistency of the calculated weights in case a problem arises, the consistency ratio (CR) is calculated as following:
    - Calculation of the eigenvector and the maximum eigenvalue for each matrix.
    - Calculation of an approximation to the consistency index (CI):

$$CI = \frac{\lambda max - n}{n-1} \quad (1)$$

Where λmax is the maximum eigenvalue for each matrix.

- To ensure the consistency of the pairwise comparison matrix, the consistency judgment must be checked for the appropriate value of n by CR that is:

$$CR = \frac{CI}{RI} \quad (2)$$

Where RI is the random consistency index. The RI values for different numbers of n are shown in [4].

If CR≤0.10, the degree of consistency is satisfactory.

If CR>0.10, there are serious inconsistencies. In this case, the pairwise comparison should be recalculated.

4. Results:

After using the AHP method, we obtained the following weights of each criterion.

Table 3. The weights of the criteria for PV.

Criteria	Weight	Sub-criteria	Weight
Location	6%	Distance to urban areas (km)	1,62
		Distance from road and railway network (km)	0,66
		Distance for electricity grid (km)	3,72
Orography	19%	Slope (°)	19,00
Climate	62%	GHI (kWh/m²)	62,00
Water source	13%	Distance from water ways (km)	10,4
		Distance from dams (km)	2,6

Table 4. The weights of the criteria for wet cooling CSP.

Criteria	Sub-criteria	Detailed weights	Global weights
Climate	Solar radiance values (DNI) (kWh/ m2/year)	57,7%	57,7%
Orography	Suggested values of slopes (°)	25,6%	25,6%
Water resources	Distance from water ways (km)	9,36%	11,7%
	Distance from dams (km)	2,34%	
Location	Distance from cities and urban areas (km)	1,4%	5,0%
	Distance from road and railway network (km)	0,5%	
	Distance from airports (km)	0,6%	
	Distance from Electricity grid (km)	2,6%	

After obtaining the weights, we applied them on the criterion of each technology. As a result, we generated the ranking maps below.

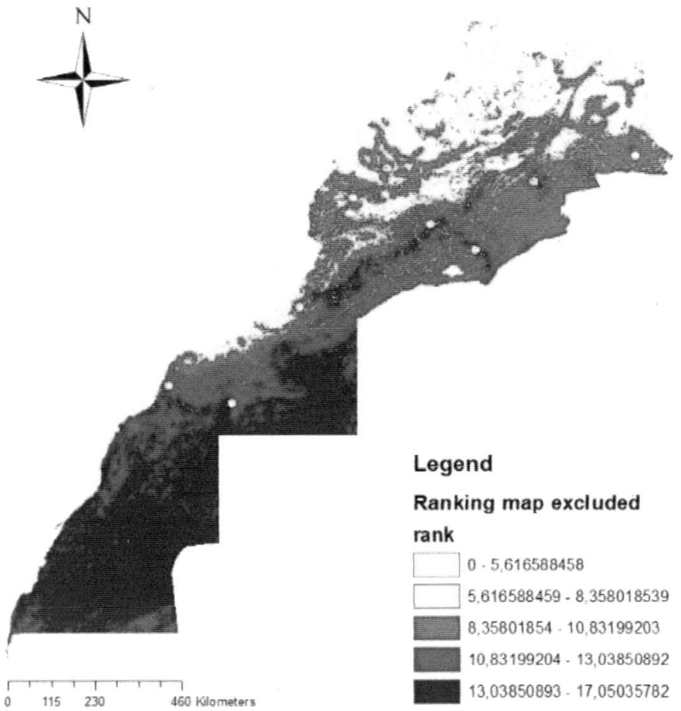

Figure 2. The ranking map of PV.

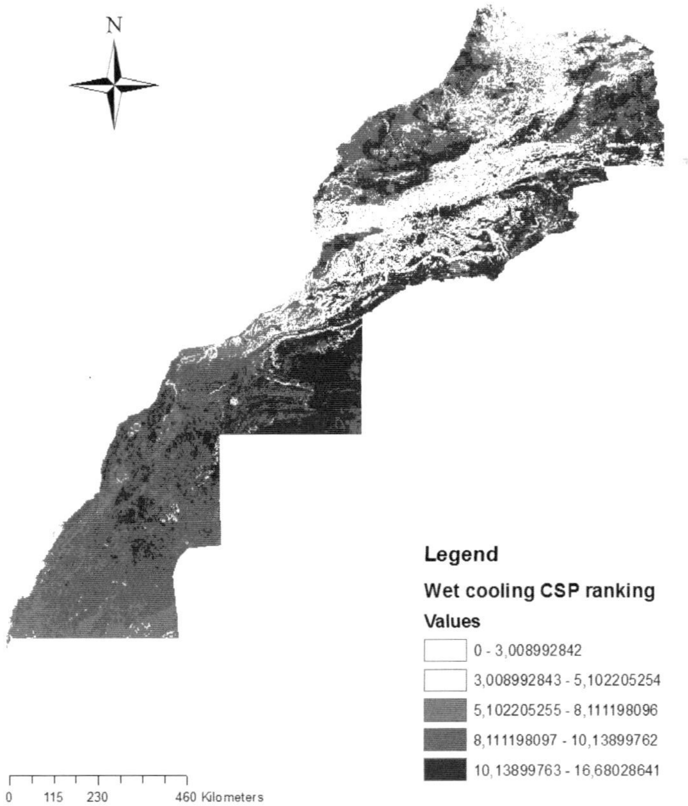

Figure 3. The ranking map of wet cooling CSP potential.

5. Conclusion

Energy is a key factor to development whether it is social, economic, or environmental. While fossil fuels are running out, renewable energy resources suggest clean alternatives to them. The sun is a powerful source of energy since solar energy can be used for generating electricity, heating, lighting and cooling buildings etc....

This paper shows an application of combining AHP with GIS for site selection of solar farms in Morocco. The AHP is used to evaluate the importance and determinate weights of criteria. The AHP methodology integrated with GIS are very important for the effective estimation of the solar farms sites. Finally, a ranking map will be created combining all criteria that will provide a decision support to choose the most suitable solar farms sites.

6. Acknowledgment

The authors are gratefully thanking IRESEN (Institut de Recherche de l'Energie solaire et Energies Nouvelles) and the Green Energy Park team for their support for this project

References

[1] "Marrakech Action Proclamation", 2016, COP22, Morocco
[2] Ettaik Z, 2017, Ministry of Energy, Mines, Water and Environment of Morocco homepage. [online]. Available: http://www.mem.gov.ma/SiteAssets/Monographie/DirectionsCentrales/DEREE.pdf
[3] Tahri M, Hakdaoui M and Maanan M, nov. 2015, "The evaluation of solar farm locations applying Geographic Information System and Mutli-Criteria Decision-Making methods: case study in southern Morocco", *Ren. and Sust. En. Rev.***51**, 1354-62
[4] Höfer T, Sunak Y, Siddique H and Madlener R, Feb. 2016, " Wind farm siting using a spatial Analytic Hierarchy Process approach: A case study of the Städteregion Aachen", *Ap. En.***163**, 222-43
[5] Alami Merrouni A, Elwali Elalaoui F, Mezrhab A, Mezrhab A and Ghennioui A, "A GIS-AHP combination for the sites assessment of large-scale CSP plants with dry and wet cooling systems. Case study: Eastern Morocco", *Sol. En. E.***166**,2-12
[6] Helioclim V3 de Armines/ Mines ParisTech/ Transvalor
[7] Uyan M, Dec. 2013, "GIS-based solar farms site selection using analytic hierarchy process (AHP) in Karapinar region, Konya/Turkey", *Ren. and Sust. En. Rev.* **28**, 11-17
[8] Wang J J, Jing YY, Zhang C F and Zhao J H, 2013,"Review on multi-criteria decision analyses aid in sustainable energy decision-making", *Ren. and Sust. En. Rev.***13**, 2263-78

Nomenclature

CSP: Concentrated Solar Power
PV: Photovoltaic
AHP: Analytical Hierarchy Process
MCDM: Multi-Criteria Decision Making
GIS: Geographic Information System
GHI: Global Horizontal Irradiation
DNI: Direct Normal Irradiation
MODM: Multi-object decision making
MADM: Multi-attribute decision making
Pii: describes the importance of the criterion i to the criterion i in the matrix
Pji: describes the importance of the criterion j to the criterion i in the matrix
Pij: describes the importance of the criterion i to the criterion j in the matrix
CI : Consistency Index

N : Number of the criteria
CR : Consistency Ratio
RI : Random Index
λmax: Eigenvalue of the pairwise comparison matrix

Thermocline Thermal Storage Using a Packed bed of Moroccan Rocks

K El Alami[1], M Asbik[1], N Zari[2] and S Rachidi[3]

[1] ERTE, Ecole Normale Supérieure de l'Enseignement Technique, Université Mohammed V, Rabat, Morocco.
[2] Moroccan Foundation for Advanced Science Innovation and Research (MAScIR)
[3] Institut de Recherche en Energie Solaire et Energies Nouvelles (IRESEN), 16, Rue Amir Sidi Mohamed, Souissi, Rabat, Morocco.

E-mail: [1]dijaelalami@gmail.com , [1] mohamed. asbik@um5.ac.ma, [2]n.zari@mascir.com, [3]rachidi@iresen.org

Abstract. This work is a contribution to the numerical study of the thermal behavior of a thermocline storage system. The energy conservation equations associated with their boundary conditions, have been solved numerically in the two filed media solid (rock) and fluid (HTF) constituting the packed bed thermal energy storage (TES). A parameter study of the packed bed dimensions, fluid flow rate, particle diameter and the ratio $\frac{H}{D}$ were carried out to evaluate the charging/discharging profiles of storage system. The results have been validated and they are in good agreement with the literature.

Keywords- Thermocline; Packed Bed; Rock; Heat Transfer Fluid.

1. Introduction

Today, concentrating solar power (CSP) technology is one of the most promising technologies for electricity production. However, the intermittency of this clean energy remains a technological challenge to be met by developing high temperature thermal storage systems (TES) [1-2]. These systems offer the possibility of retrieving the stored energy during periods of high demand and/or in the absence of the source of solar energy (nights and/or periods under adverse climatic conditions) [3].

Currently, the most mature TES system technology (Andasol, Noor 1,...) is based on the use of two tanks (hot/cold) in which a fluid (molten salt) circulates allowing the heat storage (charging/discharging) by using the principle of sensible heat [4]. This technology is quite expensive since it represents an initial investment ranging between 15 and 20% of the solar power plant cost [5-6]. Recently, a single thermal storage tank system has been developed at the laboratory scale "Sandia". Compared to this configuration, the TES single tank system can save up to 35% of the investment cost [7-8].Thus, Pacheco et al. [8] developed a packed bed TES system at pilot scale filled with rocks and showed that quartzite and silica sand are the most suitable storage materials in combination with molten salt (Hitec XL) as working fluid. In the same context, Zanganeh et al. [9] studied a packed bed of rocks at a temperature up to 650°C, using air as the heat transfer fluid (HTF), and described a storage system that could be used in a CSP. In 2016, Grirate et al. [10] compared the potential of six rock types proposed as storage materials and their compatibility with synthetic oil as HTF. The thermal properties of these rocks were evaluated experimentally in a temperature range between 25°C and 300 °C.

In this paper, a numerical simulation of the thermal behavior of a thermocline storage system was performed. A local rock "granite" is used as a storage material and "Therminol VP1" as a heat transfer fluid, the effects of different parameters namely: equivalent diameter of the rocks d, the aspect ratio $\frac{H}{D}$, mass flow rate per unit cross section G, were studied to determinate those which gives better performances of the storage system TES.

2. Problem Formulation

2.1. Physical model

The studied storage device (packed bed) is illustrated in the Fig. 1a. It is a cylindrical tank filled with Moroccan rock (Granite) containing void spaces that are occupied by a heat transfer fluid (Therminol VP1). Thermal storage is carried out via the principle of thermocline which is based on a thermal gradient to separate the hot zone from the cold zone inside the considered system. Thermal heat storage consists of one cycle (or more) which is a physical process composed of two stages, namely, the charging and the discharging periods [11]. Fig. 1b shows the operating principle of the thermocline zone, which can represents up to one third of the height of the tank. This zone moves downward during the charging stage whereas it progresses upward when discharging phase takes place. In order to store the heat energy (charging phase), the hot fluid at a temperature T_h is introduced into the upper part of the tank and it flows downward through the packed bed and hence the cold fluid escapes through the diffuser of the lower part. During this step, the hot fluid transfers the heat energy to the storage materials, initially at the temperature T_c, up to a temperature T_h. In contrast to this, the direction of HTF is reversed during the heat retrieve stage since the cold fluid penetrates via the lower part of the tank and then receives the stored energy from rocks. The recovered energy will be forwarded to the power block of a concentrating solar power plant (CSP).

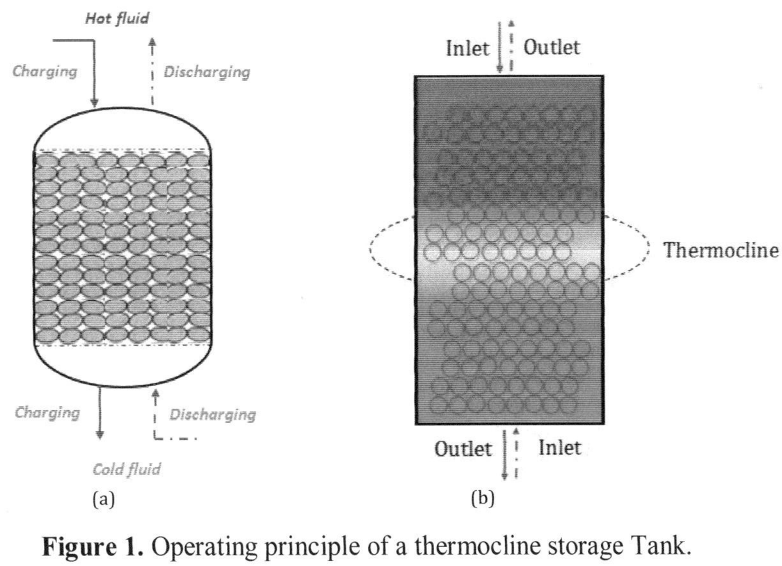

Figure 1. Operating principle of a thermocline storage Tank.

2.2. Hypotheses

In this study, the following hypotheses were made:

- the system is assumed to be one-dimensional;
- for the solid particles, lumped system model is considered (the Biot number, $B_i = \frac{h_p d_p}{k_s}$, should be less than 1);

- the packed bed has an infinite effective thermal conductivity in the radial direction;
- there is no temperature gradient inside the particles;
- the top and bottom of the system are insulated;
- radiation heat transfer is negligible;
- the heat generation inside the packed bed is not taken into consideration.

2.3. Mathematical modelling

The thermal behavior of the tank was developed and simulated by the Schumann equations [12], which describe the heat transfer between the fluid and a compact bed. Thus, the dimensionless conservation equations are written in the fluid and solid phases respectively:

$$\frac{\partial \theta_f}{\partial \tau} + \frac{\partial \theta_f}{\partial X} = \frac{1}{P_{e_f}} \frac{\partial^2 \theta_f}{\partial X^2} + S_{t_{f,s}}(\theta_s - \theta_f) + S_{t_p}(\theta_\infty - \theta_f)$$

(1)

$$\frac{\partial \theta_s}{\partial \tau} = \frac{1}{P_{e_s}} \frac{\partial^2 \theta_s}{\partial X^2} + kS_{t_{f,s}}(\theta_f - \theta_s) \quad (2)$$

Where:

$$\theta = \frac{T - T_{amb}}{T_h - T_{amb}}; \qquad X = \frac{x}{H}; \qquad \tau = \frac{t\, v_f}{H}$$

Before discretizing the conservation equations (1-2), the storage tank is divided into N vertical regular layers. On the other hand, the initial and boundary conditions associated with the above equations are:

1) The initial temperature distributions of the solid and fluid phases should be specified for the N layers.
2) The inlet temperature of the fluid phase should be specified for each time step $\Delta\tau$.
3) Adiabatic conditions for the fluid phase at the outlet of the physical domain “layer N+1”, for the charging phase are expressed by Eq. (3):

$$\frac{\partial \theta_f(X=1)}{\partial X} = 0 \quad (3)$$

4) Adiabatic conditions for the solid phase at the inlet and outlet of the physical domain “Layers 0 and N+1” respectively for the charging phase, are represented by Eq. (4):

$$\frac{\partial \theta_s(X=0)}{\partial X} = \frac{\partial \theta_s(X=1)}{\partial X} = 0 \quad (4)$$

The previous equations have been solved numerically by using the implicit finite difference method [13].

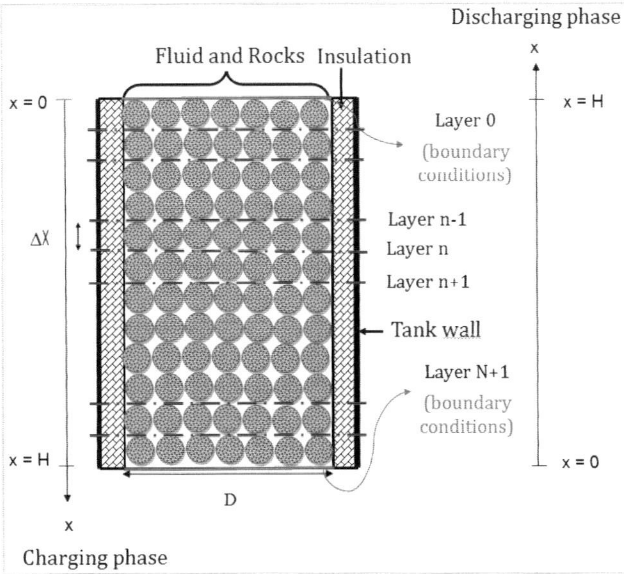

Figure 2. Packed bed storage system scheme.

3. Results and discussion

3.1. Validation of the simulation tool

The numerical simulation was carried out using the MATLAB (R2011b) software, based on the mathematical model presented above. The numerical results have been validated using the experimental data of Pacheco et al [8] which consists in discharging phase during two hours. Dimensions, operating parameters and the thermo-physical properties of the two field media fluid and solid were evaluated according to the reference of Van Lew et al. [14].

Figure 3. exhibits the evolution of numerical temperature profiles against the experimental results of Pacheco et al. [8]. It is clear that numerical and experimental results are in good agreement. Consequently, the model could be able to predict the thermal behavior of a single tank storage unit.

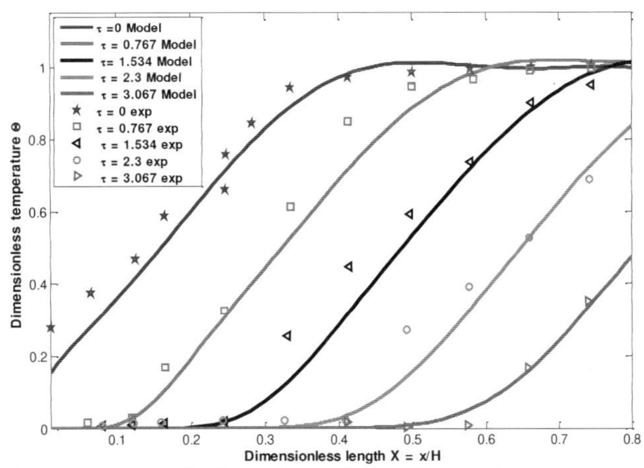

Figure 3. Comparison between numerical results and experimental measurements of Pacheco et al. [8]. Exp: experimental measurements, Model: numerical results.

3.2. Parametric study

In this work, the parametric study was carried out for two different circumstances:

- The first thermal cycle (3h charging/ 3h discharging);

- A series of five consecutives thermal cycles (3h charging/ 3h discharging).

The thermo-physical properties of HTF are presented in Table 1 whereas the thermo-physical properties of rock, the dimensions of the storage system and the operating parameters used in the simulation study are listed in Table 2.

Table 1: Thermal properties of heat transfer fluid [11]

Heat transfer fluid	Oil : Therminol VP1
Composition	73.5% diphenyl oxide and 26.5% biphenyl oxide
Temperature of solidification	12 °C
Maximum temperature of operation	400 °C
Vapor pressure	10.9 bars at 400 °C
Thermal capacity (250 °C)	1.89 $MJ/m^3.K$
Thermal conductivity (250 °C)	0.106 W/m.K
Dynamic viscosity (250 °C)	0.29 mPa.s
Availability	Low
GHG	3 kg CO_2eq kg^{-1}
Degree of Dangerousness	High

Table 2: Dimensions and operating parameters of the cylindrical TES system [15-17]

Dimensions	Value	Unit
Tank height	3	m
Tank diameter	2	m
Surface area	3.1416	m^2
Rock diameter	0.03	m
ε	0.27	--
Operational parameters		
$\dot{m}_{ch/disch}$	0.3	kg/s
$t_{ch/disch}$	3	h
T_{amb}	298	K
T_h	523	K
U	0.678	W/m^2K
ρ_s	2600	kg/m^3
C_p	820	J/ kg K
K_s	2.8	W/m K
n	100	--

Figures 4a and 4b exhibit dimensionless temperature distributions which correspond to solid and fluid field media, during a single cycle (a: charging, b: discharging). According to these figures, it is obviously evident that after each period of charging or discharging t = 0.5, 1, 2 and 3 hours, the HTF and the storage material have practically the same temperature ($T_f \approx T_s$). Moreover, in case of charging phase, inlet of the tank, these temperatures have the same value T_h ($\theta \approx 1$). Contrary to this, they take the value of ambient temperature T_{amb} ($\theta \approx 0$) at the inlet of the discharging phase. Furthermore, it is also clear that in the thermocline region, the falling temperature is very significant. The thickness of this zone depends simultaneously on the heat transfer fluid, the storage materials and time of charging/discharging of storage system TES.

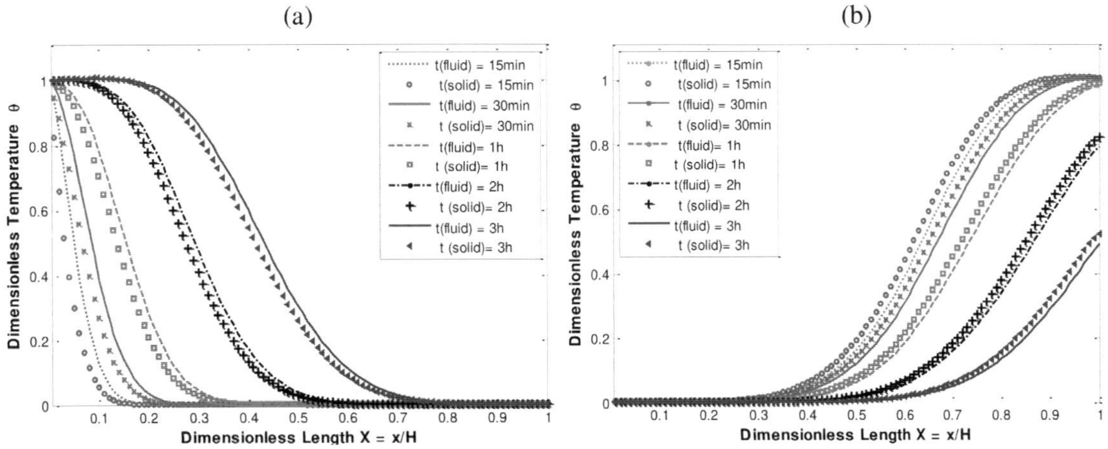

Figure 4. Dimensionless temperature profiles of the fluid and solid phases along the storage tank of the single cycle: (a) charging / (b) discharging periods.

3.3. Effects of the parameters ($\frac{H}{D}$, d, G) on the dimensionless temperature of HTF and rock

As expected, in the case of the charging process, the dimensionless temperatures evolutions of solid follow the classical behavior of thermal conduction regardless of whatever the position (Figures 5a-b: X=0.2 or X=0.8). It is clear from the curves of these figures that, at a given time, solid temperature values increase with the aspect ratio $\frac{H}{D}$, and this increase are more pronounced for its high values. Also, the variation of this parameter has a tangible effect on the thermal diffusion regime (transient regime or steady state). Thus, the time required to reach the steady state is even shorter if this ratio is high. These results can be explained by the fact that the cross section of the HTF becomes narrow since the tank height is kept constant and its diameter is varied. As for the Figures 6a-b, they show dimensionless temperature distributions of the fluid after the first (Figure 6a) and fifth (Figure 6b) thermal cycles with respect to dimensionless coordinate X. It should be noted that the augmentation of the aspect ratio accelerate the saturation of the system and hence the thermal equilibrium between rocks and the HTF (case of $\frac{H}{D} = 3$). On the other hand, it is easy to see that in the case where $\frac{H}{D} < 3$ the heat is gradually stored which generates the temperature homogenization inside rocks: $B_i < 1$ (see Figure 8a).

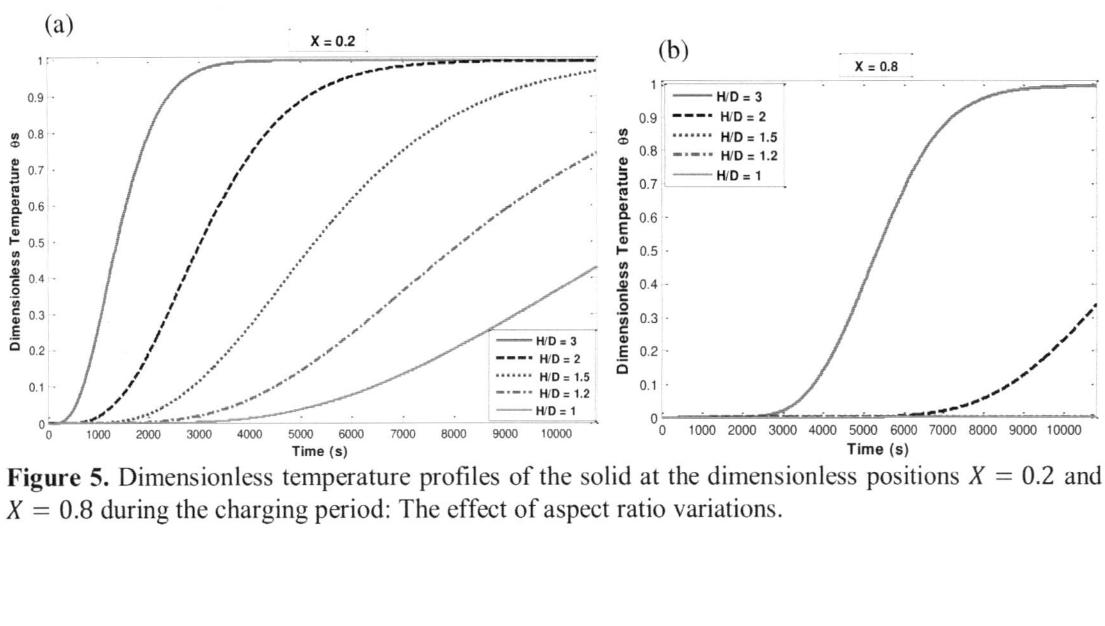

Figure 5. Dimensionless temperature profiles of the solid at the dimensionless positions $X = 0.2$ and $X = 0.8$ during the charging period: The effect of aspect ratio variations.

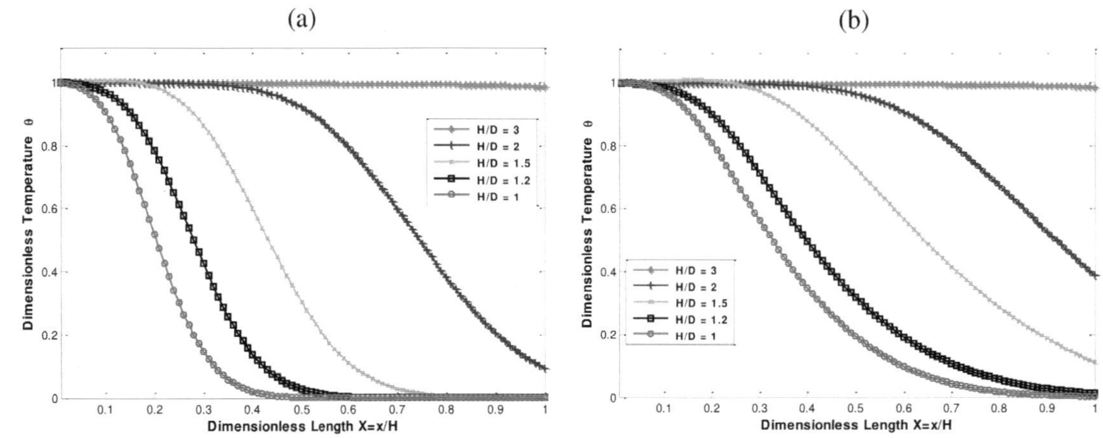

Figure 6. Dimensionless temperature distributions of the fluid along the storage tank at the end of the charging period: (a) first cycle, (b) fifth cycle.

Figures 7a and 7b show the dimensionless temperature profiles of rock and HTF (Therminol VP1) at the end of the first and fifth charging processes, for different particle diameters d = 1mm to d = 7mm. The number of Biot was compared for all diameters and it was less than 1 (Figure 8b) and therefore the assumptions mentioned above are respected. Furthermore, the difference between the storage material (Granite) and heat transfer fluid temperature (Therminol VP1) noticeably narrows for smaller particle size, the thermocline zone is sharper through the length of the tank (Figure 7a).The same behavior is observed at the end of the 5th cycle. Indeed, the small size of the particles allows a more efficient heat transfer between the solid and fluid due to the high specific exchange surface and the high coefficient of convective heat transfer.

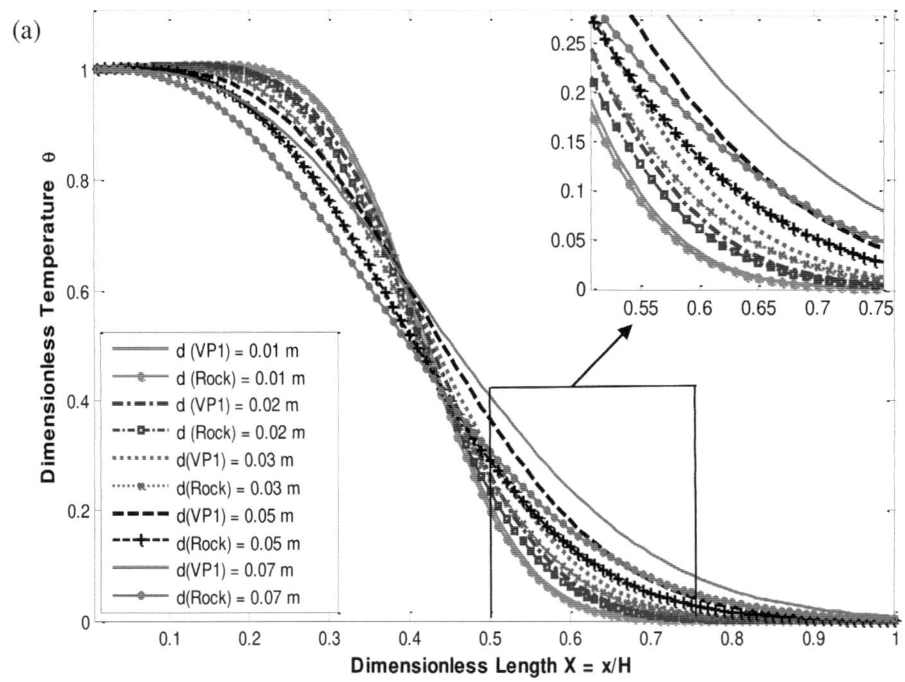

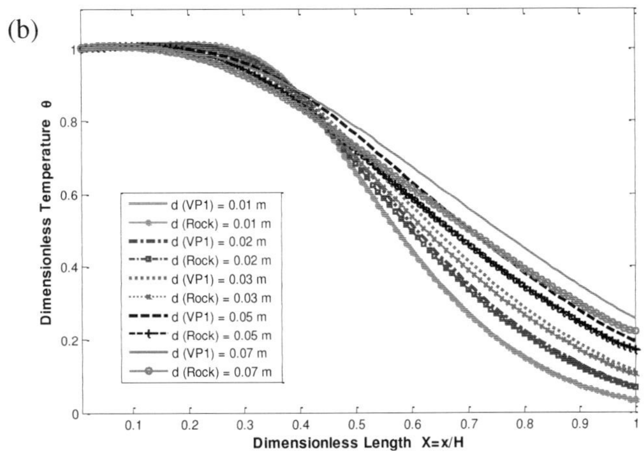

Figure 7. Effect of rock diameter variation on the dimensionless temperature profiles of the fluid and solid phases along the storage tank at the end of the charging period: (a) first cycle, (b) 5th cycle

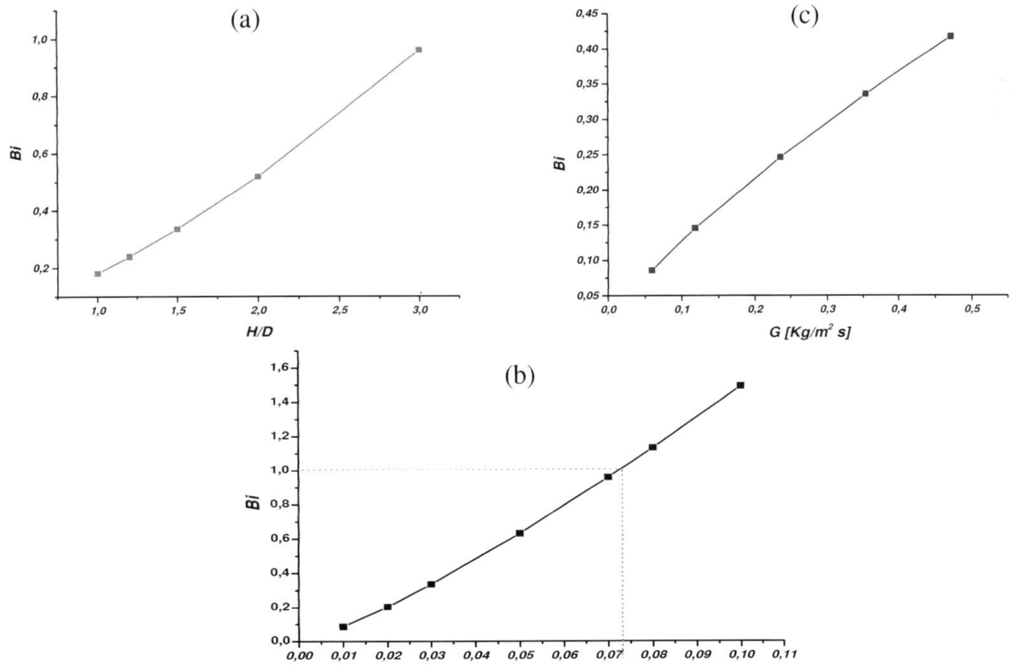

Figure 8. Variations of Biot number at the end of the charging period with respect to : (a) aspect ratio $\frac{H}{D}$, (b) particle diameter d, and (c) mass flow rate per unit cross section G.

Figures 9a and 9b exhibit the dimensionless temperature distributions inside the rock (granite) and in the oil (TerminolVP1) along the storage tank at the end of the first charging process (3h) and the 5th thermal cycle, respectively, for different values of G (0.058 kg/m^2 s ≤ G ≤0.047 kg/m^2 s). For both cases, higher flow rate leads obviously to higher temperatures at the packed bed. At the end of the first cycle, for the maximum mass flow per unit cross section, the temperature at the position $X = 0.5$ reaches 190°C ($\theta \approx 0.62$), while it remains at ambient temperature ($\theta \approx 0$), for G = 0.05 kg/m^2 s and G = 0.23 kg/m^2 s (Figure 9a). In addition, the increase of the thermocline thickness can be seen as the oil mass flow increases. Furthermore, the temperature gradient zone is larger for the 5th cycle than the first one.

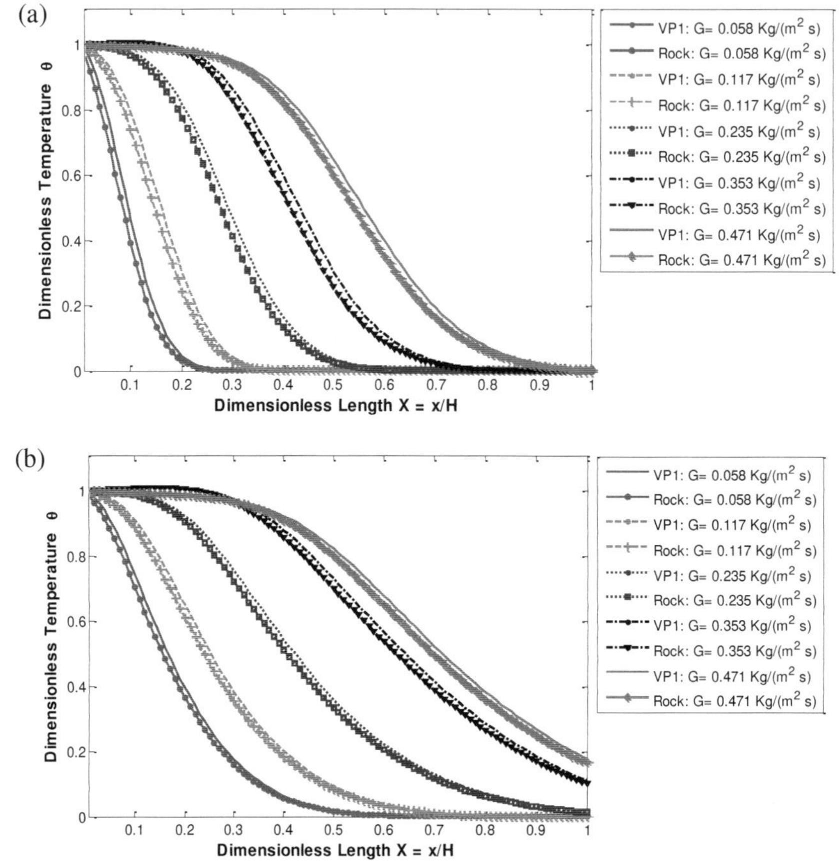

Figure 9. Effect of mass flow rate per unit cross section G on the dimensionless temperature profiles of the fluid and solid phases along the storage tank at the end of the charging period: (a) first cycle, (b) 5th cycle.

4. Conclusion

In this work, a mathematical model of heat transfer in a packed bed TES thermocline system has been developed. A simulation tool was conducted in the MATLAB software environment and validated experimentally. A parametric study of the packed bed in terms of aspect ratio, particle diameter and fluid mass flow rate was carried out during charging/discharging periods and also thermal cycling operation of a thermocline storage system.The most relevant property of the storage material is the volumetric heat capacity, while the thermal conductivity of the solid has only a minor effect. In the range of particle sizes considered, the dimensionless temperature profiles of rock and oil is found considerably better for the smallest particle size of the 5thcycle. High fluid mass flow rate and high values of aspect ratio lead to increase in temperature.

References

[1] Wang ZF 2010, Prospectives for China's solar thermal power technology development. *Energy*, **35**: 4417–20.

[2] He YL, Xiao J, Cheng ZD, Tao YB, Mcrt A, VM F, et al. 2010, A MCRT and FVM coupled simulation method for energy conversion process in parabolic trough solar collector. *Ren. Energy*; **36**: 976–85.

[3] Relloso S, Garcia E 2015, Tower technology cost reduction approach after gema solar experience. *Energy Proc*. **69**, 1660–1666.

[4] Gil A, Medrano M, Martorell I, Lázaro A, Dolado P, Zalba B and Cabeza LF 2010, State of the art on high temperature thermal energy storage for power generation:1. Renew. *Sust. Energy Rev.* **14** 31–55.
[5] Herrmann U, Kelly B, and Price 2004, Two-tank molten salt storage for parabolic trough solar power plants. *Energy*, **29**, 883–893.
[6] Angelini G, Lucchini A, and Manzolini G 2013, Comparison of thermocline molten salt storage performances to commercial two-tank configuration, *Energy Procedia*, **49**: 694–704.
[7] Kearney D, Kelly B, Herrmann U, Cable R and J. Pacheco 2004, Engineering aspects of a molten salt heat transfer fluid in a trough solar field. *Energy* **29**, 861–870.
[8] Pacheco JE, Showalter SK, Kolb WJ 2002, Development of a molten-salt thermocline thermal storage system for parabolic trough plants. *J Solar Energy Eng*; **124**:153–9.
[9] Zanganeh G, Pedretti A, Zavattoni S, Barbato M, et al. 2012, Packed-bed thermal storage for concentrated solar power-pilot scale demonstration andindustrial-scale design. *Sol. Energy* 86 (10) 3084–3098.
[10] Grirate H, Agalit H, Zari H, Elmchaouri A, Molina S and Couturier R 2016, Experimental and numerical investigation of potential filler materials for thermal oil thermocline storage, *Sol. Energy* **131**: 260-274.
[11] Hoffmann J.F 2015, Stockage thermique pour centrale solaire thermody-namique à concentration mettant en œuvre des matériaux céramiques naturels ou recyclés, *Doctoral dissertation*, Perpignan.
[12] Schumann, T.E.W. 1929, Heat transfer: a liquid flowing through a porous prism. J. Frankl. Inst. 208, 405–416. .http://dx.doi.org/10.1016/S0016-0032(29)91186-8.
[13] Dale A. Anderson et al. 1984, Computational fluid mechanics and heat transfer, Series in computational methods in mechanics and thermal sciences.
[14] Van Lew, J.T., Chan, C.L., Karaki, W., Stephens, J 2011, Analysis of heat storage and delivery thermocline tank having solid filler material. *J. Sol. Energy Eng.* **133**, 021003-1–021003-10.
[15] Eppelbaum L, Kutasov I, Pilchin A 2014, Thermal Properties of Rocks and Density of Fluids. *Appl. Geotherm.*, Springer, Berlin, Heidelberg, pp. 99–149.
[16] Miao S.Q, Li H.P, Chen G 2014, Temperature dependence of thermal diffusivity, specific heat capacity, and thermal conductivity for several types of rocks. *J.Therm. Anal. Calorim.* 115 1057–1063.
[17] Grirate H, Zari N, Elamrani I, Couturier R, Elmchaouri A, Belcadi S and P.Tochon 2014, Characterization of several Moroccan rocks used as filler material for thermal energy storage in CSP power plants. *Energy Procedia* **49** 810–819.

Nomenclature

Symbols		$S_{t_{f,s}}$	Stanton number fluid/solid (–)
A	Surface area (m^2)	P_{e_i}	Peclet number (i is f for fluid or s for solid)
d	Rock diameter (m)		
D	Diameter (m)	*Greek letters*	
H	Height (m)	θ	Dimensionless temperature (–)
t	Time (s)	k	Ratio of solid to fluid thermal conductivity (–)
τ	Dimensionless time (–)	ε	Porosity of the storage tank (–)
T	Temperature (K)		
v	Velocity (m/s)	*Subscripts*	
Q_m	Mass flow rate (kg/s)	amb	Ambient
G	Mass flow rate per unit cross section (Kg/m^2 s)	s	Solid
U	Overall heat transfer coefficient (w/m^{2}k)	f	Fluid
x	Variable length (m)	w	Wall
X	Variable dimensionless length (–)	h	Hot
Bi	Biot number (--)	c	Cold
S_{t_w}	Stanton number of tank wall (–)		

IOP Conf. Series: Earth and Environmental Science **161** (2018) 012017 doi:10.1088/1755-1315/161/1/012017

ANN-based modelling and prediction of daily global solar irradiation using commonly measured meteorological parameters

M Marzouq[1], Z Bounoua[2], A Mechaqrane[2], H E Fadili[1], Z Lakhliai[1] and K Zenkouar[3]

[1]Laboratory of Computer Sciences and Interdisciplinary Physics, National School of Applied Sciences, USMBA, Fez, PO. Box 72 Fez Morocco

[2]Laboratory of Renewable Energies and Intelligent Systems Electrical Engineering department, Faculty of Sciences and Technologies, USMBA, Fez, PO. Box 2202 Fez Morocco

[3]Intelligent System and Application Laboratory, Faculty of Sciences and Technologies, USMBA, Fez, PO. Box 2202 Fez Morocco

E-mail: manal.marzouq@usmba.ac.ma

Abstract. The aim of this work is to develop an artificial neural network (ANN) based model for accurately predicting the daily global solar irradiation in the city of Fez. The potential of the developed model is verified and appraised through the local collected database for the period 2009-2015 from the radiometric station of the Faculty of Sciences and Technology of Fez. The obtained model is MLP with feed forward back-propagation algorithm containing three input parameters and a single hidden layer with nine neurons. Coefficient of determination R^2, the mean absolute percentage error MAPE and the relative root mean square error RRMSE are respectively equal to 97.16%, 21.77% and 18.79%.

1. Introduction

Reliable knowledge of solar irradiation is very important for the design and deployment of solar energy systems [1, 2]. Since it is difficult to obtain measurements of solar irradiation and its components from most meteorological stations, several models have been developed to estimate these necessary data.

ANNs has been shown to be more suitable to predict solar irradiation than other empirical models [3]. In particular, the prediction of daily solar irradiation with ANN's has been developed in the last two decades. Recently, in the work of Chiteka and al. [4], altitude, latitude, longitude, clearness index, average temperature, humidity and pressure have been used as input parameters. Other authors used sunshine duration with commonly used parameters [5, 6]. Some additional inputs were considered for estimation like the amount of suspended particulate matters [7].

In this paper, we try to develop an ANN-based model using the most commonly accessible parameters in meteorological stations: relative humidity, temperature, solar irradiation on the top of the atmosphere and so on. Our second contribution is to study the relevance of the aforementioned input parameters individually and combined. Afterwards, we will evaluate the feasibility of the incremental combination method and test the best model using different number of neurons in the hidden layer.

Nomenclature			
G_s	Global solar irradiation (Wh/m2)	$S\Delta T$	Square temperature (°C)
ANN	ANN Artificial Neural Networks	Rh_{max}	Maximum daily relative humidity (%)
MLP	Multi-Layers Perceptron	Rh_{min}	Minimum daily relative humidity (%)
N	Number of hidden neurons	MAPE	Mean absolute percentage error (%)
Rh_{moy}	Mean relative humidity (%)	RRMSE	Relative root mean square error (%)
WS	Wind speed (m/s)	R^2	Coefficient of determination (%)
WD	Wind direction (°)	D_y	Day of the year
Rf	Rainfall (mm)	T_{max}	Maximum daily Temperature (°C)
SA	Solar altitude angle (°)	T_{min}	Minimum daily Temperature (°C)
ΔT	Difference between daily maximum and minimum temperatures (°C)	G_{toa}	Solar irradiation at the top of the top of the atmosphere (Wh/m^2)

This paper is organized as follows: the second section is dedicated to the presentation of the site and the used database followed by our ANN model description. Section 4 deals with the selection of the input parameters. Section 5 and 6 present the selection of the ANN architecture and concluding remarks.

2. Site and Database

In this study, the measurement of global horizontal solar irradiation at ground surface and other meteorological parameters was performed by a radiometric station placed on the roof of the Faculty of Sciences and Technology building at sidi Mohamed Ben Abdellah University, Fez, Morocco (latitude 33° 56' N, longitude 4° 59' W, altitude 579 m). The device collecting the solar data used in this work is a Kip & Zonen model CM-11 pyranometer. For measuring the precipitation, we use the rain gauge while air temperature and relative humidity was measured by means of a thermohygrometer. The wind speed and direction are measured with an anemometer. The measured data between the years 2009 and 2015 were used with 75% for model training and 25% for testing. Before applying our model, a quality control procedure was performed on dataset [8].

3. ANN model description

In this work, a multilayer perceptron (MLP) using Levenberg-Marquardt back-propagation training algorithm with one hidden layer is used for modelling daily values of global solar irradiation on a horizontal surface in the city of Fez.
Hyperbolic tangent and linear functions were used as activation functions in the hidden and output layers respectively. To evaluate the ANN prediction performance, many well-known prediction accuracy indices are adopted in the literature [9]. In this work, we use the coefficient of determination R^2, the mean absolute percentage error MAPE and the relative root mean square error RRMSE defined in [9, 10].

4. Selection of input parameters

In this section, we start by assessing the relevance of each input of our model. Then, we adopt an incremental method by adding progressively the most relevant inputs obtained in the first step in order to show the impact of these combinations on the accuracy of our prediction model. After that, we explore other combinations in order to evaluate the feasibility of the incremental method: the aim is to find the best combinations. In all these experimented models we used an ANN with 5 nodes in the hidden layer. Finally, the best obtained combination is tested using different number of hidden neurons. In this study, we use only one hidden layer which allows achieving several runs for each combination of inputs and hidden neurons within an acceptable time.

Table 1: Individual input parameters performances obtained for the test daily dataset on 10 runs.

Input variables	Performance indicators		
	R^2 (%)	MAPE (%)	RRMSE (%)
G_{toa}	93.01	45.72	28.26
T_{max}	92.72	39.15	28.40
SA	92.45	100.62	24.98
Rh_{min}	91.80	36.63	30.84
ΔT	89.46	39.16	34.85
Rh_{moy}	89.28	94.26	28.15
$S\Delta T$	89.15	40.06	35.40
Rh_{max}	87.57	65.51	36.34
T_{min}	87.05	39.55	32.21
Rf	85.21	54.62	40.12
WS	81.77	72.84	43.67
WD	77.62	92.41	47.82

4.1. Individual input parameters performances

The input parameters selection is the first step in developing our ANN model. The measured and calculated input data are: temperature, minimum temperature, maximum temperature, relative humidity, minimum relative humidity, maximum relative humidity, solar altitude angle, solar irradiation at the top of the atmosphere, rainfall, wind speed, wind direction and day of the year. In the variable selection process, we choose to evaluate individually the average performance of each input variable for 10 runs to avoid random effects because of the initial weight selection and to ensure the reliability of the prediction model [11]. Table 1 summarizes the obtained performance indicators for the test dataset ranked in deceasing order.

4.2. Incremental combination of input parameters

In order to take into account the effect of each input variable evaluated individually in Table 1, we suggest an incremental method by combining progressively the inputs regarding their individual performances. Six combinations of input parameters were developed and dressed in Table 2. From Table 2, it can be noted that the addition of input parameters with the best

Table 2: Average performances of the incremental combinations of input parameters obtained for the daily dataset test on 10 runs.

Combinations	Performance indicators		
	R^2 (%)	MAPE (%)	RRMSE (%)
G_{toa}, T_{max}	95.89	31.45	22.01
G_{toa}, T_{max}, SA	95.65	30.98	22.92
G_{toa}, T_{max}, SA, Rh_{min}	95.99	24.55	22.11
G_{toa}, T_{max}, SA, Rh_{min}, Rf	95.56	24.84	23.42
G_{toa}, T_{max}, SA, Rh_{min}, WS	95.70	30.75	22.89
G_{toa}, T_{max}, SA, Rh_{min}, WD	92.90	29.55	26.25

performances, taken individually, does not necessarily improve the global performance of our model.

Moreover, it is clearly shown in this Table that increasing the number of input parameters does not automatically increase the performances of the ANN model: we obtain, for example,

Table 3: Average statistical indicators obtained with various combinations of inputs for the test dataset on 10 runs.

Combinations												Performance indicators		
ΔT	Rh_{moy}	SA	G_{toa}	T_{max}	T_{min}	D_y	Rh_{max}	Rh_{min}	$S\Delta T$	WS	WD	R^2 (%)	MAPE (%)	RRMSE (%)
			X	X								95.69	31.83	22.55
X	X		X									96.77	22.73	19.99
X	X		X			X						96.79	22.31	20.01
X	X	X										96.98	21.22	19.35
	X		X						X			97.02	21.18	19.31
	X	X							X			96.79	22.53	19.93
				X	X	X	X	X				96.68	22.04	20.36
				X	X		X	X				94.57	27.60	25.53
				X	X	X	X	X				96.68	22.04	20.36
		X		X	X							96.46	23.77	21.02
	X		X			X			X			96.64	20.47	22.78
		X	X	X	X							96.37	24.93	21.30
X	X	X									X	96.95	22.98	19.41
X		X	X								X	96.74	22.25	20.05
X	X	X								X	X	96.87	23.69	19.78
X	X	X									X	96.95	22.98	19.41

comparable performances with two and five input parameters presented in the first and third raw respectively.

4.3. Choice of the best combination of input parameters

The remaining of the experimental tests aims to explore other combination possibilities even with input parameters having lower performances as reported in Table 1. Table 3 shows the experimental results for different combinations. As for the other results the average performances are obtained on 10 runs. Table 3 shows that the best combination is obtained using $S\Delta T$, G_{toa} and Rh_{moy} as inputs, with R^2= 97.02%, RRMSE =19.31% and MAPE of 21.18%. Another good combination is given by the inputs (ΔT, SA, Rh_{moy}) with very similar performances, even if the corresponding input parameters are not relevant individually.

5. Selection of the number of neurons in the hidden layer

In this section, the best previously obtained combination is tested with different numbers of neurons in the hidden layer in order to find the best architecture for our model. Table 4 and Figure 1 show the corresponding results.

Table 4: Average statistical indicators obtained with various combinations of inputs for the test dataset on 10 runs.

N	R^2 (%)	MAPE (%)	RRMSE (%)	N	R^2 (%)	MAPE (%)	RRMSE (%)
1	96.41	23.30	21.06	11	96.67	22.28	20.25
2	96.73	21.50	20.17	12	96.67	22.28	20.25
3	96.84	22.53	19.75	13	96.45	24.29	21.00
4	96.81	22.22	19.75	14	96.08	25.56	22.10
5	97.02	21.18	19.31	15	96.02	23.93	22.28
6	96.98	21.32	19.30	16	96.20	23.65	21.86
7	96.93	21.55	19.51	17	96.39	24.48	21.19
8	96.63	23.37	20.47	18	96.31	25.16	21.33
9	97.16	21.77	18.79	19	96.49	22.79	20.94
10	96.51	22.66	20.78	20	96.51	23.43	20.84

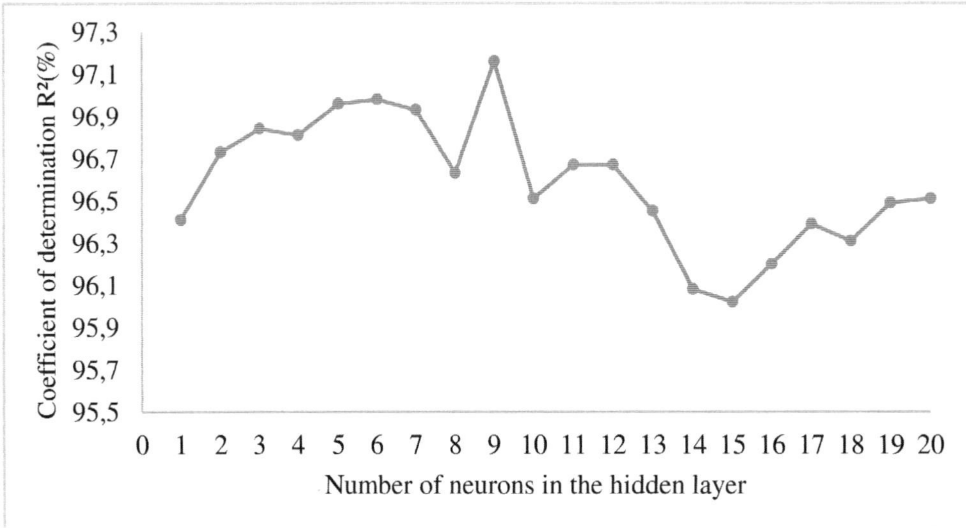

Figure 1: Coefficient of determination of the selected ANN model with different number of neurons in the hidden layer

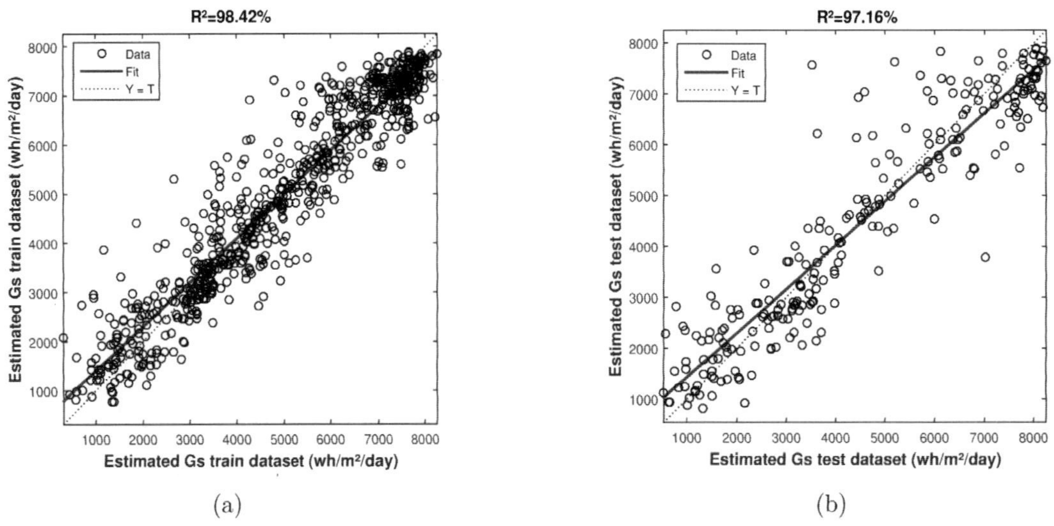

Figure 2: Predicted versus measured daily global irradiation values for our model for (a) training dataset and (b) test dataset.

Figure 2 shows the predicted values of our model with 9 neurons in the hidden layer versus the measured values. It indicates good agreement between measured and estimated values.
The results show that the suitable configuration for our model contains 9 neurons in the hidden layer. This configuration gives as performance 97.16% , 18.79% and 21.77% for R^2 , RRMSE and MAPE respectively.

6. Conclusion

This study shows the results of an effort to predict the daily global solar irradiation according to measured values such as relative humidity, temperature and solar irradiation at the top of the atmosphere which are commonly accessible parameters in the meteorological stations. During this work, we have studied the relevance of the input parameters individually and explored the feasibility of the incremental method. Finally, we have tested the effect of hidden neurons on the performances of our model.

An important outlook for this study is to propose a methodology for automatically determining the relevance of input parameters to enhance the overall performances of the ANN model.

References

[1] Iqbal M, 1979 A study of Canadian diffuse and total solar radiation data-ll Monthly average horizontal radiation *Solar Energy* vol **22** pp 87-90

[2] Sukamongkol Y, Chungpaibulpatana S and Ongsakul W 2002 A simulation model for predicting the performance of a solar photovoltaic system with alternating current loads *Renewable Energy* vol **27** pp 237-258

[3] Reddy K S and Ranjan M 2003 Solar resource estimation using artificial neural networks and comparison with other correlation models *Energy Convers. Manage.* vol **44** pp 2519-30

[4] Chiteka K and Enweremadu C C 2016 Prediction of global horizontal solar irradiance in Zimbabwe using artificial neural networks *Journal of Cleaner Production* vol **135** pp 701–711

[5] Lam J C, Wan K K W and Yang L 2008 Solar radiation modelling using ANNs for different climates in China *Energy Conversion and Management* vol **49** pp 1080-90

[6] Yadav A K, Malik H and Chandel S S 2014 Selection of most relevant input parameters using WEKA for artificial neural network based solar radiation prediction models *Renewable and Sustainable Energy Reviews* vol **31** pp 509-519

[7] Vakilia M, Sabbagh-Yazdi S-R, Kalhorb K and Khosrojerdi S 2015 Using artificial neural networks for prediction of global solar radiation in Tehran considering particulate matter air pollution *Energy Procedia* vol **74** pp 1205-12

[8] Bounoua Z and Mechaqrane A 2017 Quality control of hourly global and diffuse solar irradiation on horizontal surface at Fez city, Morocco (Fez) Proc. du 2ème colloque fronco-marocain sur les Energies Renouvelables et leur intégration aux réseaux de transport et de distribution pp 122-123

[9] Marzouq M, El Fadili H, Lakhliai K and Zenkouar K 2017 A review of solar radiation prediction using artificial neural networks Inter. Conf. on Wireless Technologies, Embedded and Intelligent Systems (Fez) IEEE pp 1-6

[10] Voyant C, Notton G, Kalogirou S, Nivet M L, Paoli C, Motte F and Fouilloy A 2016 Machine Learning methods for solar radiation forecasting: a review *Renewable Energy* vol **105** pp 569-82

[11] Bosch L, López G and Batlles F J 2008 Daily solar irradiation estimation over a mountainous area using artificial neural networks *Renewable Energy* vol **33** pp 1622-28

IOP Conf. Series: Earth and Environmental Science **161** (2018) 012018 doi:10.1088/1755-1315/161/1/012018

A low-cost PV Emulator for testing MPPT algorithm

A Chalh [*1], **S Motahhir**[1], **A El Hammoumi**[1], **A El Ghzizal**[1] **and A Derouich**[1].

[1]*Laboratory of Production engineering, Energy and Sustainable Development, Smart Energy Systems and Information Processing Research Team, Higher school of technology, SMBA University, Fez, Morocco.*

*abdelilah.chalh@usmba.ac.ma

Abstract. The high cost of commercial PV emulators requires finding new solutions for building low-cost system having similar behavior of PV panel. In this work, we present a low-cost emulator using a variable DC supply with a series variable resistor; based the maximum power transfer theorem. On the other hands, a study of the behavior of this emulator is done. The results of experiments test are shown that our PV emulator can provide a simple, efficient and a low-cost way for researchers to test and verify their MPPT algorithms in a laboratory environment.

1. Introduction

The demand for energy in different forms is constantly increasing, particularly renewable energies, namely: hydraulic, wind, solar, geothermal and bioenergy energy. Today, solar energy has taken a large part of the market due to the continued development of PV system technology and lower prices. The power provided by the PV panel is maximum only when the latter operates at its maximum power point (MPP). As a result, several studies have been presented in the literature aimed at monitoring this point, such as: Perturb & Observe (P & O) [1], [2], incremental conductance (INC) [3] - [6], fractional open circuit voltage [7], fractional short-circuit current [8], fuzzy logic control [9], and neural network [10], etc.

On the other hands, the conversion of solar energy into electricity varies aleatory according to the level of solar radiation and the atmospheric temperature [11], [12]. Therefore, proposed MPPT algorithms must be tested under different conditions so as to validate their robustness. In general, there are some methods for testing MPPT algorithms in a laboratory environment. For example, we found one which is based on using an actual outdoor installed PV panel/array in the laboratory [13]. But, the disadvantages of this method are the requirement of external space, dependence on the weather condition, and the expensive solution of the PV system. To solve all these limitations, PV emulators [14], [15] have been proposed to replace PV systems, which make easy the set of a test profile desired by the researcher. Recently, there have been several types of commercial PV emulators, among them, the Agilent Solar Simulator used in [16], and other PV emulators based on a programmable DC power supply, but they are very expensive. Therefore, this work proposes a low-cost PV emulator to test the real implementation of MPPT algorithm for PV systems. This PV emulator is based on a simple DC supply and resistor, thus it allows a variety of I-V and P-V curves for different values of solar radiation to be simulated inside laboratory regardless of actual solar radiation levels. This PV emulator is simple and easy to build in a conventional laboratory environment. The design of the proposed emulator is based on the 'maximum power transfer theorem'.

This paper is organized as follows. The next section exposes design of PV emulator. Section 3 presents the experimental results. Finally, the conclusion is given in section 4.

2. Design of PV emulator

2.1. Maximum power transfer theorem

PV emulator must give a power curve as this of PV panel that exhibits a peak. Therefore, the principle of the maximum power transfer theorem is used for the PV emulator design which is composed of a variable DC supply in series with a variable resistor. This DC supply can provide its maximum power when load resistor equals to the series resistor and as presented in Fig. 3, this emulator can give a P-V curve that exhibits a peak in the power curve which can be tracked by the MPPT algorithm to be tested. The 'maximum power transfer theorem' can be explained as follows [17]:

Based on Fig. 1, the following equations can be found:

$$\boldsymbol{P}_{load} = \boldsymbol{I} * \boldsymbol{V} = \boldsymbol{I}^2 * \boldsymbol{R}_{load} \tag{1}$$

$$\boldsymbol{I} = \frac{\boldsymbol{E}}{\boldsymbol{R}_{series} + \boldsymbol{R}_{load}} \tag{2}$$

Replacing (2) in (1), equation (3) is found:

$$\boldsymbol{P}_{load} = \frac{\boldsymbol{R}_{load}}{(\boldsymbol{R}_{series} + \boldsymbol{R}_{load})^2} * \boldsymbol{E}^2 \tag{3}$$

$$\frac{\boldsymbol{dP}_{load}}{\boldsymbol{dR}_{Load}} = \boldsymbol{E}^2 \frac{\boldsymbol{R}_{series} - \boldsymbol{R}_{load}}{(\boldsymbol{R}_{series} + \boldsymbol{R}_{load})^3} \tag{4}$$

The power is at its maximum when its derivative is equal to zero and from equation (4) this can be reached when the load resistor is equal to the series resistor as follows:

$$\boldsymbol{R}_{load} = \boldsymbol{R}_{series} \tag{5}$$

Therefore, in this case, maximum power transfer condition (5) will appear when the voltage of the load is half of that generated by the DC supply.

$$\boldsymbol{V} = \frac{\boldsymbol{E}}{2} \tag{6}$$

2.2. Description of the set-up

Fig. 1 presents the basic schematic of the proposed PV emulator, which is constituted of a DC power supply and a series resistor to give a P-V curve that exhibits a peak. In Fig. 2 the proposed PV emulator gives the I-V curves with a negative slope for a fixed $\boldsymbol{R}_{series}$ and $\boldsymbol{R}_{load}$ is varied from its minimum to maximum, this is experimentally validated in section 3. Fig. 3 illustrates the power goes through the maximum when the load voltage is half of the DC supply voltage. On the other side, the P-V curve generated by the PV module shows that the peak occurs at 0.7 V_{oc}. Therefore, the maximum power is transferred to the charge when the equivalent resistor of DC/DC converter and the load is equal to the value of series resistor

IOP Conf. Series: Earth and Environmental Science **161** (2018) 012018 doi:10.1088/1755-1315/161/1/012018

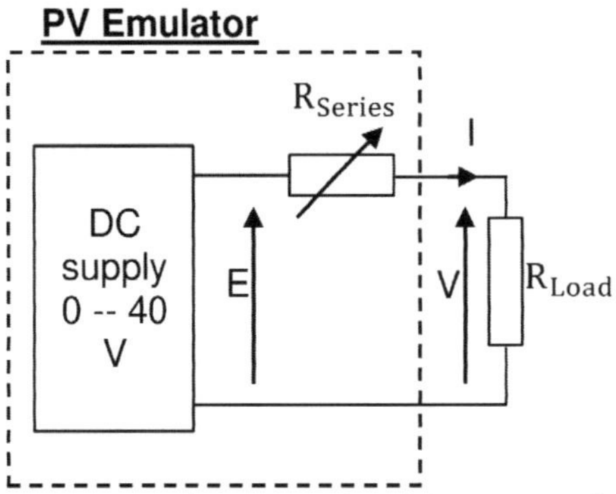

Figure 1. Basic schematic of the PV emulator.

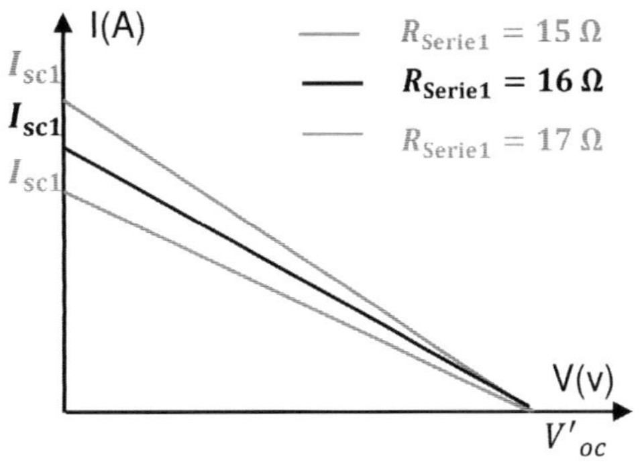

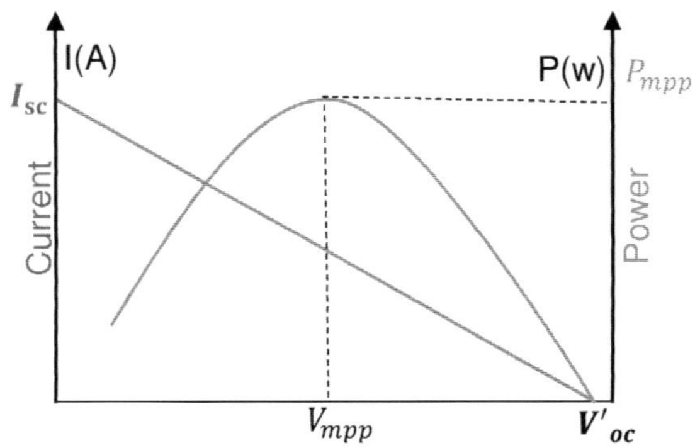

Figure 2. I–V curves for different series resistor. **Figure 3.** P-V and I-V characteristics of emulator

2.3. Dimensioning of the PV emulator

The maximum power is transferred to the load resistor by the PV emulator when the equivalent resistor (between PV emulator and load) is equal to the value of the series resistor. In order to dimension the series resistor of our emulator that will translate the actual behavior of the point of maximum power, the voltage and current values of the PV panel at MPP are required. Thereafter, the series resistor can be calculated by the following equation:

$$R_{serie} = R_{mpp} = \frac{V_{mpp}}{I_{mpp}} \tag{7}$$

Generally, and as known in the literature, it is a single resistor value can lead the PV panel to operate at MPP [17], [18]. Therefore, a DC/DC converter controlled by an MPPT algorithm must be placed between the panel and the load to operate the PV panel at MPP. In Fig. 4, a complete schematic view of the PV emulator control is shown.

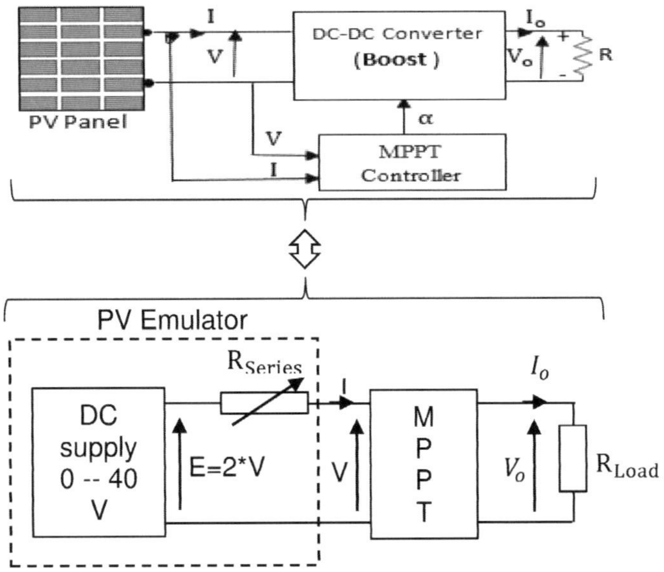

Figure 4. A complete schematic of the PV emulator control.

3. Experimental results

3.1. Validation of the proposed PV emulator

The circuit model and the experimental setup of the developed system are presented in Fig. 5 and Fig. 6 respectively in order to validate the performance of the proposed PV emulator.

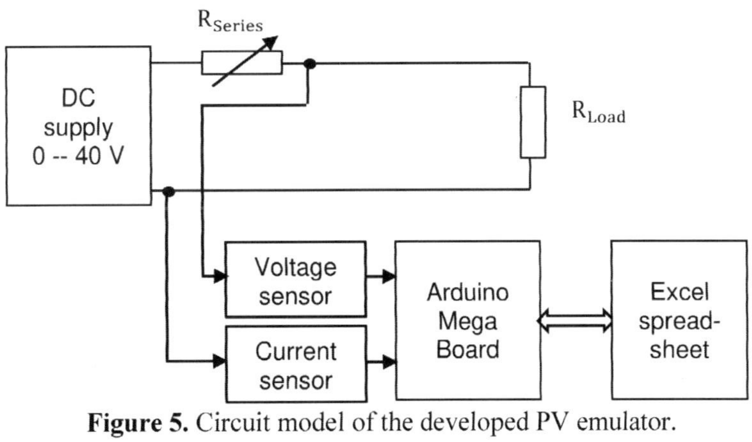

Figure 5. Circuit model of the developed PV emulator.

The voltage and current data of the PV emulator obtained by voltage and current sensors are transmitted directly in real-time in Excel. That by using the PLX-DAQ data acquisition Macro which allows communication between the microcontroller of the Arduino Mega and Excel spreadsheet by the intermediary of a serial communication [19]. Fig. 7 presents the experimental P-V and I-V curves characteristics of the PV emulator. We then experimentally notice that the peak in the power curve occurs when the output voltage of PV emulator equals half of the DC power supply voltage.

IOP Conf. Series: Earth and Environmental Science **161** (2018) 012018 doi:10.1088/1755-1315/161/1/012018

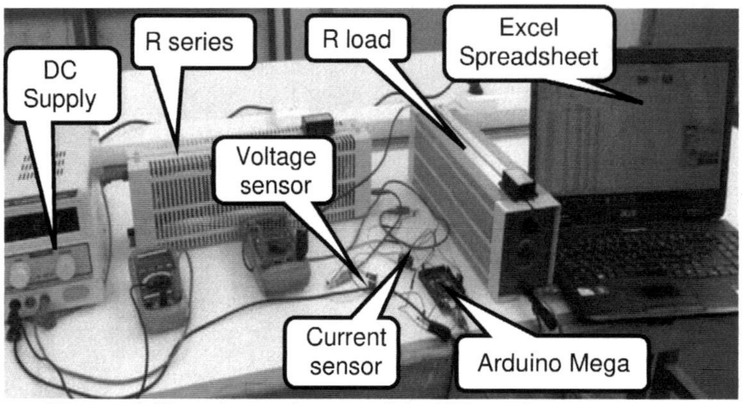

Figure 6. Experimental setup of the PV emulator system.

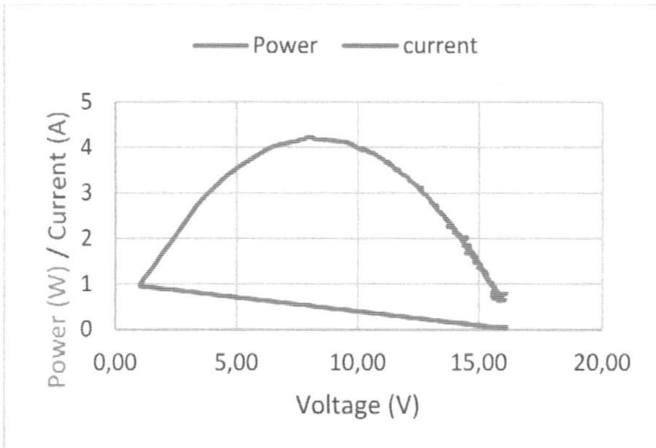

Figure 7. Experimental P-V and I-V curves characteristics of the PV emulator.

3.2. Effect of sun radiation variation

Fig. 8 shows the experimental I-V curves for a fixed V'_{oc} and various series resistors. According to this figure, the short-circuit current changes on the PV emulator when the value of R_{serie} varies.

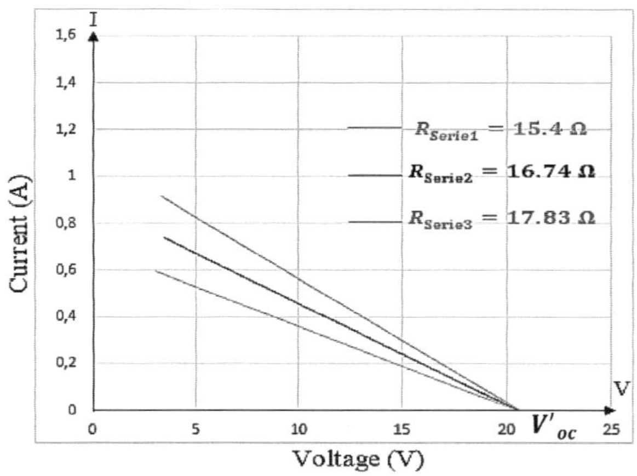

Figure 8. Experimental I–V curves for different series resistance at fixed V'_{oc}.

When the solar insolation varies, the short-circuit current changes linearly. Fig. 9 presents the experimental P–V curves for a fixed V'_{oc} and various series resistances. This situation is similar to variation of the insolation in case of PV panel.

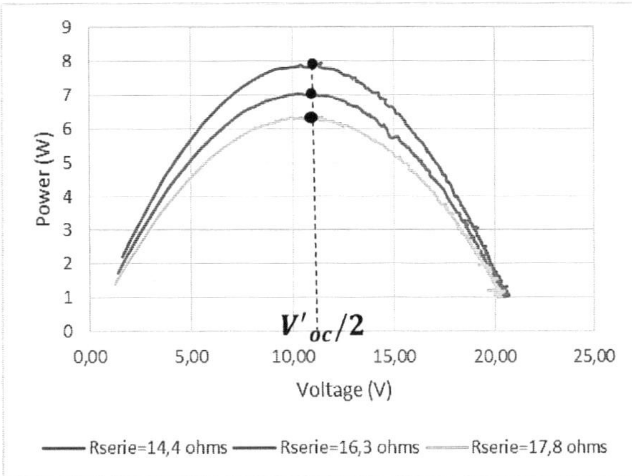

Figure 9. Experimental P–V curves for many series resistance at fixed V'_{oc}.

3.3. *The effect of temperature variation*

When the cell temperature changes, the open-circuit voltage changes significantly. So, this case is simulate the principle of the maximum power transfer theorem is used for the PV emulator in the present system by guarding the R_{series} constant and varying the DC supply voltage. Fig. 10 presents the experimental P-V curves for a fixed R_{series} and various V'_{oc}.

The V_{mpp} point moves to the right as the Maximum value of the voltage DC supply is increased. This is similar to changing temperature conditions in PV systems.

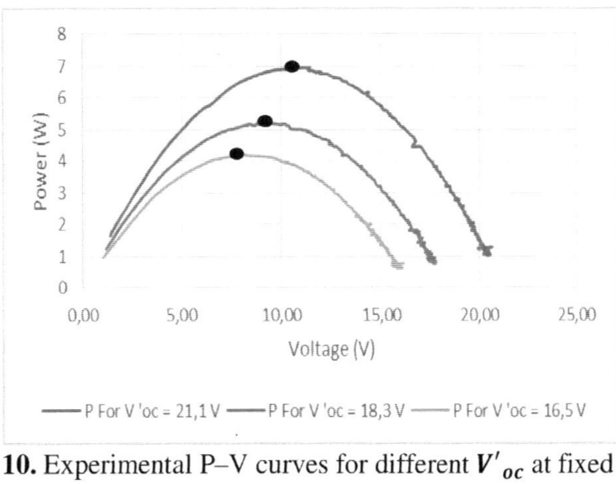

Figure 10. Experimental P–V curves for different V'_{oc} at fixed R_{series}.

Consequently, we can simulate the effects of solar radiation and temperature variations by varying, respectively, the value of the series resistor and the voltage of the DC supply.

4. Conclusion

In this paper, a low-cost PV emulator to emulate the behavior of PV panels device has been designed and implemented. The principle of the maximum power transfer theorem is used for the PV emulator design, which is composed of a variable DC supply in series with a variable resistor. Thus, various experimental tests to confirm the effectiveness of the developed PV emulator are presented in this study. It has been found also from the P-V curve that it exhibits a peak in the power, which can be tracked by an MPPT algorithm. Hence, the obtained results suggest that the proposed system is suitable and can be profitable for researchers to test their MPPT algorithms in a laboratory environment with a reduced budget.

References

[1] Pandey A, Dasgupta N and Mukerjee A K 2008 High-performance algorithms for drift avoidance and fast tracking in solar MPPT system. IEEE Transactions on Energy Conversion, 23(2), 681-689.
[2] Femia, N, Petrone G, Spagnuolo G and Vitelli, M. 2005 Optimization of perturb and observe maximum power point tracking method. IEEE transactions on power electronics, 20(4), 963-973.
[3] Motahhir S, El Ghzizal A, Sebti S, & Derouich A 2018 Modeling of photovoltaic system with modified incremental conductance algorithm for fast changes of irradiance, International Journal of Photoenergy.
[4] Safari A and Mekhilef S 2011 Simulation and hardware implementation of incremental conductance MPPT with direct control method using cuk converter. IEEE transactions on industrial electronics, 58(4), 1154-1161.
[5] Hsieh, G C, Hsieh H I, Tsai C Y and Wang C H 2013 Photovoltaic power-increment-aided incremental-conductance MPPT with two-phased tracking. IEEE Transactions on Power Electronics, 28(6), 2895-2911.
[6] Motahhir S, El Ghzizal A, Sebti S, and Derouich A 2017 MIL and SIL and PIL tests for MPPT algorithm, Cogent Eng, 4(1), 18.

[7] Ahmad J 2010 A fractional open circuit voltage based maximum power point tracker for photovoltaic arrays. In Software Technology and Engineering (ICSTE), 2010 2nd International Conference on (Vol. 1, pp. V1-247).
[8] Singh G K 2013 Solar power generation by PV (photovoltaic) technology: A review. Energy, 53, 1-13.
[9] Bianconi E, Calvente J, Giral R, Mamarelis E, Petrone G, Ramos-Paja C A and Vitelli, M 2013 A fast current-based MPPT technique employing sliding mode control. IEEE Transactions on Industrial Electronics, 60(3), 1168-1178.
[10] Dolan D, Durago J and Crowfoot J 2010 Simulation of a photovoltaic emulator. In North American Power Symposium (NAPS), 2010 (pp. 1-7). IEEE.
[11] Mei Q, Shan M, Liu L and Guerrero J M 2011 A novel improved variable step-size incremental-resistance MPPT method for PV systems. IEEE transactions on industrial electronics, 58(6), 2427-2434.
[12] Schofield D M K, Foster, M P and Stone, D A 2011 Low-cost solar emulator for evaluation of maximum power point tracking methods. Electronics Letters, 47(3), 208-209.
[13] Hohm D P and Ropp, M E 2000 Comparative study of maximum power point tracking algorithms using an experimental, programmable, maximum power point tracking test bed. In Photovoltaic Specialists Conference, Conference Record of the Twenty-Eighth IEEE (pp. 1699-1702).
[14] Khouzam K, Ly C, Koh C K and Ng, P Y 1994 Simulation and real-time modelling of space photovoltaic systems. In Photovoltaic Energy Conversion, Conference Record of the Twenty Fourth. IEEE Photovoltaic Specialists Conference-1994, 1994 IEEE First World Conference on (Vol. 2, pp. 2038-2041).
[15] Mukerjee A K and Dasgupta, N 2007 DC power supply used as photovoltaic simulator for testing MPPT algorithms. Renewable Energy, 32(4), 587-592.
[16] Tey K S and Mekhilef S 2014 Modified incremental conductance MPPT algorithm to mitigate inaccurate responses under fast-changing solar irradiation level. Solar Energy, 101, 333-342.
[17] Kong C S 1995 A general maximum power transfer theorem. IEEE Transactions on Education, 38(3), 296-298.
[18] Motahhir S, El Ghzizal A, Sebti S and Derouich A 2015 Une ressource pédagogique pour l'enseignement par simulation: cas des panneaux photovoltaïques. In Workshop International sur les Approches Pédagogiques & E-Learning.
[19] El Hammoumi A, Motahhir S, Chalh, El Ghzizal A and Derouich A 2018 Low-cost virtual instrumentation of PV panel characteristics using Excel and Arduino in comparison with traditional instrumentation Renewables: Wind, Water and Solar 5 (1), 3.

Nomenclatures

- E : DC supply output voltage [V]
- I : PV emulator output current [A]
- I_0 : MPPT output current [A]

- I_{mpp} : panel output maximum power point current [A]
- P_{load} : Load power [W]
- R_{load} : Load resistance [Ω]
- R_{Series} : series resistance [Ω]
- R_{mpp} : resistor value at MPP [Ω]
- V : PV emulator output voltage [V]
- V_{mpp} : voltage at MPP [V]
- V_0 : Boost output voltage [V]
- V'_{oc} : PV emulator open-circuit voltage [V]
- V_{oc} : panel open-circuit voltage [V]

Greek Letters

- α : duty cycle

Abbreviations

- CV: constant voltage
- DC: Direct Current
- INC: Incremental Conductance
- MPP: maximum power point
- MPPT: maximum power point tracking
- PV: Photovoltaic
- P & O: Perturb and Observe

Real-time virtual instrumentation of Arduino and LabVIEW based PV panel characteristics

A El Hammoumi*[1], S Motahhir[1], A Chalh[1], A El Ghzizal[1] and A Derouich[1].

[1] *Laboratory of Production engineering, Energy and Sustainable Development, Smart Energy Systems and Information Processing Research Team, Higher school of technology, SMBA University, Fez, Morocco.*

*Corresponding author, E-mail: aboubakr.elhammoumi@usmba.ac.ma

Abstract. This paper describes a virtual instrument based on a low-cost embedded board to monitor and plot the PV panel characteristics under real operation condition. The system design is based on a low-cost Arduino acquisition board in which the ATMega328 microcontroller is integrated. The acquisition is made through a low-cost current and voltage sensors and the data are transmitted in LabVIEW by using LIFA Interface for Arduino. Hence, the I-V (current-voltage) and P-V (power-voltage) characteristics for PV panel, which processed under actual conditions, can be obtained and plotted directly on a monitoring platform in LabVIEW. The proposed instrument can be used for educational or research purposes using a low-cost and easily hardware without having extensive knowledge about electronic engineering. The present instrumentation technique provides easy access to the collected data for further analysis.

1. Introduction

The increasing demand for energy in line with the world's challenge to reduce the use of fossil fuels, which pollute the environment, has greatly increased interest in renewable energy sources (RES) [1], [2]. PV solar power is one of the RES for electricity generation and it is the most popular in the world because of the continuous evolution of PV technology and lower prices [3]. As a result of the popularity of PV solar power, many researchers around the world are working in this field of science to contribute to the current knowledge about this technology and improve its cost-effectiveness. Therefore, they must have scientific instruments to conduct the required experiments. Tracing of PV characteristics (current-voltage (I-V) and power-voltage (P-V) curves) of a PV device is the most important and basic experiment test, that should be performed by any researcher or anyone interested in PV technology, to obtain information about its behavior and performance [4]. Different traditional instruments are used in order to acquire the PV panel characteristics such as Multimeters. Nevertheless, by using this kind of instrument, there is a high probability that the characteristics obtained and traced manually do not give enough information about the PV device state because it can not be taken directly in real-time [5]. In addition, the experiments test are tiring and require a lot of time.

On the other hand, different commercial data loggers are available for purchase on the internet for virtual instrumentation of PV panel characteristics [6], [7], such as the DAQPro 5300 [8] and the CAMPBELL Scientific CR800 [9]. Generally, the most data loggers are too expensive and demand to be coupled with special software, which increased cost and requiring additional specific skills. In addition, they have many disadvantages that make them customized for specific applications, limited in terms of the input voltage and the number of input/ output channels, inadequate input channels for the connection of some sensors. For these reasons, we must design and developed new data instruments that can monitor PV systems at low-cost and with flexible design. For that, many virtual instrument systems for the PV panel have been proposed in the literature, such as virtual instrumentation in LabVIEW for real-time PV system monitoring [10]-[12] and virtual instrumentation by the standard simulation software Matlab [13]. In this paper, we present a real-time virtual instrument system based on low-cost hardware to monitor and trace the P-V and I-V characteristics of the PV panel in real operating conditions. The system design of the proposed instrumentation is based on LabVIEW and Arduino. A current and voltage sensors sense the output current and voltage from PV panel. The measured data of the current, voltage and power are plotted directly on a monitoring platform developed in LabVIEW. The main advantages of the present technique are its simplicity and utilization of the low-cost components such as Arduino UNO board, low-cost current and voltage sensors. Moreover, the current, voltage, power and the characteristics I-V and P-V of the PV panel are plotted directly in LabVIEW. That is done by building a simple block program compared to other LabVIEW instruments in the literature that using multiple and complex blocks in the LabVIEW program. This work allows the user to acquire, supervise and trace the solar panel characteristics in real time with a simpler, cheaper and faster way.

The paper is arranged as follows: The design and description of the proposed instrument system are presented in Section 2. The performed experiment and the results are done in Section 3. Finally, Section 3 summarizes this paper.

2. The PV panel instrumentation system

Figure 1 shows the schematic diagram of the PV panel instrumentation system. The current and voltage sensors sense the output current and voltage from PV panel. Then, the readings of the two sensors are transmitted to the microcontroller of the Arduino UNO board. While the LabVIEW Interface for Arduino (LIFA) serves as an interface between the Arduino microcontroller and computer to make communication between the microcontroller and LabVIEW through a serial connection. During the acquisition process, the measured data of the current, voltage and power are plotted directly in LabVIEW. As shown in figure 1, the Arduino board is the heart component of the real-time acquisition system. The design of the proposed instrument is divided into two parts: hardware system and software system.

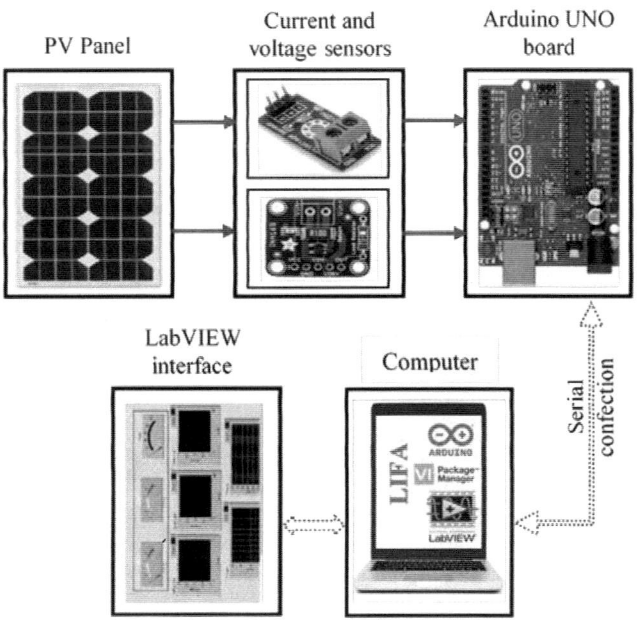

Figure 1. Schematic diagram of the PV panel instrumentation system.

IOP Conf. Series: Earth and Environmental Science **161** (2018) 012019 doi:10.1088/1755-1315/161/1/012019

2.1. Hardware

The proposed PV panel instrument system consists of a PV panel, Arduino UNO board, current sensor, and voltage sensor.

A. PV Panel

PV panels are intended to recover the energy of solar radiation to transform it into continuous electrical energy [14]. A monocrystalline PV panel M20-36 is used in this work, and its features at STC (Standard Test Conditions) are presented as follow [15]:

- Maximum power, Pmax: 20 W;
- Voltage at Pmax, Vmp: 18.76 V;
- Current at Pmax, Imp: 1.07 A;
- Short-circuit current, Isc: 1.17 A;
- Open- circuit voltage, Voc: 22.70 V;
- Temperature coefficient of Voc, Kv: 0.35%/°C;
- Temperature coefficient of Isc, Ki: -0.043%/°C;
- Number of cells: 36;
- Light-generated current, Iph: 1.173 A;
- Diode saturation current, Is: 2.6797e-11 A;
- Ideality factor: 1.0036;
- Shunt Resistance Rsh: 405.962 Ω;
- Series Resistance Rs: 1.0547 Ω.

B. Arduino UNO

The data acquisition board used in this work is Arduino UNO, it is a low-cost board [5]. Figure 2 shows the Arduino UNO board, and table 1 presents its technical specifications [16].

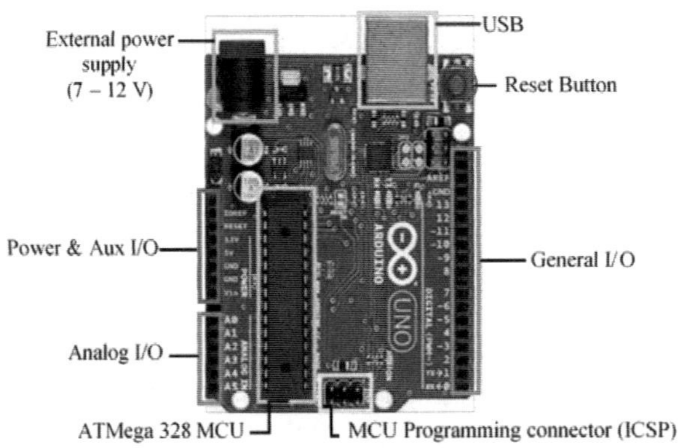

Figure 2. Arduino UNO board.

Table 1. Technical specifications of Arduino Uno board

Microcontroller	ATMega328
Operating Voltage	5 V
Input Voltage (recommended)	7 – 12 V
Input Voltage (limits)	6 – 20 V
Digital Input / Output Pins	14 (of which 6 provide PWM output)
Analog Input Pins	6
DC Current per I/O Pin	40 mA
DC Current for 3.3V Pin	50 mA

Flash Memory	32 KB
SRAM	2 KB
EEPROM	1 KB
Clock Speed	16 MHz

C. Voltage sensor

For measuring the PV panel's output voltage, a voltage sensor F031-06 is used. Figure 3 shows the F031-06 voltage sensor module, figure 4 presents its connectors with the Arduino UNO board and the specifications of this sensor are shown in table 2.

Figure 3. F031-06 voltage sensor module.

Table 2. Features of the F031-06 voltage sensor module

Features	**F031-06 voltage sensor module**
Input Voltage Range	DC 0-25 V
Voltage Detection Range	DC 0.02445 V - 25 V
Analog Voltage Resolution	0.00489 V

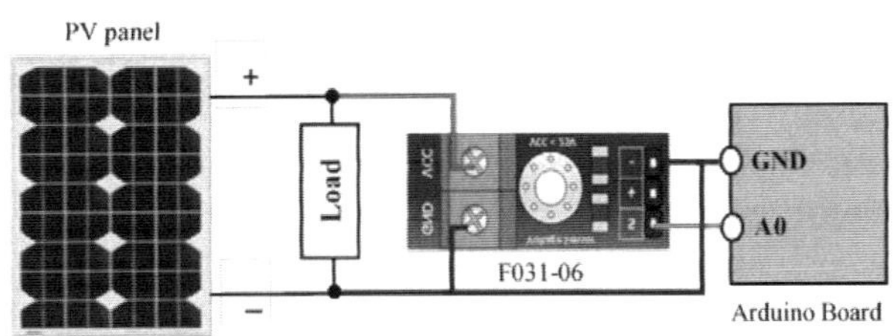

Figure 4. Schematic view of the F031-06 sensor connected to the Arduino board.

The F031-06 voltage sensor uses the principle of the voltage divider, it is essentially a voltage divider using a resistance of 30 kΩ and a resistance of 7.5 kΩ as shown in figure 5 [17]. It is used to reduce the input voltage up to 5 times compared to the original voltage because the maximum analog input voltage of the Arduino microcontroller is 5 V. The voltage sensor module will be installed in parallel with the PV panel load as shown in figure 4. The sensor reading is a digital value (V_{out}) which varies between 0 and 1023. Since the ADC of the microcontroller is coded in 10 bits, the resolution of the PV panel module voltage is 0.00489 V (5/1023) and the input voltage of this module must be more than 0.02445 V (0.00489 V× 5). Hence, the output voltage of the PV panel can be formulated as the following equation:

$$V = 5 \times V_{out} \times \frac{5}{1023} \tag{1}$$

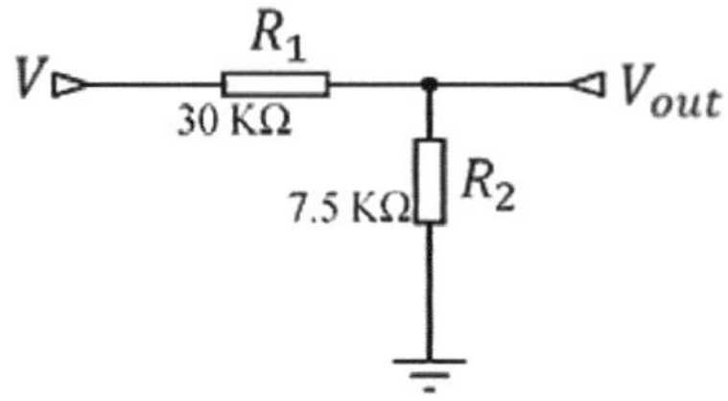

Figure 5. Voltage divider circuit.

D. Current sensor

The current sensor used to sense the PV panel output current is the INA169 module (figure 6), it can measure a continuous current up to 5 A [18]. In table 3, the specifications of the selected current sensor are presented.

Figure 6. INA169 current sensor module.

Table 3. Features of the INA169 current sensor module

Features	**INA169 current sensor module**
Common Mode Voltage Range	DC 2.7 V - 60 V
Full-Scale Sense Input Voltage	500 mV
Input Offset Voltage (Max)	±1000 µV/V
Input Offset Drift (Max)	±1 µV/°C
Nonlinearity Error (Max)	±0.1%
Total output Error (Max)	±2%
Common Mode Rejection Ratio (Typ)	120 dB
Bandwidth	440 kHz
Supply Voltage Range	DC 2.7 V - 60 V
Operating Temperature Range	-40 °C - 85 °C
Quiescent Current	60 µA

Figure 7 shows the INA169 current sensor circuit [18]. The INA169 is a high-side current monitor that measures the voltage drop across a sense resistor (R_S). Then, a voltage level (V_o) is generated at the output resistor (R_L) [18]. Therefore, as shown in figure 7, the output current of sensor module can be obtained by the following equation:

$$I_s = \frac{V_o \times 1K}{R_s \times R_L} \tag{2}$$

Where:

V_o: is the voltage we measured at the output of the INA169.
R_S: is the value of the shunt resistor, this is set at 0.1 Ω.
R_L: is the value of the output resistor, this is set at 10 KΩ.

Hence, the output current of the PV panel can be formulated as the following equation:

$$I = \frac{V_o \times 5}{1023} \quad (3)$$

The output voltage of the INA169 module is the input voltage at an Arduino analog pin (A1), which varies between 0 and 1023. The current sensor module must be connected in series to the positive side of the PV panel and that of the load as shown in figure 8.

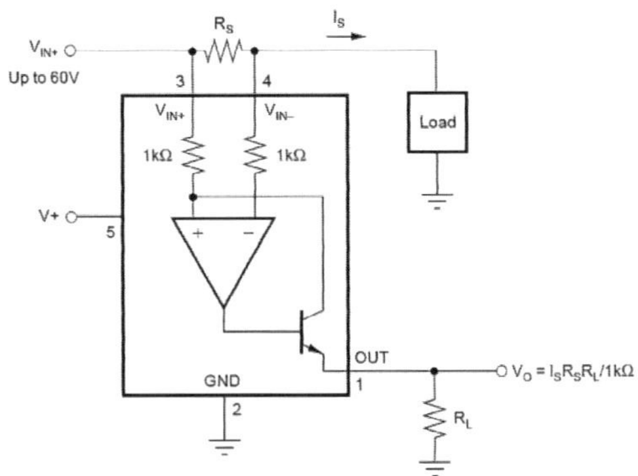

Figure 7. INA169 current sensor circuit [18].

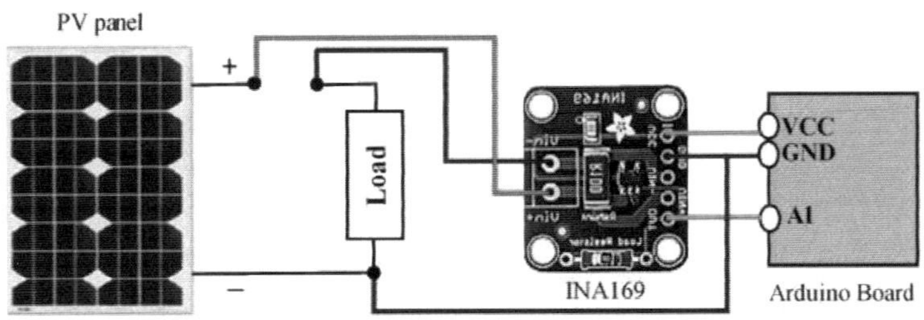

Figure 8. Schematic view of the INA169 sensor connected to the Arduino board.

2.2. *Software*

The real-time virtual instrumentation system of PV panel characteristics was designed using a LabVIEW and Arduino based controller. For that, in the software part, we present all tools requested for sending data from Arduino board to LabVIEW, hence, the steps for obtained the I-V and P-V curves of PV panel in real-time.

A. LabVIEW Interface For Arduino (LIFA)

To send data from Arduino to LabVIEW, the LIFA is necessary to make the connection between the Arduino microcontroller and computer. For this, to make the interface as shown in figure 9, you will need to follow the following steps:

- Install the LabVIEW software (LabVIEW 2011 or later);
- Download the NI-VISA drivers and install it. Since the Arduino appears as a serial instrument device for LabVIEW, the NI-VISA drivers must be downloaded and installed to communicate with Arduino board in LabVIEW;
- Download the JKI VI Package Manager (VIPM) and install it, it allows to install and update LabVIEW libraries;
- Install the LIFA by following the next steps:

Lunch LabVIEW → click on tool → go to VIPM → browse to LIFA in the list of packages → click on the Install Upgrade Packages button;

- Download the open source Arduino IDE software and install it;
- The LIFA provides a sketch program that must be uploaded to the Arduino before you can use the virtual instruments to communicate with it. For this, Open the Arduino IDE application and following the next steps to uploading LIFA firmware to the Arduino board:
  - ✓ Go to File → Click on Open → go to the sketch which is located at: C:\ProgramFiles(x86)\National Instruments\LabVIEW 2015\vi.lib\LabVIEW Interface for Arduino\Firmware\LIFA_Bas;
  - ✓ Select the type of used Board (Arduino UNO);
  - ✓ Connect the board to computer via USB cable and select the used serial port;
  - ✓ Verify and Upload the sketch program LIFA to the Arduino microcontroller;
- After that, you can start building your code in Arduino with LIFA Block Diagram.

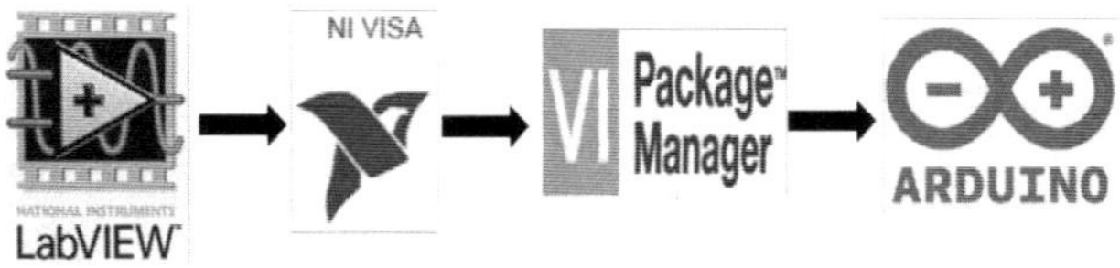

Figure 9. Process of installing LIFA and its firmware to Arduino board.

B. LabVIEW program

In figure 10 and figure 11, the overview of the block diagram and the software interface of the real-time virtual instrumentation system of PV panel characteristics are shown. For enriching understanding, the block diagram is separated into four part. The first one illustrates the blocks of initialization and configuration of the serial connection between the hardware (Arduino UNO) and software (LabVIEW), that by defining the type of used Arduino (Arduino UNO), type of serial communication (USB) and the Baud Rate (9600 bits per second). In the second part, the voltage and current readings will be obtained from the microcontroller by using the analog read function in LIFA's palette in which the initialization of the selected Arduino pins as inputs was made, A0 pin for voltage reading and A1 pin for current reading. The third part shows that the current, voltage and output power readings are carried out and will be plotted directly according to time by using scrolling graphs in LabVIEW palette. Note that the reading of the voltage is multiplied by 5 because as it is demonstrated in the hardware part that the voltage reading by the microcontroller varies between 0 and 5 V, it is necessary to multiply it by 5 to obtain the PV panel's output voltage. The PV panel's output power reading is done out by the product of the current and voltage reading. Finally, in the last part, the I-V and P-V curves of PV panel characteristics will be plotted in software interface of the system by using x/y graphs in LabVIEW palette.

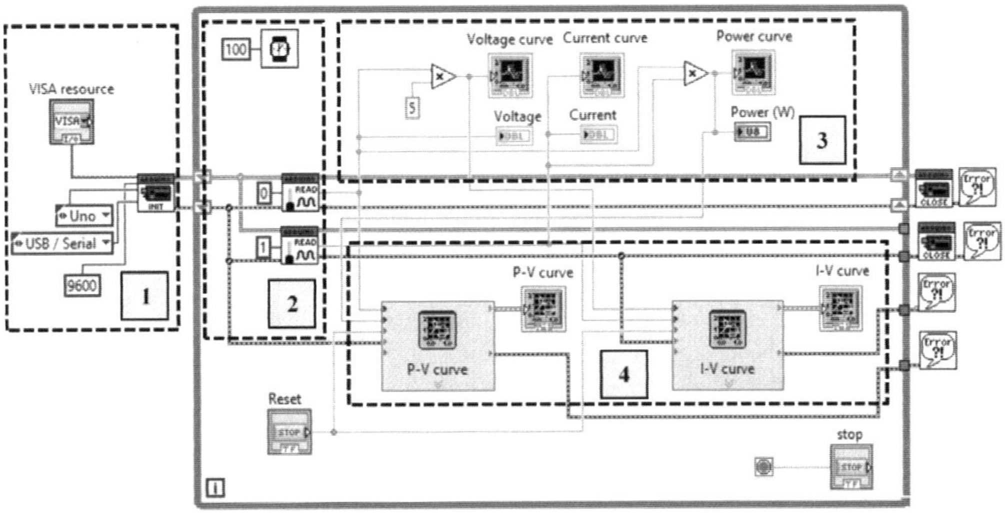

Figure 10. Block diagram of the real-time virtual instrumentation system under LabVIEW.

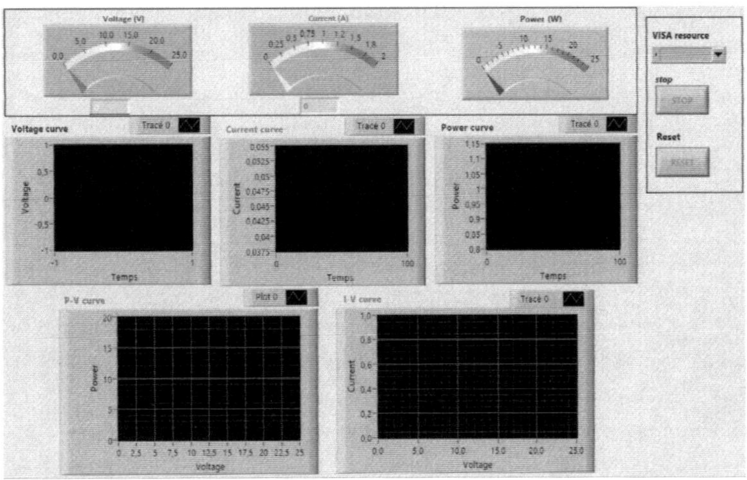

Figure 11. The software interface of the real-time virtual instrumentation system under LabVIEW.

3. Experimental results

In figure 12, the experimental setup of the real-time virtual instrumentation system is shown. Apart PV panel, Arduino UNO board, voltage and current sensor, different components are used in the experimental setup such us lamps of 100 W that acts as a solar simulator, a variable resistance between 0 and 300 Ω as a load and acting as a light intensity driving. The microcontroller's operation is programmed in the block diagram to measure the values relative to the current and the voltage captured by the sensors successively every 100 ms. Once the Arduino board is connected to the computer through USB cable, we launch the software interface of our system under LabVIEW. Then, we defined in the visa block the serial port where the Arduino board is connected to the computer. Thereafter, after the execution of the program, the output data will be plotted in real-time on the software interface under LabVIEW. The results of direct monitoring and the I-V/ P-V curves of the PV panel are shown in figure 13.

IOP Conf. Series: Earth and Environmental Science **161** (2018) 012019 doi:10.1088/1755-1315/161/1/012019

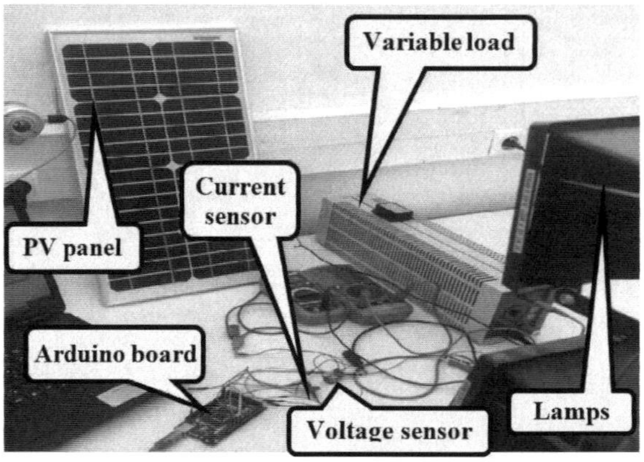

Figure 12. Experimental setup of the measurement system

In figure 14, a comparison between the P-V characteristic of the PV panel which is obtained by the proposed solution and that obtained by an experimental test through calibrated Millimeters is shown. The results presented in figure 14 indicate that the developed real-time virtual instrument in this work is valid. The price of the hardware components used for data acquisition in the proposed instrumentation is presented in table 4.

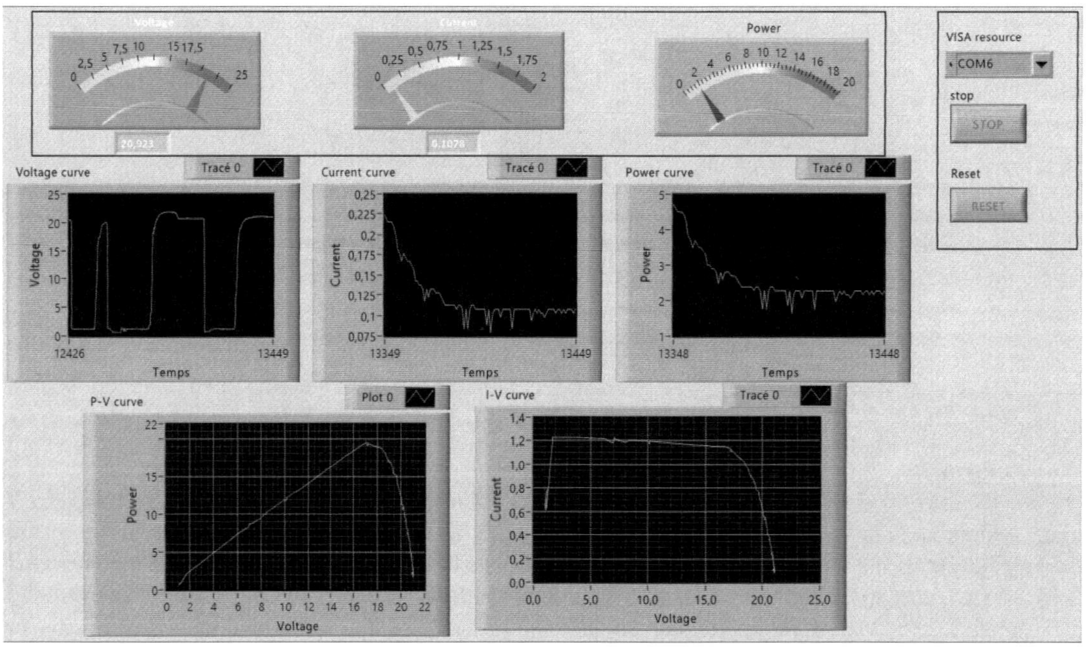

Figure 13. Obtained results in the software interface of the real-time virtual instrumentation system.

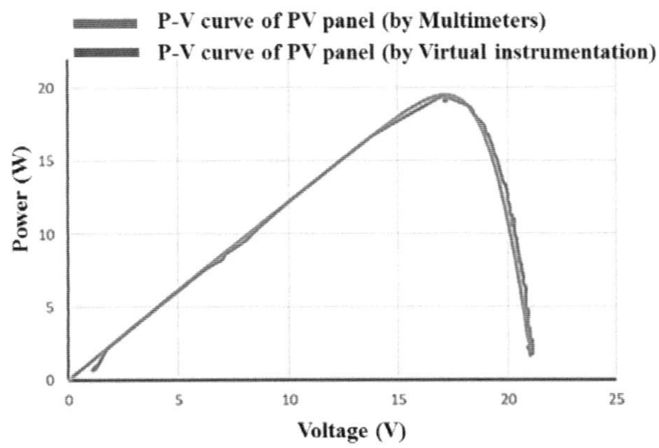

Figure 14. P-V curve of PV panel.

Table 4. Price of hardware components used for data acquisition

Components	Price ($)	
Arduino UNO board	4.75	(www.aliexpress.com)
F031-06 voltage sensor module	0.74	(www.aliexpress.com)
INA169 current sensor	3.30	(www.aliexpress.com)
Total price	8.79	

4. Conclusion

This article describes a real-time virtual instrument system of PV panel characteristics based on a low-cost embedded board, which used to monitor and trace the characteristics I-V and P-V of a PV panel under the real condition. For that, PV panel, Arduino UNO board, voltage and current sensors have been used as hardware components coupled to the open-source Arduino IDE, LabVIEW and LIFA as the software for the system design. The present system is validated by comparing the obtained data in LabVIEW interface with that obtained with the experimental test using a calibrated Multimeters. Results show that the data of our instrument are in accordance with experimental data.

References

[1] Corcelli F, Ripa M and Ulgiati S 2017 End-of-life treatment of crystalline silicon photovoltaic panels. An emergy-based case study *Journal of Cleaner Production* **161** pp. 1129-1142.
[2] Singh G K 2013 Solar power generation by PV (photovoltaic) technology: A review *Energy* **53** pp. 1-13.
[3] Breyer C, Bogdanov D, Gulagi A, Aghahosseini A, Barbosa L S Koskinen O and Farfan J 2017 On the role of solar photovoltaics in global energy transition scenarios *Progress in Photovoltaics: Research and Applications* **25** No. 8 pp. 727-745.
[4] Montes-Romero J, Piliougine M, Muñoz J V and Fernández E F 2017 Photovoltaic Device Performance Evaluation Using an Open-Hardware System and Standard Calibrated Laboratory Instruments *Energies* **10** No. 11 p. 1869.
[5] El Hammoumi A, Motahhir S, Chalh, El Ghzizal A and Derouich A 2018 Low-cost virtual instrumentation of PV panel characteristics using Excel and Arduino in comparison with traditional instrumentation *Renewables: Wind, Water and Solar* **5** No. 1 p. 3.
[6] Munoz, M. A., Alonso-García M C, Vela N and Chenlo F 2011 Early degradation of silicon PV modules and guaranty conditions *Solar energy* **85** No. 9 pp. 2264-2274.
[7] Hernday P 2011 Field Applications for I-V Curve Tracers *SolarPro* pp. 76–106.
[8] Fourtec Fourier Technologies 2018 http://www.fouriersystems.com/ Accessed 25 March 2018.
[9] Campbell Scientific 2018 http://www.campbellsci.com/ Accessed 25 March 2018.
[10] Koutroulis E and Kalaitzakis K 2003 Development of an integrated data-acquisition system for renewable energy sources systems monitoring *Renewable Energy* **28** No. 1 pp. 139-152.

[11] Ulieru V D, Cepisca C and Ivanovici T R 2010 Measurement and analysis in PV systems *In Proceedings of the International Conference on Circuits, ICC,* pp. 22-24.
[12] Hemza A, Abdeslam H, Rachid C, Pasquinelli M and Barakel D 2015 Tracing current-voltage curve of solar panel Based on LabVIEW Arduino Interfacing *Bilişim Teknolojileri Dergisi* **8** No. 3 pp. 117–123.
[13] Chouder A, Silvestre S, Sadaoui N and Rahmani L 2012 Modeling and simulation of a grid connected PV system based on the evaluation of main PV module parameters *Simulation Modelling Practice and Theory* **20** No. 1 pp. 46-58.
[14] Motahhir S, El Ghzizal A, Sebti S and Derouich A 2018 Modeling of Photovoltaic System with Modified Incremental Conductance Algorithm for Fast Changes of Irradiance *International Journal of Photoenergy* **2018** pp. 1–13.
[15] TDC-M20-36 solar panel 2018 https://tdcsolar.en.alibaba.com/product/60018196685-220763581/20w_mono_small_solar_panel_from_chinese_factory.html?spm=a2700.8304367.rect38f22d.16.47185198OcAXoD Accessed 10 March 2018.
[16] Arduino Uno 2018 https://store.arduino.cc/arduino-uno-rev3 Accessed 20 March 2018.
[17] Voltage Sensor Module 2017 http://www.emartee.com/product/42082/Voltage%20Sensor%20/ Accessed 2 October 2017.
[18] INA169 Datasheet 2017 http://www.ti.com/lit/ds/symlink/ina169.pdf?HQS=TI-null-null-alldatasheets-df-pf-SEP-wwe.pdf Accessed 20 October 2017.

STUDY OF THE DIELECTRIC AND MICROSTRUCTURE PROPERTIES OF (1-x) $CaCu_3TI_4O_{12}$ – (x) $PbZr_{0.65}Ti_{0.35}O_3$ COMPOSITE

N Hadi[*,1], T Lamcharfi[1], F Abdi[1], N S Echatoui[1], F Z Ahjyaje[2], A Harrach[3] and S Chouikh[1]

[1] Laboratory of Signals, Systems and Components, USMBA. FST Fez, B.P. 2202, Morocco.
[2] Laboratory of Materials Natural Substances, Environment and Modeling, LMSNEM, USMBA. FP-Taza -. Morocco,
[3] Laboratory of Condensed Matter Chemistry, USMBA. FST-Fes, B.P. 2202, Morocco.

*Email: salehmabed@gmail.com

Abstract. In this paper, structural, dielectric and electrical properties of calcium copper titanate and $PbZr_{0.65}Ti_{0.35}O_3$ ceramic composite with composition (1-x) $CaCu_3Ti_4O_{12}$ – (x) $PbZr_{0.65}Ti_{0.35}O_3$ (where x = 0.00, 0.30, 0.50, 0.70 and 1.00) has been distinguished. The $CaCu_3Ti_4O_{12}$ sample was prepared by employing a high temperature solid state reaction technique and $PbZr_{0.65}Ti_{0.35}O_3$ synthesized by sol-gel route. X-ray diffraction (XRD), scanning electron micrograph (SEM), FT-IR spectrum (IR) was used to reach the structural properties of the studied compounds. The XRD and IR studies confirm the formation of pure phase for x = 0.00 and 1.00 concentrations i.e cubic and tetragonal, respectively. The SEM show change in the grain distribution, size and shape at addition of $PbZr_{0.65}Ti_{0.35}O_3$. Dielectric measurements were made by using an impedance analyzer, the results show that the dielectric constant of the CCTO/PZT composite is greater than that of pure CCTO and than that of pure PZT. The temperature dependence of the ac conductivity indicated that the conduction process is due to singly and doubly ionized, and frequency dependence of the ac conductivity of composites follows Jonscher's power law.

1. Introduction

Calcium Copper Titanate ($CaCu_3Ti_4O_{12}$, CCTO) based perovskite materials have been widely used for various applications, such as multilayer capacitor (MLCC), dynamic random access memory (DRAMs), microwave devices, electronic devices in automobiles and aircrafts such as multilayer capacitor and electronic devices, due to their excellent dielectric properties [1-3]. However, CCTO ceramic has high dielectric constant ($\varepsilon_r \sim 10^4 - 10^5$) independent of temperature (100-400K) which is desired for microelectronic applications. On the other hand, the CCTO ceramic has higher dielectric loss that limits its practical applications [4-6]. Recently, extensive studies of the formation of perovskite oxides of the systems $Ca_{1-x}M_xTi_{1-x}M_xO_3$ (M = Cu, Y, Sr, Ba, Pb; M¢ = Co, Al, Fe, Cr) were performed. Almeida et al. [7] investigated the properties of the composite BaTiO3–CCTO film and its potential use as antennas and microdevices. In an analogous previous work, N. Hadi et al.[8] investigated the stability of the perovskite phase and the dielectric properties in the (1-x)CCTO-x$BaTiO_3$ ceramic (CCTO and $BaTiO_3$ perovskites), the value of dielectric constant was about 56,000 when x = 0.10. In this work, we report the effect of the presence of the PZT on the dielectric and microstructure properties of the CCTO. We compare these properties for the (1-x) $CaCu_3Ti_4O_{12}$–(x) $PbZr_{0.65}Ti_{0.35}O_3$ composites.

2. Experimental

The ceramic of (1-x)$CaCu_3Ti_4O_{12}$-x$PbZr_{0.65}Ti_{0.35}O_3$(where x = 0.00, 0.30, 0.50, 0.70 and 1.00) was synthesized by a modified solid-state route via three steps. **Firstly**, $CaCu_3Ti_4O_{12}$(CCTO) was synthesized by a solid state process using $CaCO_3CuO$ and TiO_2 were used as starting materials. In this route, stoichiometric ratios of the regents were mixed in agate motor for 1h, after that stirring in the medium of acetone for 3h. The mixed powder was calcined in air at 1050 ^{0}C for 4h. **Secondly**, $PbZr_{0.65}Ti_{0.35}O_3$ (PZT) ceramic powder was also synthesized by a Sol-Gel process using Lead (II) acetate trihydrate, $Pb(CH_3COO)_2.3H_2O$ zirconium acetate, $Zr(CH_3COO)_4$ and titanium isopropoxide, $Ti(OCH(CH_3)_2)_4$,. The metal acetate were dissolved in distilled water to obtain standard aqueous solutions of Pb^{2+} and Zr^{4+}, and mixed in separate beakers along with stoichiometric amounts of $Ti(OCH(CH_3)_2)_4$. The resulting dispersion was heated on a hot plate under stirring at 70–80 °C to evaporate the water and allow self-ignition. By controlling the hydrolysis condition of the complex solution, a PZT gel was formed. The dry gel was calcined at 700 °C for 4h in atmosphere PZT powders were obtained. **Finally**, the (1-x)$CaCu_3Ti_4O_{12}$-x$PbZr_{0.65}Ti_{0.35}O_3$ prepared of $CaCu_3Ti_4O_{12}$ and $PbZr_{0.65}Ti_{0.35}O_3$were carefully weighed in stoichiometric proportion and mixed thoroughly in agate motor for 1h then stirring in acetone for 2h. The powder was dried at 400 °C for 2h. The dried powder was then pressed into disks using polyvinyl alcohol (PVA) as a binder. The final sintering of the pellets was done at 1000 °C for 8h using a heating rate 3 °C/min.

3. Results and discussion

X-ray diffraction patterns of the CCTO and PZT powder are shown in **Fig. 1**.The CCTO and PZT powders show a single phase which is highly crystalline in nature where the main peaks of the ceramics are comparable to those of the standard ceramic XRD patterns of CCTO (JCPDS 75-2188) and PZT (JCPDS 33-0784), which was identified to be cubic and tetragonal phase respectively.

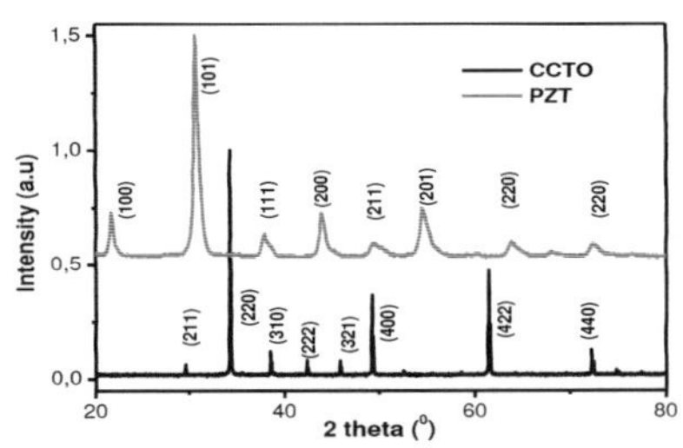

Fig. 1 X-ray diffraction pattern of CCTO and PZT pure

The phase formation of CCTO and PZT was further confirmed from FT-IR spectrum shown in **Fig.2.** The IR spectra of the CCTO ceramic in the region of 400–1200 cm^{-1} is dominated by three broad absorptions centered at 566, 516 and 447 cm^{-1} [7]. A similar spectrum is observed for the ceramic CCTO prepared in this work, with absorptions at 573, 523 and 451 cm^{-1} (see **Fig. 2**). The absorptions associated to CCTO are: 573 cm-1 related to Ca-O, 447 cm-1 related to vibrational mode of the Ti-O-Ti and 523 cm^{-1} which is assigned to bending of Cu-O. For the PZT ceramic, two main absorptions were observed at 570 and 410 cm^{-1} which is consonant for the PZT ceramic, according to Singh et al. [6].

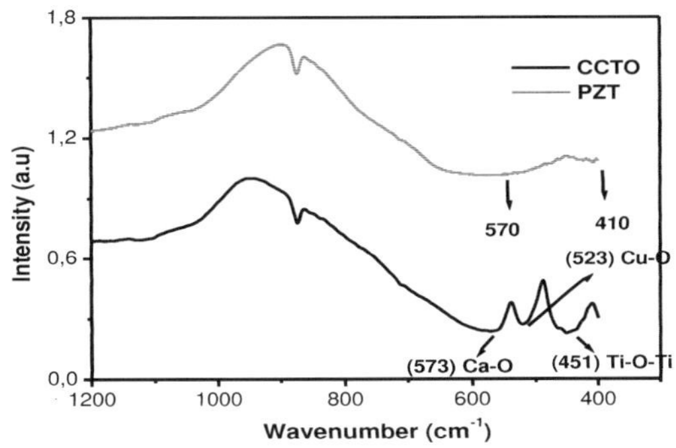

Fig. 2 FT-IR spectrum of CCTO and PZTpure.

The microstructure of CCTO, PZT and CCTO-PZT samples sintered at 1000 ^0C observed by SEM is shown in **Fig. 3.** It is found that the shape, the size and the distribution of grain of the microstructures of all the composites change with addition of PZT in CCTO. The size, distribution and shape of grain of the microstructures of all the samples confirmed the polycrystalline nature of the sample and well developed grains were observed in all the compositions. The grain size increases on increasing PZT contents compared with grains of pure PZT. The variation of average grain size with composition is shown in **Table 1**.

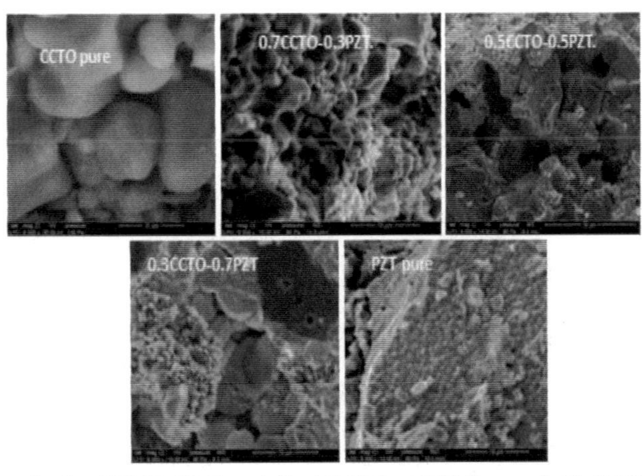

Fig. 3. SEM micrographs for CCTO, PZT and (1-x)CCTO-xPZT for $x = 0.00, 0.30, 0.50, 0.70$ and 1.00

Table 1: Variation average grain size of (1-x) CCTO–xPZT ceramics (x = 0.00, 0.30, 0.50, 0.70 and 1.00).

(1-x) CCTO–xPZT	grain size (μm)
x= 0.00	8.750
x= 0.30	5.747
x= 0.50	8.552
x= 0.70	5.000
x= 1.00	1.750

Fig. 4a and b show the behaviour of dielectric constant and dielectric loss of these ceramics as a function of temperature at 10 kH_Z. From Fig. 4a, the dielectric constant is almost constant up to 150, 200 and 300 °C for CCTO, CCTO-PZT and PZT, respectively. After this range, ε_r an increase gradually attaining a maximum value εmax at Curie temperature and thereafter it decreases with the further increase in temperature. This dielectric anomaly indicates a phase transition from ferroelectric to paraelectric phase at that particular temperature (Curie temperature, T_m). The Curie temperatures of CCTO, (1-x)CCTO-xPZT and PZT ceramics are 321, 337 (x = 0.30), 348 (x = 0.50), 360 (x = 0.70) and 370 °C (see table 2), respectively. The T_m shifts towards higher temperatures (T_m~360 °C) with increasing PZT content upon x=0.70. As can be noted from Fig. 4a and Table 2, the values of ε_r increased in the CCTO-PZT samples in temperature range of this study. The dielectric peak sharp around Tm, decreased with the increase of PZT content, which is one of the characteristics of the disordered perovskite structure with diffuse phase transition. The effect of the PZT content on the temperature dependence of CCTO dielectric loss or tan (δ) is shown in Fig. 4(b). The dielectric loss of pure CCTO is maximum with respect to other samples in the temperature range of this study. The PZT doped CCTO samples show small dielectric loss in which is dependent on PZT content. Variation of the values of dielectric loss at high temperature may be attributed to the conduction mechanism [10].

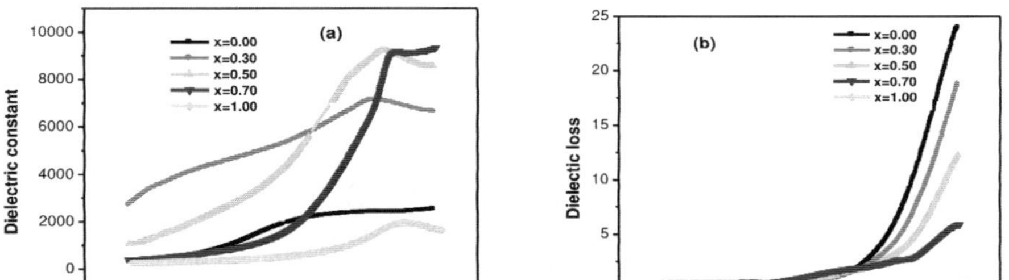

Fig. 4. Plot of real part of permittivity ε_r and dielectric loss tan (δ) of (1-x)CCTO-xPZT as a function of temperature at 10kHz (**a**) real part of permittivity ε_r (**b**) dielectric loss

Table 2: Variation of ε_r maximum and transition temperature (T_m) of (1-x) CCTO–xPZT ceramics.

(1-x) CCTO–xPZT	T_m	$\varepsilon_{r\ max}$
x= 0.00	321	1813.26
x= 0.30	337	5756.75
x= 0.50	348	6652.32
x= 0.70	360	6079.00
x= 1.00	369	1686.75

Fig. 5 shows the ac conductivity (σ_{ac}) as a function of frequency at 160 ^{0}C for all (1-x)CCTO-xPZT composites. The σ_{ac} is calculated from the dielectric data by using the relation [11];

$$\sigma_{ac} = \omega\varepsilon_0\varepsilon'' \quad \textbf{(1)}$$

where σ_{ac} is the ac conductivity, ω is the frequency, ε_0 the permittivity of space free ($8.854.10^{-12}$ F/m), and ε'' is imaginary part of the dielectric constant. The frequency dependence of conductivity which obtained it in this study can be described by the Jonscher's universal power law [11, 12] given by:

$$\sigma_{ac}(\omega) = \sigma_{dc} + A\omega^n \quad \textbf{(2)}$$

where σ_{dc} is the frequency independent conductivity, A is the dispersion parameter and n is the dimensionless frequency exponent, with the values 0 and 1 which is a measure of the degree of interaction or an ideal Debye dielectric dipolar-type and an ideal ionic-type crystal respectively. The term $A\omega^n$ contains the ac dependence and characterizes all dispersion phenomenons. The plots showed a dispersive nature at 220 ^{0}C due to the typical RC network. Low frequencies region is dominated by conductive process in the ceramic while the high frequencies one is controlled by capacitive region [12]. The conduction difference between these compounds is due to the high defect concentration, mainly due to the oxygen vacancy resulting from the simple interaction between the components. The high conductivity in the composites at large frequencies is believed to be due to the hopping of the mobile species. Its proportionally to ω^n given in the equation 2, indicates that the electrical network (RC) response is qualitatively similar to the universal dielectric response (UDR) [9].

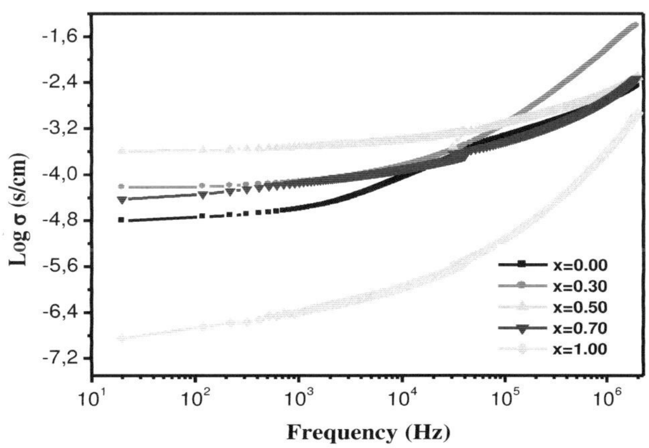

Fig. 5. Variation of ac conductivity with frequency at 220 ^{0}C of (1-x) CCTO–xPZT ceramics (x = 0.00, 0.30, 0.50 0.70 and 1.00).

4. Conclusion

In the present work, we have prepared the (1-x)CCTO-xPZT system with compositions of x = 0.00-1.00 by using solid state reaction route. X- ray diffraction pattern shows the formation of a pure phase for x = 0.00 and 1.00 compositions. The FT-IR studies confirm the formation of pure phase for CCTO and PZT powder and showed the presence of three and two main peaks, respectively. The dielectric study shows that, for CCTO-PZT, an enhanced of the dielectric constant compared with pure CCTO and PZT samples.

References

[1] Rai U S, Singh L, Mandal K D, Singh N B 2014 An Overview on Recent Developments in the Synthesis, Characterization and Properties of High Dielectric Constant Calcium Copper Titanate Nano-Particles. *Nanosci Technol* **1(2)** 1-17.

[2] Deschanvers A, Raveau B, Tollemer F 1967 Substitution of Copper for a divalent metal in perovskite-type titanates. *Bull. Soc. Chim.* Fr.4077-407

[3] Laxman S, Rai U S, Mandal K D and Singh N B 2014 Progress in the growth of $CaCu_3Ti_4O_{12}$ and related functional dielectric perovskites *progress in Crystal Growth and Characterization of Materials* **60** 15–62. http://dx.doi.org/10.1016/j.pcrysgrow.2014.04.

[4] Hadi N, Lamchafi T, Abdi F, Elbasset A, and Sayouri S 2017 Effect of Sintering Temperature on the Dielectric and Microstructure Properties of $CaCu_3Ti_{3,90}$ $Co_{0,10}$ O_{12} Ceramic *International Journal of Development Research*, 07, **(04)** 12432-12436.

[5] Laxman S, Rai U S, Mandal K D, and Rai A K 2014 Effect of site selection on dielectric properties of Fe doped $CaCu_3Ti_4O_{12}$ electro-ceramic synthesized by citrate nitrate gel *Indian J Phys,* **88 (7)** 665–670.

[6] Laxman S, Rai U S, Mandal K D, and Rai A K 2013 Effects of partial substitutions of transition metal cations on dielectric properties of $Ca_{1-x}La_xCu_{3-x}MnxTi_{4-x}Ni_xO_{12}$ (x= 0.05, 0.10) electro-ceramics synthesized by semi-wet chemical method. *Journal of Advanced Ceramics* **2** 119–127.

[7] Almeida A F L, Fechine P B A, Góes V, Valente M A, Miranda M A R and Sombra A S B 2004 Dielectric properties of $BaTiO_3$-$CaCu3Ti_4O_{12}$composite screen-printed thick films for high dielectric constant devices in the medium frequency (MF) range, *Mater. Sci. Eng*, 111-113.

[8] Hadi N,Lamchafi T, Abdi F Gouitaa N, Echatoui N S, and Zouhairi M 2017 Structural, Dielectric and Electrical Properties of (1-x) $CaCu_3Ti_4O_{12}$-(x) $BaTiO_3$ Ceramics *Journal of chemistry*: vol. **29**. No, (8)1811-1816.

[9] Raju K, Rashmi R and Seema S**,** 2014 Influence of $BiFeO_3$ addition on the electrical properties of (Na, K) (Nb, Ta)O_3 ceramic system using impedance spectroscopy *Adv. Mat. Lett.* **5(11)**, 658-665.

[10] Arun C, Hemant S, Naithani U C, Shubhash S, Uday. P, Pratiksha D, Anuradha M, Alok Srivastava, Sharma R.K. 2011 Structural, dielectric and electrical properties of Lead zirconate titanate and $CaC_{u3}T_{i4O12}$ ceramic composite *Adv. Mat. Lett.*, **2(1)**, 26-3.

[11] Vijay k T, Chary S A, Suresh Bhardwaj, Awasthi A M, Narender S R 2013 Dielectric relaxation, Ionic conduction and Complex impedance studies on $NaNO_3$ fast ion conductor. *International Journal of Materials Science and Applications*, **2(6)** 173-178.

[12] Almond D P. and Bowen C.R. 2004 Anomalous Power Law Dispersions in ac Conductivity and Permittivity Shown to be Characteristics of Microstructural Electrical Networks *Physical Review Letter*, **92**, 157601.

IOP Conf. Series: Earth and Environmental Science **161** (2018) 012021 doi:10.1088/1755-1315/161/1/012021

The computational study of bridge effect in D-π-A photosensitive dyes, based on triphenylamine

M Lazrak[1*], H Toufik[1*], S M Bouzzine[1,2], H Bih[1] et F Lamchouri[1]

[1] Laboratoire Matériaux, Substances Naturelles, Environnement et Modélisation (LMSNEM), Université Sidi Mohamed Ben Abdellah de Fès, Faculté Polydisciplinaire de Taza, BP 1223 Taza Gare, Taza, Maroc
[2] Centre Régional des Métiers de l'Education et de la Formation, BP 8, Errachida, Maroc

*E-mails: malak.lazrak@usmba.ac.ma / hamid.toufik@usmba.ac.ma

Abstract. For an efficient electron transfer from donor to acceptor unit, the heterocycles units play a crucial role as π-bridges in the Donor-π-Acceptor (D-π-A) organic sensitizers for dye-sensitized solar cells (DSSCs). A series of organic D-π-A dyes, based on triphenylamine donor (D), cyanoacrylic acid acceptor (A)and bridged by different π-conjugated spacer systems, are studied by density functional theory (DFT) and time-dependent DFT (TD-DFT) with the Becke 3-Parameter-Lee-Yang-Parr (B3LYP) functional and the 6-31G (d, p) basis set.
Generally, the fundamental parameters, such as the HOMO and LUMO energies, of studying dyes have been discussed. The calculated results show that the narrow energy gaps (1.98 to 2.48 eV) and the large visible absorption bands (from 556.72 to 707.10 nm), suggest the studied compounds as promising light harvesters materials for organic solar cell applications.
Furthermore, our theoretical results demonstrate that sensitizer D5 with the benzothiadiazole spacer has a broader absorption (707.10 nm) and the lower energy gap (1.98 eV) compare to other dyes; this opens a new way to design new efficient organic sensitizers for dye sensitized solar cells (DSSCs).

1. Introduction

The vast majority of global energy consumption is provided by the exploitation of non-renewable fossil resources. Although, the unavoidable depletion of this type of resources and the environmental impacts resulting from their consumption poses, due to the growing global energy demand, important questions concerning the guarantee of access to energy for future generations. In this context, renewable energies, such as wind, geothermal, biomass or solar energy, are today the subject of many important researches. Among these energies, solar energy is considered one of the promising alternatives as it is available aplenty from the sun. This energy could therefore represent a viable alternative to fossil fuels. Nevertheless, there is still a lot of progress to be made in this field, particularly in terms of the production cost of the modules.

Among the new concepts studied in the field of photovoltaics, dye-sensitized solar cells (DSSC) have been the subject of much research since the early work of O'Regan and Grätzel in 1991[1], involving less expensive production methods; they consist of semiconducting metal oxides photosensitized by a photosensitive dye. This type of photovoltaic cells has already shown high power conversion efficiency (PCE), making it an excellent alternative to marketed silicon-based solar cells [2,3].

One of the major research challenges for the production of high performance photovoltaic cells is the development of new photosensitizers. The Ruthenium (Ru)-based sensitizers have been commonly used as efficient sensitizers, such as N3 and N719, and their PCEs (power conversion efficiencies) reach above 10% [4]. However, seeing that the Ru-based sensitizers have some limitations as the rareness and cost of Ru metal, many researchers are focusing on organic dyes as promising sensitizers, this is attributed to their low production cost, their flexibility manipulates and their environment friendly applications [5,6]. The organic dyes, almost all have an electron donor (D) and electron acceptor (A) connected with π-conjugated spacer, namely D-π-A dyes.

On the basis of studies carried out on certain D-π-A organic dyes, we have identified that triphenylamine (TPA) is among the best donors used in the construction of effective organic sensitizers due to its non-planar molecular configuration, resistant to aggregation [7,8]. Cyanoacrylic acid is chosen as the electron-accepting unit because of its high electron-withdrawing capacity and its high TiO_2 bonding capacity, which facilitates the injection of electrons [9]. In addition to the molecular engineering of donors and acceptors, the development of new types of π-spacers are one of the essential requirements to improve the optoelectronic parameters in D-π-A organic dyes, as it is crucial for the electron transfer from the donor to acceptor fragments and beneficial to the light harvesting ability.

In this study, we were systematically investigated theoretically five D-π-A sensitizers containing the TPA donor (D) unit and the cyanoacrylic acid acceptor (A) unit connected by different π-spacers such as phenyl, pyridine, thiophene, bithiophene and benzothiadiazole. The choice of these bridges units is justified by their performance in the same type of DSSCs. They are considered to be promising π-conjugated spacers in D-π-A organic dyes and have enhanced the efficiency of solar devices [10,11,12]. Therefore, we were interested in exploring the effect of this bridging on the geometries and optoelectronic properties. The presented study of these compounds depicted in Figure 1 could help to design more efficient photovoltaic organic materials, for aim to find the best material which is used as a π-spacer electron in dyes applied in dye-sensitized solar cells (DSSCs).

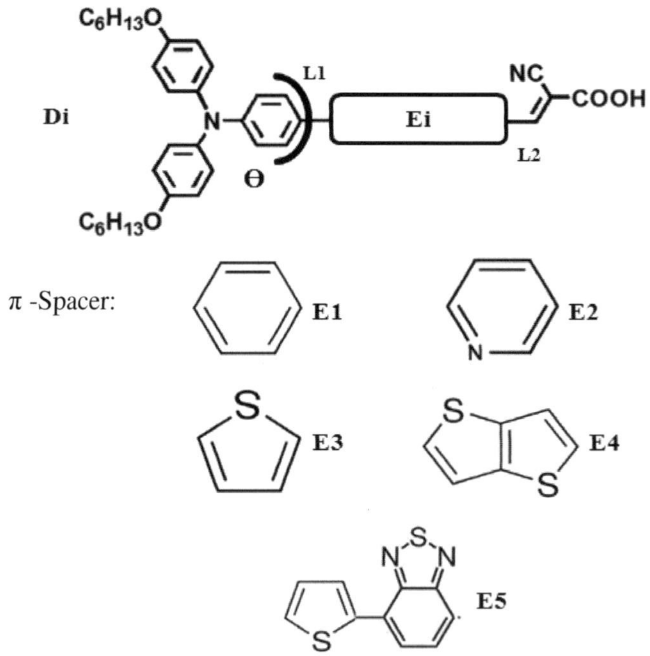

Figure 1. Molecular structures of study dyes Di (i = 1 to 5).

2. Computational Details

In this work, the geometrical and electronic properties are studied by the Density Functional Theory (DFT) using the B3LYP correlation exchange functional (Becke's three-parameter hybrid functional, and Lee-Yang-Parr's correlation functional), known to be the most hybrid functional used for a wide variety range of organic molecules, combined with 6-31G(d, p) basis set of neutral and doped forms [13-16].

The excitation energies, the wavelengths and oscillator strengths were simulated using Time Dependent Density Functional Theory method (TD-DFT). These calculations were carried out including solvent effects (tetrahydrofuran, THF) using the conductor-polarizable continuum model (CPCM).All calculations were performed with the Gaussian 09 packages [17].

3. Results and discussions

3.1 Geometrical properties

The geometrical configuration of all sensitizers Di (i=1-5) is optimized by DFT calculations at B3LYP/6-31G (d, p) level. The calculated dihedral angles θ between the donor group and π-spacer, the bond lengths (L1) between the donor group and π-spacer and those (L2) between the acceptor group and π-spacer for studying compounds are listed in Table 1.

Table 1. Geometrical parameters of the studied compounds calculated at the B3LYP/6-31G(d,p) level.

Dyes	Forms	L1(Å)	L2(Å)	θ(°)
	Neutral	1.47	1.45	30.51
D1	Cation	1.47	1.46	32.46
	Anion	1.46	1.42	23.75
	Neutral	1.47	1.44	8.67
D2	Cation	1.48	1.46	10.66
	Anion	1.46	1.41	2.17
	Neutral	1.45	1.42	18.19
D3	Cation	1.45	1.43	13.26
	Anion	1.44	1.39	1.72
	Neutral	1.46	1.42	22.41
D4	Cation	1.45	1.43	9.63
	Anion	1.44	1.39	1.25
	Neutral	1.47	1.42	30.78
D5	Cation	1.46	1.43	25.25
	Anion	1.47	1.40	22.69

For each dye, the calculated bond lengths L1 and L2 are in the range of 1.42-1.47 Å for the neutral forms. We can conclude that they don't change significantly which implies that the change of spacer doesn't have an effect on the bond lengths. However, during the doping process, the bond lengths of the negatively doped dyes become shorter, while the bond lengths for those positively doped become generally longer 'compared to the work done by our research group' [18,19,20].

On the other hand, the dihedral angles for D1, D2, D3, D4 and D5 are 30.51°, 8.67°, 18.19°, 22.41° and 30.78° respectively, which indicates that the dihedral angle of the D2 is significantly smaller; this can be ascribed to the nitrogen atom (N) of pyridine ring which induces a coplanar structure by hydrogen bonding interaction between the nitrogen atom and phenyl hydrogen. In addition, the decrease of bond lengths, for the negatively doped dyes, is accompanied by a huge decrease of dihedral angles.

3.2. Electronic properties

The highest occupied molecular orbital (HOMO) and the lowest unoccupied molecular orbital (LUMO) of study dyes are calculated by DFT method, the band gap is evaluated as the difference between the HOMO and LUMO energies. Table 2 summarizes the calculated electronic parameters obtained in the neutral and doped forms. The results were compared with available experimental data [7].

Table2. Electronic properties of the studied compounds calculated at DFT/B3LYP/6-31G(d, p) level.

Dyes	Forms	HOMO (eV)	LUMO (eV)	Egap (eV)	$Egap^{exp}(eV)^{a}$
	Neutral	-4.87	-2.42	2.45	2.56
D1	Cation	-8.18	-4.61	3.57	
	Anion	-0.15	1.41	1.56	
	Neutral	-4.91	-2.54	2.37	
D2	Cation	-8.20	-4.82	3.38	-
	Anion	-0.32	1.46	1.78	
	Neutral	-4.94	-2.46	2.48	
D3	Cation	-8.07	-4.96	3.11	-
	Anion	-0.33	1.47	1.8	
	Neutral	-4.9	-2.49	2.41	
D4	Cation	-7.84	-4.89	2.95	-
	Anion	-0.46	1.32	1.78	
	Neutral	-4.82	-2.84	1.98	
D5	Cation	-7.65	-5.11	2.54	-
	Anion	-0.90	0.76	1.66	

[a]Experimental Gap Energies of the studied dyes (eV)[7].

The calculated Egap are in the range of 1.98-2.48 eV for neutral form, and increase in the following order: D5 <D2 <D4 <D1 <D3. We note that the energy gap decreases from the neutral to the negatively doped form, while increasing to the positively doped form. The weaker band gap of D5 indicates a significant effect of intramolecular charge transfer, which would cause an absorption spectrum shifted to red. This is attributed to thiadiazole compound inserted at the π-spacer.

On the other side, as shown in table 2, the obtained value at B3LYP/6-31G(d,p) level of energy gap for D1 is close to these obtained experimentally, this is attributed to the appropriate method used (DFT). In addition, this theoretical method will help us to identify the dyes with the most interesting optoelectronic properties for solar cell applications, and will guide us in the synthesis of these dyes and the experimental study of their optoelectronic properties.

As depicted in Figure 2, the HOMO orbital of D5 is delocalized over the donor (D), whereas the LUMO orbital shows distributions through the acceptor and π-spacer units, which indicates that there are good electron–separated states between HOMO and LUMO. These distributions enhance the Intra Charge Transfer (ICT) by facilitating the charge transfer from the electron donor to the acceptor.

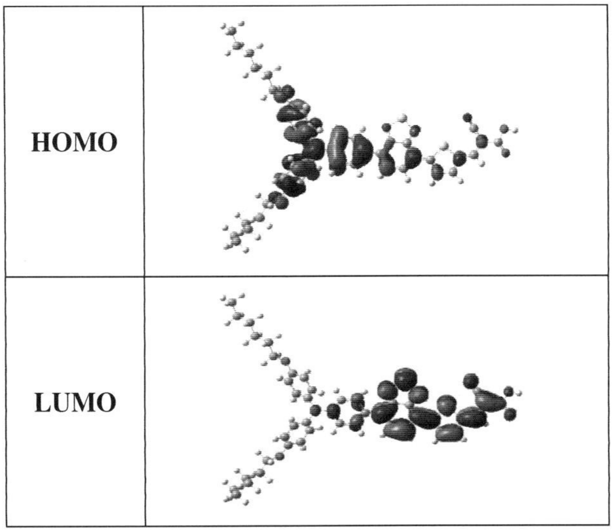

Figure 2. Representation of HOMO and LUMO molecular orbitals of the D5 compound.

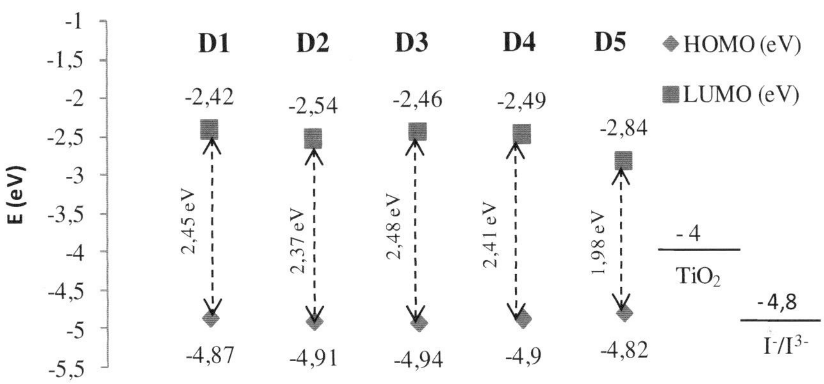

Figure 3. Sketch of the HOMO, LUMO and Egap levels of the studied molecules.

For DSSC application the HOMO of the dye must be below the redox couple of I^-/I_3^- electrolyte and the LUMO of the dye must be situated above the TiO_2 conduction band (-4.0 eV) [21]. We note through the Figure 3 that all the five dyes are supposed to be injected very easily from the excited state S1 to the TiO_2 conduction band. On the other hand, these molecules that lose electrons could get electrons from the electrolyte.

3.3. Optical properties

The functional theory of time-dependent density (TD-DFT) has emerged as a powerful tool for theoretical processing of absorption spectra [22,23]. TD-DFT/B3LYP/6-31G (d, p) method was used to simulate the optical property of the studied dyes in the gas phase. Table 3 lists the maximum absorption wavelength (λ), oscillator strengths (f) and excitation energies (ΔE) of each dye. The solvent effects were included in the calculation in order to predict the experimental spectra with a reasonable accuracy, seeing that most chemical processes occur in the solution phase [24]. Based on the previous experimental work, the tetrahydrofuran (THF) was used as solvent.

Table 3. Optical properties of the studied compounds calculated at the TD-DFT/B3LYP/6-31G(d,p) level.

Dyes	Solvent	ΔE (eV)	oscillator strength(f)	λ (nm)
D1	With	2.23	0.58	556.72
	Without	2.06	0.66	601.15
D2	With	2.19	0.73	565.67
	Without	2.03	0.84	610.35
D3	With	2.29	0.81	541.59
	Without	2.07	0.91	598.34
D4	With	2.22	0.88	558.28
	Without	2.02	1.00	612.06
D5	With	1.75	0.67	707.10
	Without	1.62	0.78	763.54

The analysis in Table 3 shows that the excitation energies are about 2eV for all the studied dyes. In addition, all compounds have an absorption maximum in the visible region of approximately 542 to 763 nm, which can be attributed to an Intramolecular Charge Transfer (ICT) between the different units [25].

The calculated wavelength λ of study compounds decreases in the following order: D5> D4> D2> D1> D3, which is almost the same order of the band gaps. Furthermore, we remark that a near infra-red is realized (707 nm) of D5; its absorption spectrum covers the whole visible region then extends to the near infra-red (NIR) region, which indicates the strongest ICT interaction between units in D5. As a result, the calculated spectral data confirm that the introduction of benzothiadiazole as a π-spacer in D-π-A system could favorable the long wavelength light harvesting, beneficial for increasing the power conversion efficiency of the solar cells. The absorption maxima of all dyes are shifted by the solvent effect (THF): bathochromic displacement. The magnitudes of the spectral offset are about 50 nm for all dyes, this clearly indicated the significant influence of the solvent that lead to enhance the conjugation on the studied structures [26].

4. Conclusion

In summary, a series of D-π-A organic dyes were theoretically studied by introducing different π-conjugated spacers: phenyl, pyridine, thiophene, bithiophene and benzothiadiazole combined with triphenylamine and cyanoacrylic acid respectively as donor and acceptor, to understand the bridge impact on the optoelectronic properties of dyes. All geometric structures, electronic and optical properties were studied using DFT and TDDFT methods.

In view of the HOMOs, LUMOs, band gaps and wavelengths, we can conclude that the selected dyes can be used as units in dye sensitized solar cells. Overall, D5, with the benzothiadiazole π-spacer, should be the best photosensitizer for use in DSSC; this is due to his low band gap and his favorable absorption which allow a better light harvesting. Furthermore, the DFT study clearly reveals that the electron injection into the CB of TiO_2 and regeneration of dyes by I^-/I_3^- where feasible.

This theoretical approach is a guiding tool to understand the structural relationships of these compounds and to design another new promising organic sensitizers for the future development of DSSCs.

References

[1] O'Regan B and Grätzel M 1991 *J. Nature* **353** 737–740
[2] Hagfeldt A and Grätzel M 2000 *J Acc. Chem. Res* **33** 269–277
[3] Grätzel M 2001 *J. Nature* **414** 338–344

[4] Aghazada S, Gao P, Yella A, Marotta G, Moehl T, Teuscher J, Moser J E, De Angelis F, Grätzel M and Nazeeruddin M K 2016 *Inorg. Chem* **55** 6653–6659
[5] Hagfeldt A, Boschloo G, Sun L C, Kloo L and Pettersson H 2010 *J. Chem. Rev* **110** 6595–6663
[6] Mishra A, Fischer M K R, Bauerle P and Angew 2009 *J. Chem. Int* **48** 2474–2499
[7] Eom Y K, Hong J Y, Kim J and Kim H K 2017 *J. Dyes. Pigm* **136** 496-504
[8] Fahim Z, Bouzzine S M, AitAicha Y, Bouachrine M and Hamidi M 2018 *Res. Chem. Intermed* **44** 2009–2023
[9] Mahmood A 2016 *Sol. Energy* **123** 127–144
[10] Wang S, Guo J, He L, Wang H, Zhao J, and LuC 2013 *Synthetic metals* **168** 1-8
[11] Hasanein A A, Elmarassi Y R and Kassem E N 2016 *J. Mol. Model* **22** 115
[12] Misra R, Maragani R, Arora D, Sharma A and Sharma G 2016 *J. Dyes. Pigm* **126** 38-45
[13] Chou H H, Hsu C Y, Hsu Y C, Lin Y S, Lin J T and Tsai C 2012 *J. Tetrahedron* **68** 767-773
[14] Becke A D 1993 *J. Chem. Phys* **98** 1372–1377
[15] Becke A D 1988 *J. Phys Rev.* A **38** 3098–3100
[16] Lee C, Yang W and Parr R G 1988 *J. Rev.* B **37** 785–789
[17] Frisch M J, Trucks G W, Schlegel H B, Scuseria G E, Robb M A and Cheeseman J R 2009 *Gaussian 09.* A p02
[18] Toufik H, Bouzzine S M, Lamchouri F, Nawdali M, Hamidi M and Bouachrine M 2012 *J. Mater. Environ. Sci. 3* **2** 286-293
[19] Toufik H, Bouzzine S M, Ninis O, Lamchouri F, Aberkane M, Hamidi M and Bouachrine M 2012 *Res. Chem. Intermed* **38** 1375-1388
[20] Toufik H, Bouzzine S M, Ninis O, Aberkane M, Lamchouri F, Hamidi M and Bouachrine M 2011 *J. Physical Studies* **16** 1702-1708
[21] Li R, Lv X, Shi D, Zhou D, Cheng Y and Zhang G 2009 *J. Phys. Chem.* C **113** 7469-7479
[22] Asbury J B, Wang Y Q, Hao E, Ghosh H and Lian T 2001 *J. Res. Chem. Intermed* **27** 393-406
[23] Luo C, Bi W, Deng S, Zhang J, Chen S and Li B 2014 *J. Phys. Chem.* C **118** 14211-14217
[24] Mao M, Zhang X, Zhu B, Wang J, Wu G and Yin Y 2016 *J. Dyes. Pigm* **124** 72-81
[25] Wu W, Yang J, Hua J, Tang J, Zhang L and Long Y 2010 *J. Mater. Chem* **20** 1772-1779
[26] Kafafy H, Wu H, Peng M, Hu H, Yan K, El-Shishtawy R M and Zou D 2014 *Int. J. Photoenergy* **2014** 1-9

A comparative study of ANN and Kalman Filtering-based observer for SOC estimation

H Ben Sassi[1], F Errahimi[1], N Es-Sbai[1] and C Alaoui[2].

[1]Laboratory of Renewable Energies and Intelligent Systems (LERSI), Faculty of Sciences and Technology, Sidi Mohamed Ben Abdellah University Fez.
Box 2202, Fez, Morocco

[2]University EUROMED, Fez, Morocco

E-mail: hicham.1bensassi@gmail.com

Abstract. Electrical Vehicle Batteries (EVB) study has gained a lot of interest in recent years, with the aim of better managing their use, due to the high changes in the electric vehicle dynamics and operational modes, which could cause severe damages to the battery if not properly managed. Recently lithium-ion (Li-ion) batteries have become the most suitable technology for electric vehicles, because of their interesting features such as a relatively long cycle life, lighter weight and high energy density. However, there is a lot of work that is still needed to be done in order to ensure safe operating lithium-ion batteries, starting with their internal status monitoring, cell balancing within a battery pack and thermal management. In this paper, a comparative study of two different methods for state of charge estimation techniques are presented: Kalman filtering observers and artificial neural network based observers. The respective results are compared in terms of accuracy, implementation requirement, and overall performances.

1. Introduction

Nowadays, electric mobility is starting to define society and is becoming more and more essential to daily activities. Safe and durable battery is of a great significance for this type of mobility, hence the increasing interest of research activity oriented to battery studies in order to ensure safe operating mode and to control the battery in case of any abnormal functioning conditions that could damage the battery, especially in the case of lithium-ion technology. Li-ion has presented itself as the most suitable existing technology for electrical storage, but still suffers from safety issues. An overcharge or overdischarge of the battery could lead to overheating and sometimes causes fire runway, which is why state of charge monitoring is a necessity. However, being seen that the state of charge is an unmeasurable variable, several approaches and algorithms have been proposed and implemented for lithium-ion batteries SOC estimation [1-5]. Out of these approaches, two have stood out due to their high performances, and are widely implemented especially in applications that require high precision. The first one is Artificial Neural Network (ANN) which is a data-oriented approach, it doesn't require a detailed physical knowledge of Li-ion batteries (internal resistance, capacity…). This method is based on the black box battery models which describe the relationship between the SOC and the factors that influence its value. The implementation of this strategy was realized in [6] on a field-programmable gate away (FPGA) platform for speed estimation of the two-mass drive System. The second most popular method is Kalman filters which are online-based and sometimes considered as the optimal mean to predict and correct time-varying "state" of a dynamic system repeatedly as the

system operates [4]. A detailed etended kalman filter (EKF) hardware/software implementations on Altera FPGA is proposed in [7] for robot mapping applications. The EKF algorithm was coded in C and then implemented on Altera Nios II FPGA. In [8], a detailed design and implementation of a SOC observer based on EKF on FPGA platform was presented, comparing the performances of an FPGA implemented EKF and those of the same observer implemented on Matlab/Simulink environment. In [9], a combination of both methods was investigated for higher accuracy. In this paper, a comparative investigation of these two approaches is presented in terms of their performances.
The present paper is organized as follows: the second paragraph deals with the design of both strategies. Simulation results and discussion are presented in the third paragraph. This manuscript ends with a conclusion.

2. SOC estimation

2.1 Kalman filters

Kalman filters are online and sometimes considered as the optimal mean to predict and correct time-varying system in a way that minimizes the mean of the squared error. They are model based observers that require a prior knowledge of the battery's model and accurate internal parameters identification. Hence in this paper the battery is modeled by using a shunt resistor R_0, an *RC* branch and open circuit voltage source U_{oc}. The model is presented in figure.1. The internal parameters of the model, which were identified using Levenberg Marquardt nonlinear least square algorithm are given in Table (1). Equations (1) and (2) are the state space representation of the battery model. In equation (1) the state variables are the state of charge and the voltage across the RC branch: $X=[SOC,V_{RC}]^T$, T is the sampling time, I_{b_k} represents the current across the battery terminals, while w_k and C_b are respectively the system noise covariance matrix and the nominal capacity of the battery. In equation (2) Y_k is the output voltage of the battery, whereas v_k represents the measurement noise covariance matrix.

$$X_{k+1} = f(X_k, I_{b_k}) = \begin{bmatrix} 1 & 0 \\ 0 & e^{-\frac{T}{RC}} \end{bmatrix} X_k + \begin{bmatrix} \frac{-T}{C_b} \\ \mathrm{R}(1 - e^{-\frac{T}{\mathrm{RC}}}) \end{bmatrix} I_{b_k} + w_k \quad (1)$$

$$Y_k = h(X_k, I_{b_k}) = U_{oc}(SOC) - X(2)_k - R_0 I_{b_k} + v_k \quad (2)$$

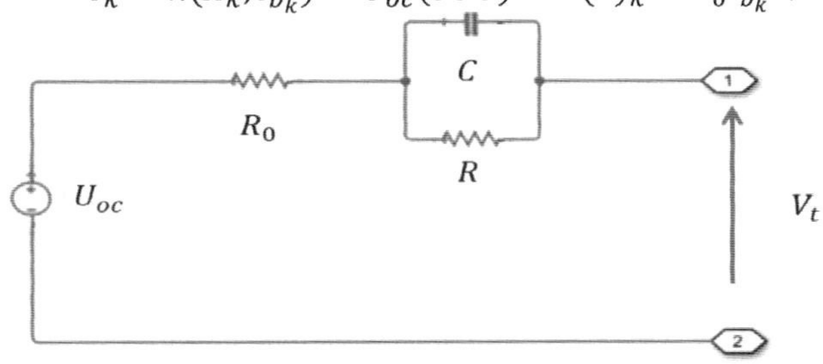

Figure 1: Lithium-ion battery model.

Table 1. Battery internal parameters.

Parameter	**R0(Ω)**	**C (F)**	**R(Ω)**
Value	0.0016	24000	0.0024

In literature there exists a variety of kalman filters, from classical linear Kalman filter to extended Kalman filter [4] and sigma point Kalman filters (SPKF) [10]. They all share the same principle and only differ in linearization process. In this paper, the performances of the unscented Kalman filter (UKF) are presented, since the state of charge estimation is a nonlinear problem and UKF is nonlinear Kalman filter which is based on the unscented transformation as linearization method. UKF algorithm is presented below, for more background theory refer to reference [11].

Step 1. Initialization

IOP Conf. Series: Earth and Environmental Science **161** (2018) 012022 doi:10.1088/1755-1315/161/1/012022

$$Q_k = \mathrm{E}(w_k w_k^T);\ R_k = \mathrm{E}(v_k v_k^T)\ ;\ \ \mathrm{E}(w_k v_k^T) = 0 \tag{3}$$

$$P_0 = \mathrm{E}[(x_0 - \bar{x}_0)(x_0 - \bar{x}_0)^T] \tag{4}$$

$$\bar{x}_0 = \mathrm{E}(x_0) \tag{5}$$

Where Q_k and R_k are diagonal matrices that represent respectively the process noise (w_k) covariance and the observation noise (v_k) covariance, P_0 is the state covariance matrix and $\overline{X}_0$ represents the main of the state variable.

Step 2. Compute sigma points

A set of weighted sigma points are deterministically computed, as shown in the following equation:

$$\mathcal{X}_{k-1} = [\bar{x}_{k-1},\ \bar{x}_{k-1} + \left(\sqrt{(n+\lambda)P_{x,k-1}^{-}}\right), \bar{x}_{k|k-1} - \left(\sqrt{(n+\lambda)P_{x,k-1}^{-}}\right)] \tag{6}$$

Where λ is a scaling parameter that represents how far the sigma points are separated from the mean and n is the system dimension, in our case n=2. For more background theory refer to reference [8].

Step 3. Prediction phase

1) Propagating sigma points through the system/state function $f(x, i_{bk})$:

$$\mathcal{X}_{k|k-1}^{(i)} = f(\mathcal{X}_{k-1}^{(i)}, i_{bk}) \tag{7}$$

2) Calculating the propagated mean and covariance:

$$\bar{x}_{k|k-1} = \sum_{i=0}^{2n} Wm.\mathcal{X}_{\boldsymbol{k|k-1}}^{(i)} \tag{8}$$

$$P_{x,k|k-1}^{-} = \sum_{i=0}^{2n} Wc[\mathcal{X}_{k|k-1}^{(i)} - \bar{x}_{k|k-1}][\mathcal{X}_{k|k-1}^{(i)} - \bar{x}_{k|k-1}]^T + Q_{k-1} \tag{9}$$

Where i stands for the number of sigma points, while Wm and Wc are respectively mean calculation and covariance calculation weights of sigma points.

3) Computing the new sigma points matrix using $\bar{x}_{k|k-1}$ and $P_{x,k|k-1}^{-}$:

$$\mathfrak{x}_{k|k-1} = \left[\bar{x}_{k|k-1},\ \bar{x}_{k|k-1} + \left(\sqrt{(n+\lambda)P_{x,k|k-1}^{-}}\right), \bar{x}_{k|k-1} - \left(\sqrt{(n+\lambda)P_{x,k|k-1}^{-}}\right)\right] \tag{10}$$

4) Propagating the computed sigma points through the output/state function $h(x, i_{bk})$:

$$y_{k|k-1}^{(i)} = h(\mathfrak{x}_{k|k-1}^{(i)}, i_{bk}) \tag{11}$$

5) Calculating the mean of the output variable:

$$\bar{y}_{k|k-1} = \sum_{i=0}^{2n} Wm.y_{k|k-1}^{(i)} \qquad \text{for i=1...2n} \tag{12}$$

Step 4.Measurement update

1) Calculating the covariance of the measurement vector and the cross covariance respectively.

$$P_{y.k} = \sum_{i=0}^{2n} Wc[y_{k|k-1}^{(i)} - \bar{y}_{k|k-1}][y_{k|k-1}^{(i)} - \bar{y}_{k|k-1}]^T \tag{13}$$

$$P_{xy.k} = \sum_{i=0}^{2n} Wc[\mathfrak{x}_{k|k-1}^{(i)} - \bar{x}_{k|k-1}][y_{k|k-1}^{(i)} - \bar{y}_{k|k-1}]^T \tag{14}$$

2) Calculating the Kalman gain, in order to correct the estimations :

$$K_k = P_{xy.k}.P_{y.k}^{-1} \tag{15}$$

3) Update the estimated state with Kalman gain:

$$\hat{X}_k = \hat{X}_{k|k-1} + K_k(y_k - \bar{y}_{k|k-1}) \tag{16}$$

4) Update the propagated state covariance using kalman gain:

$$\hat{P}_k = P_{x.k|k-1}^{-} - K.P_{y.k}.K^T \tag{17}$$

2.2 Artificial neural network

An ANN comprises a set of interconnected neurons that mimic the information processing of the human brain. It is very accurate and powerful tool that can model almost any nonlinear system if it is trained correctly. In literature there are several types of NN, and the choice of one over the other depends on the complexity of the system and the desired applications. In this paper, multilayer feed-forward neural network (FFNN) is chosen due to its high performance in estimation problems. The FFNN is composed of three types of layers: an input layer with nodes to represent the input variables, hidden layers to model the nonlinearity between the system input and output, and an output layer to represent the system output variables. Based on an amperic choice after several tests and simulations, the chosen neural network is composed of 3 neurons in the first layer and 6 neurons in the hidden layers. The network has as input the voltage across the battery terminals, the discharge and the charge current, and temperature. The output of the ANN is the state of charge. Figure. 2 show the structure of the chosen ANN.

The ANN is trained offline using data collected from charging and discharging process of 2Ah lithium-ion battery in Matlab/Simulink environment. The battery was charged with fixed current profile until it has reached 61%, and then switched to fixed voltage profile and the current decreases exponentially until the battery is fully charged. For the discharging process the battery was fully discharged with variable amplitude current profile. A sample of the current profile applied to the battery is presented in Fig. 3. Temperature variation was taking into consideration during all the charging/discharging process, because it affects the internal parameters of the battery such as the internal resistance and the SOC, as demonstrated in reference [11].

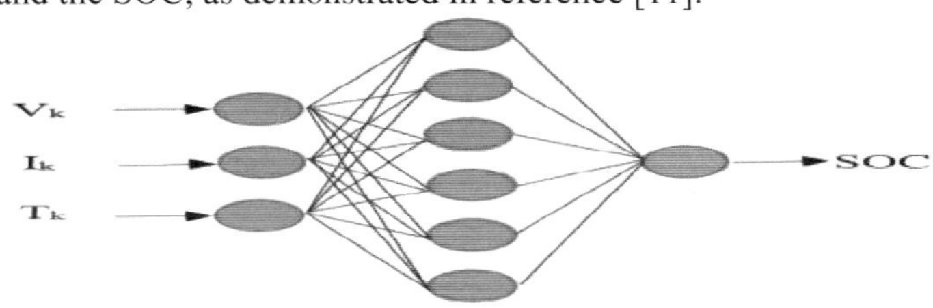

Figure 2: Structure of the chosen multilayer feed-forward neural network.

3. Simulation and results discussion

Following the design processes, both the UKF based observer and the ANN are simulated and tested under the same current profile presented in Figure. 3. Figures. 4 and 5 represent simulation results of the UKF, and Figures. 6 and 7 are those of the ANN.

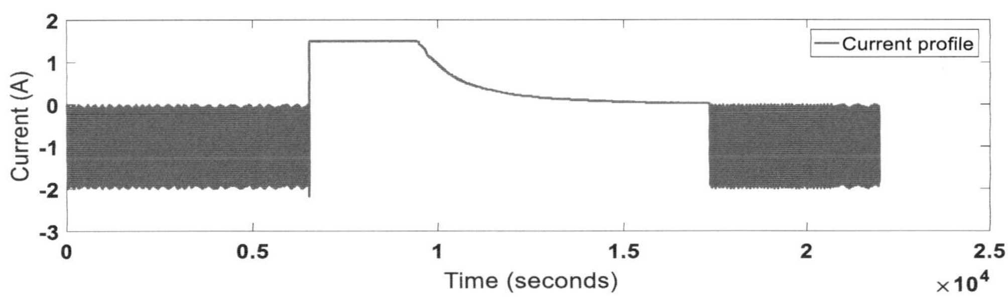

Figure 3: Current profile applied to both observers.

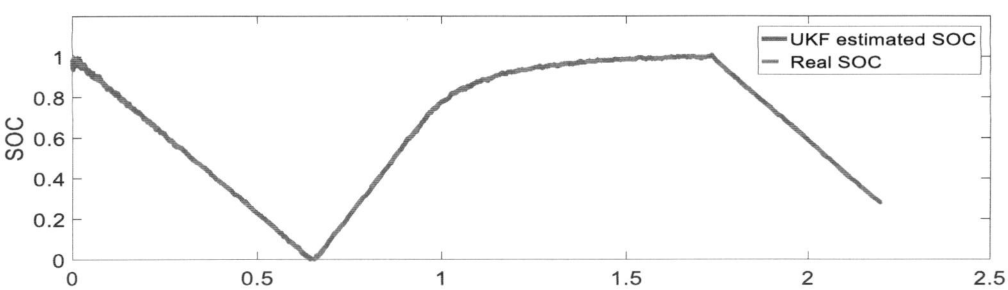

Figure 4: UKF estimated SOC vs real SOC.

IOP Conf. Series: Earth and Environmental Science **161** (2018) 012022 doi:10.1088/1755-1315/161/1/012022

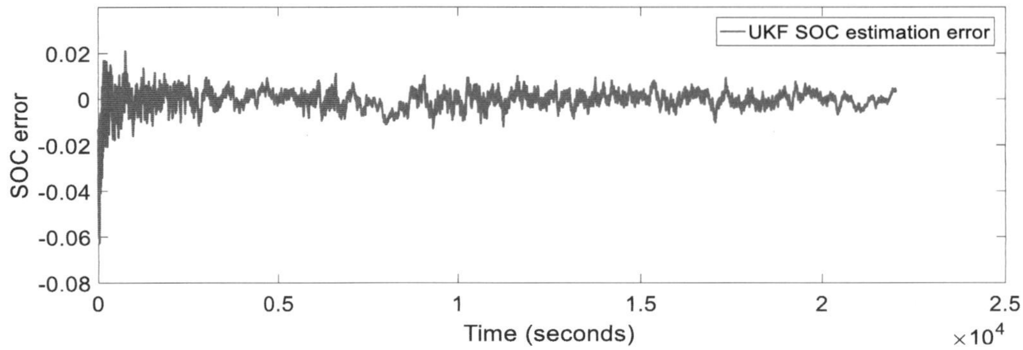

Figure 5: SOC estimation using UKF.

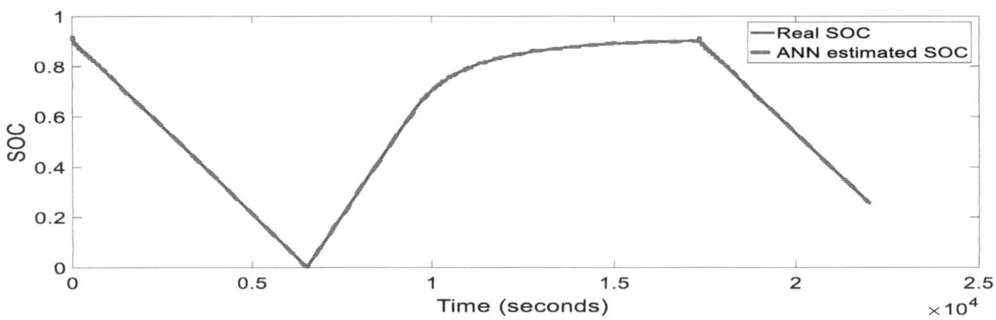

Figure 6: ANN based SOC vs real SOC.

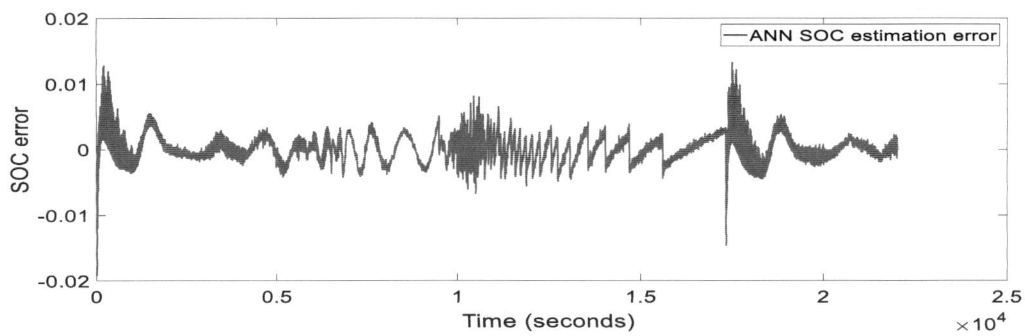

Figure 7: ANN based SOC estimation error.

From the above simulation results, it is clear that both strategies perform well in SOC estimation, although the ANN, which has maximal error value less than 1.8% and average error less than 0.5%, is slightly more accurate than the UKF, which has maximal error value less than 5% and average error less than 1%. This slight advantage of the ANN over UKF in terms of precision is due to modelling errors in the design process of the UKF and internal parameters uncertainty. On the other hand, the UKF has the advantage of being more accurate than the ANN for functioning condition outside the range covered by the training data. Once the battery is modelled and the internal parameters are accurately determined, Kalman filters produce more accurate estimations for any functioning condition.

From design point of view, both strategies require certain amount of data. In the case of the ANN, the data is required to train the network, while for Kalman filters, it is used to identify the battery's internal parameters. Nevertheless, the amount of data required for ANN is much more important (350000 sample of each input data: current voltage and temperature, in addition to the state of charge, compared to 3000 in the case of UKF) in order to cover the real-life loading conditions as much as possible in terms of SOC span, which is time consuming and extremely difficult to do. The design of the ANN observer is much easier than the UKF, which requires an accurate model and internal parameters identification, in addition to an accurate tuning of its covariance matrices [13-14].

From implementation point of view both strategies have been implemented in several works [8, 15, 16, 17]. The UKF requires more computing power [8,15], due to the complexity of the equations, and less memory than the ANN, which requires a huge memory to store the training data [16, 6].

4. Conclusion

In this paper, a comparative study of Kalman filters and the artificial neural network was carried out, from design challenges to estimation accuracy and ending with resources required for implementation. From the obtained results, it can be concluded that both strategies are competitive in terms of precision, and the choice of one over the other depends on the available resources; if sufficient data to cover a wide operating range is available, together with a powerful platform for implementation, the ANN is powerful and recommended, otherwise, Kalman filters are preferable, due to their capability of operating with less data and under unpredicted scenarios.

5. References

[1] Hannan M A, Lipu M S H, Hussain A and Mohamed A 2017 A review of lithium-ion battery state of charge estimation and management system in electric vehicle applications: Challenges and recommendations *Renewable and Sustainable Energy Reviews* **78** 834.

[2] Ng K S, Moo C S, Chen Y P and Hsieh Y C September 2009 Enhanced coulomb counting method for estimating state-of-charge and state-of-health of lithium-ion batteries *Applied Energy* **86** 1506-11.

[3] Chen X, Shen W, Dai M, Cao Z, Jin J, and Kapoor A April 2016 Robust adaptive sliding mode observer using RBF neural network for Lithium-ion battery state of charge estimation in electric vehicles *IEEE Transactions on Vehicular Technology* **65** 1936 – 47.

[4] Plett G L 2004 Extended Kalman filtering for battery management systems of LiPB-based HEV battery packs part 3. State and parameter estimation *Journal of Power Sources* **134** 277–292.

[5] Burgos C, Sàez D, Orchard M E and Càrdenas R 2015 fuzzy modelling for the state-of-charge estimation of lead-acid Batteries" *Journal of Power Sources* **274** 355-366.

[6] Orlowska K T, and Kaminski M August 2011 FPGA implementation of the Multilayer Neural Network for the speed estimation of the two-mass drive system" *IEEE transactions on industrial informatics* **7** 436 – 445.

[7] Bonato V, Peron R, Wolf F D, De Holanda J A M, Marques E and Cardoso J M P 2007 An FPGA implementation for a Kalman Filter with application to mobile robotics *International Symposium on Industrial Embedded Systems. SIES '07 IEEE.*

[8] Otero N, Eichi R H, Rodriguez Andina J J and Chow M Y July 3-5, 2014 FPGA implementation of an observer for state of charge estimation of lithium-Polymer Batteries *International Conference on Mechatronics and Control ICMC Jinzhou China.*

[9] He W, Williard N, Chen C and Pecht M 2014 State of charge estimation for Li-ion batteries using neural network modeling and unscented Kalman filter-based error cancellation *Internationnal Journal of Electrical Power and Energy Systems IJEP* **62** 783–791.

[10] Plett G L 2006 Sigma-point Kalman filtering for battery management systems of LiPB-based HEV battery packs: Part 2: simultaneous state and parameter estimation *Journal of Power Sources* **161** 1369–84.

[11] Gao L, Liu S, and Dougal R A september 2002 Dynamic Lithium-Ion Battery Model for System Simulation *IEEE transactions on components and packaging technologies,* **25** 495 - 505.

[12] Julier S. et Uhlmann J, 1997 New extension of the Kalman filter to nonlinear systems, *The 11th International Symposium on Aerospace/Defence Sensing, Simulation and Controls, Orlando*, pp. 182-193.

[13] Leonardo A S and Da Cruz J J june 2016 Automatic tuning of the unscented Kalman filter and the blind tricyclist problem *IEEE control systems magazine* **36** 70 – 85.

IOP Conf. Series: Earth and Environmental Science **161** (2018) 012022 doi:10.1088/1755-1315/161/1/012022

[14] Sun F, Hu X, Zou Y and Li S 2011 Adaptive unscented Kalman filtering for state of charge estimation of a lithium-ion battery for electric vehicles *Energy* **36** 3531-40.
[15] Dhaouadi R, Mohan N and Norum L july 1991 Design and Implementation of an Extended Kalman Filter for the State Estimation of a Permanent Magnet Synchronous Motor *IEEE transactions on power electronics* **6** 491 – 497.
[16] Botros N M and Abdul-Aziz M december 1994 Hardware implementation of an artificial neural network using field programmable gate arrays (FPGA's) *IEEE transactions on induslrial electronics* **41** 665 – 667.

IOP Conf. Series: Earth and Environmental Science **161** (2018) 012023 doi:10.1088/1755-1315/161/1/012023

Comparative study of different MPPT methods for wind energy conversion system

O Zebraoui[1] and M Bouzi[1]

[1] Laboratory of Mechanical Engineering, Industrial Management and Innovation
Faculty of Science and Technical, Hassan 1 University, Settat, Morocco
E-mail: otmanezebraoui@gmail.com

Abstract. In this paper we present a comparative study of different MPPT methods for a wind energy system based on permanent magnet synchronous generator (PMSG). These techniques are: Hill Climbing Search (HCS), Optimal Torque Control (OTC), Power Signal Feedback (PSF) and Fuzzy Logic control (FLC). These methods are developed with the aim of comparing their performance on the wind system energy efficiency, according to the response time, the maximum power to be achieved and the system behaviour during the optimal power point tracking MPP. And for a better analysis study, we simulated the system under variable wind speed conditions. The results show that intelligent control based on fuzzy logic is more efficient and presents better performance compared to other MPPT algorithms.

1. Introduction

A wind energy system is one of the most widely used renewable energy sources in isolated sites, thanks to its economic and environmental advantages compared to other systems based on fossil energy sources. Several research papers present modeling and optimization studies of wind energy systems for small installations [1]-[3], using permanent magnet synchronous generators (PMSG) because of its high reliability, requires less maintenance, no need for gearboxes and more efficiency. Also to get maximum power at all times and whatever the weather conditions, the use of an MPPT controller is required to operate the system in its maximum power point MPP. In the literature, there are two types of MPPT methods: direct and indirect methods. The direct techniques are independent of the knowledge of generator characteristics and the climatic conditions, namely, Hill Climbing Search (HCS) [5] [6] and Fuzzy Logic controller (FLC) [4]-[8]. The indirect methods are based on knowledge of the generator characteristics such as Optimal Torque Control (OTC) and Power Signal Feedback (PSF) [3] [11].

In this work, we present a modeling and simulation study of a wind energy system consists of wind turbine, PMSG generator, ac/dc converter, dc/dc booster converter and a DC load. Thus, an MPPT controller was developed to optimize the system energy efficiency by tracking the optimum operating point, using different MPPT algorithms that are analyzed and compared in order to optimize their performance on our system.

2. Modelling of Wind Energy System

In this work, we propose a wind energy conversion system consists of wind turbine, PMSG generator which converts the mechanical energy of the turbine into electrical energy, a ac/dc diode rectifier to convert AC voltage to DC Voltage, dc/dc boost converter that allows to be achieved the desired output voltage and power by controlling the MPPT controller, and the complete system supplies a DC electrical load using a DC bus line. Fig. 1 shows the proposed system architecture.

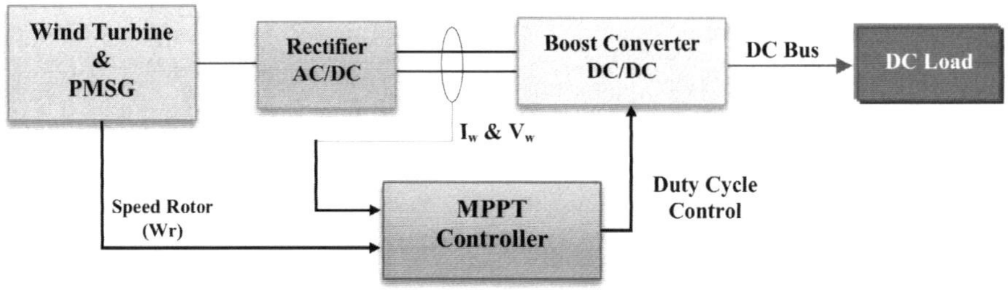

Figure 1. Wind energy system architecture.

2.1. Wind Turbine Modelling

The mechanical power P_m generated by the turbine is given by (1):

$$P_m = \frac{1}{2}\rho S C_p(\lambda, \beta) V_w^{\ 3} \tag{1}$$

Where: C_p is the turbine power coefficient which depends on the tip speed ratio of the rotor blades λ and the blade pitch angle β. V_w is the wind speed (m/s), ρ is the air density (Kg/m^3), S is the area of swept of turbine.

The power coefficient C_p is as follows:

$$C_p(\lambda, \beta) = C_1\left(C_2\frac{1}{\lambda_i} - C_3\beta - C_4\right)e^{\frac{-C_5}{\lambda_i}} + C_6\lambda \tag{2}$$

Where: C_1=0.5176, C_2=116, C_3=0.4, C_4=5, C_5=21, C_6=0.0068 and the parameter λ_i is defined by:

$$\frac{1}{\lambda_i} = \frac{1}{\lambda+0.08\beta} - \frac{0.035}{\beta^3+1} \tag{3}$$

The input mechanical torque T_m for the PMSG generator is obtained by:

$$T_m = \frac{P_m}{\omega_r} \tag{4}$$

Where, ω_r is the rotor speed of a wind turbine.

The tip speed ratio TSR of the wind turbine is defined as following:

$$\lambda = \frac{\omega_r R}{V_w} \tag{5}$$

Where, R is the radius of wind turbine rotor in meters (m).

Tables 1 and 2 summarize the parameters of the wind turbine and the PMSG generator used in this study:

Table 1. Parameters of the wind turbine.

Parameter	Value
Nominal Mechanical Output Power	8.5 kW
Base Wind Speed	12 m/s
Air Density **ρ**	1.225kg/m^3
Pitch Angle **β**	0°

IOP Conf. Series: Earth and Environmental Science **161** (2018) 012023 doi:10.1088/1755-1315/161/1/012023

Table 2. Parameters of PMSG.

Parameter	Value
Number of Pole Pairs	5
Stator Phase Resistance	0.425 Ω
Armature Inductance	0.000835 H
Friction Factor	0.001189 Nms
Inertia constant	0.01197 kg.m^2

Fig. 2 shows the wind turbine mechanical power as a function of the rotor speed P_m (ω_r) for various wind speed values:

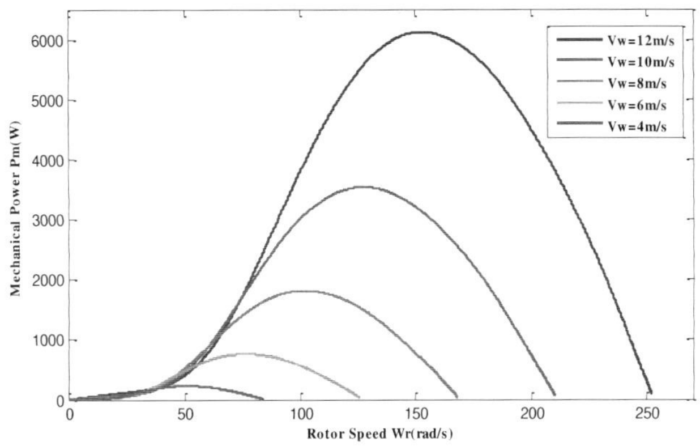

Figure 2. Pm(ω_r) Wind turbine mechanical power for various wind speed values.

Fig. 3 shows the wind turbine mechanical torque as a function of the rotor speed T_m (ω_r) for different wind speed values.

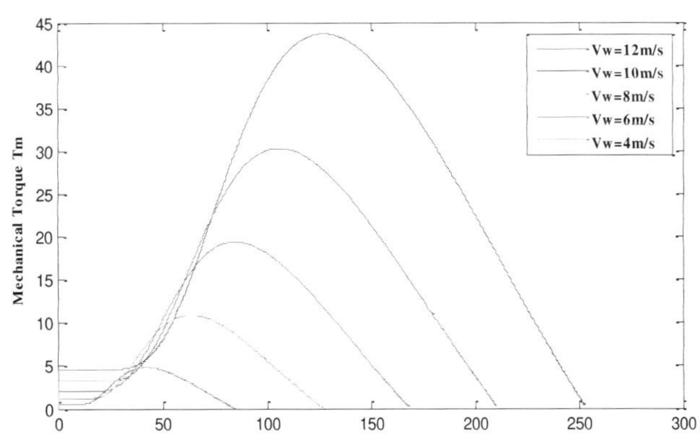

Figure 3. Tm(ω_r) Wind turbine mechanical torque for various wind speed values.

Fig. 4 shows the power coefficient $C_p(\lambda)$ for various pitch angle values β. the optimal value of C_p is corresponding to β=0 and to the tip speed ratio optimal value (λ_{opt}=8.144) which equal to 0.48.

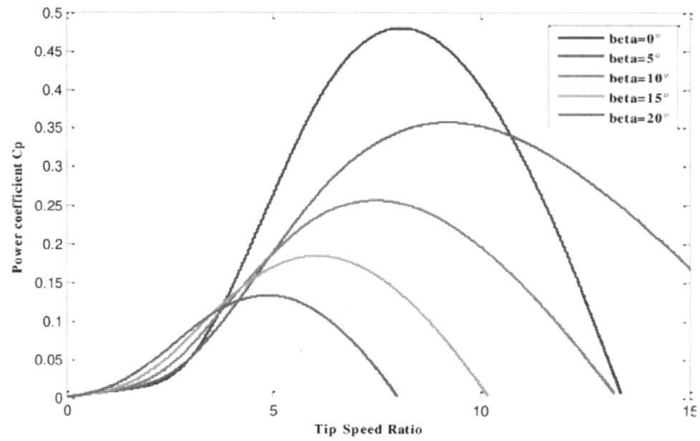

Figure 4. Power coefficient for various values of the pitch angle β.

According to these characteristics of the wind turbine, it can be seen that for the different wind speed values, there is an operating point where these characteristics P_m, T_m and C_p are optimal. This shows the need for the use an MPPT controller to reach and track this maximum power point whatever the variation in wind speed.

2.2. Modelling of DC/DC Boost Converter

A DC/DC boost converter is one of the most widely used converters in a wind energy conversion chain, which converts a low input voltage into a high output voltage. It has an adaptation interface between the PMSG generator and the electrical load in order to operate the system in its maximum power point by controlling its MOSFET transistor using a PWM signal and the duty cycle of the MPPT controller. Fig. 5 shows the equivalent circuit of boost converter:

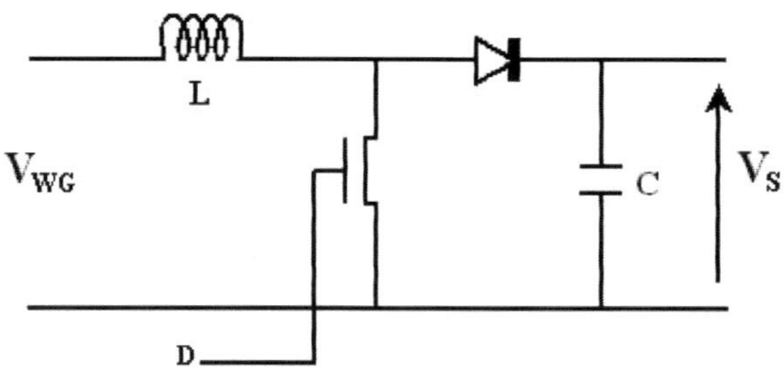

Figure 5. Boost converter.

The output voltage of the boost converter is expressed as a function of the input voltage by:

$$V_s = \frac{V_{WG}}{1-D} \quad (6)$$

Where: D is the duty cycle.

3. MPPT Algorithms

In wind energy conversion system, the use of an MPPT controller is essential, seen its important role to extract the maximum energy produced by the PMSG generator whatever wind speed, its principle based on the execution of MPPT algorithm which allows to reach and track the maximum power point MPP. There are several MPPT techniques used in wind systems, that they can be distinguished according to two types: Direct methods and other indirect. Indirect methods such as Optimal Torque Control (OTC) and Power Signal Feedback (PSF) are based on knowledge of the wind turbine characteristics $C_{p\text{-}opt}$ and λ_{opt} [1] [3]. However, direct methods do not require prior knowledge of these characteristics, namely: Hill Climbing Search (HCS) [8] [9] and Fuzzy Logic control (FLC) [9] [10].

In this work, we propose the development of its algorithms and compare their performance on our system efficiency.

3.1. Optimal Torque Control (OTC) Method

The principle of the OTC control is to adjust the torque of PMSG generator to its reference torque for the different wind speeds, it requires knowledge of the optimal turbine characteristics $C_{p\text{-}opt}$ and λ_{opt}.

In the optimal case ($C_{p\text{-}opt}$ and λ_{opt}), and according to the equations (1), (4) and (5), wind speed, mechanical power and optimum torque can be written as follows:

$$V_w = \frac{\omega_{r.opt} R}{\lambda_{opt}} \tag{7}$$

$$P_{m.opt} = \frac{1}{2}\rho\pi R^2 C_{p.opt} V_w^3 = K_{opt}.\omega_{r.opt}^3 \tag{8}$$

$$T_{m.opt} = \frac{P_{m.opt}}{\omega_{r.opt}} = K_{opt}.\omega_{r.opt}^2 \tag{9}$$

Where K_{opt} is a constant determined by the wind turbine characteristics and is defined by:

$$K_{opt} = \frac{1}{2}\cdot\frac{\rho\cdot\pi\cdot R^5\cdot C_{p.opt}}{\lambda_{opt}^3} \tag{10}$$

Fig. 6 shows the block diagram of the OTC method:

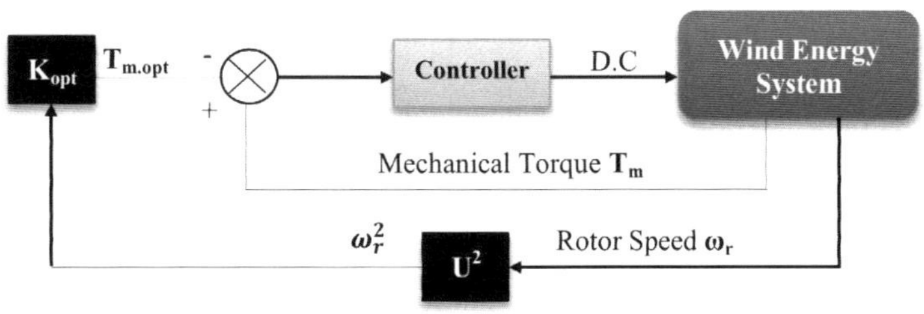

Figure 6. OTC method.

3.2. Power Signal Feedback (PSF) Method

The PSF technique allows to maximize the output power of a PMSG generator by calculating the reference power for a particular wind speed, using the equation (8) of the power optimal. Then a comparison is made between this calculated power and the current power to generate a control signal which is applied to the boost converter, in order to ensure that the power coefficient C_p and the tip speed ratio λ are always at its optimum values. Fig. 7 shows the principle of the PSF method:

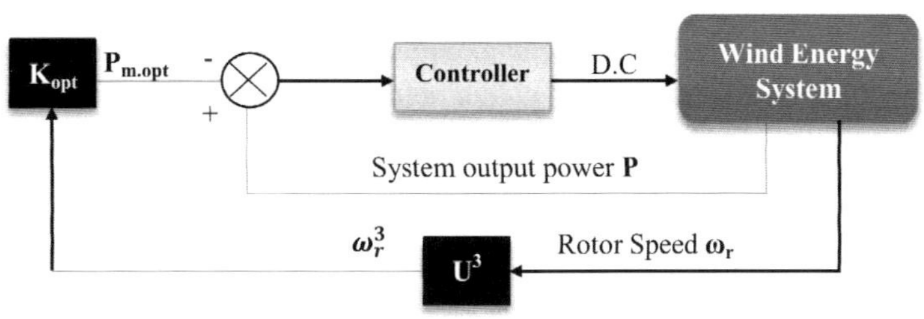

Figure 7. PSF method.

3.3. Hill Climbing Search (HCS) Method

The HCS method is the most technique MPPT used in a wind system because of its simplicity and independence to wind turbine characteristics, the power coefficient maximal $C_{p\text{-}opt}$ and the tip speed ration optimal λ_{opt}. Its principle is based essentially on a periodic perturbation (incrementing or decrementing) of the wind turbine rotor speed in small step, then observing the effect of this perturbation on the output power P of the PMSG generator, by comparing the current value of the power with its previous value. And as explained by the flow chart of this algorithm in Fig. 8, If a small increase of the rotor speed results in an increase of output power (case $\Delta P/\Delta\omega>0$), the perturbation will continue in the same direction, Otherwise (case $\Delta P/\Delta\omega<0$), the perturbation will be in the opposite direction, this strategy is repeated until reaching the maximum power point corresponds to the zero slope $\Delta P/\Delta\omega=0$.

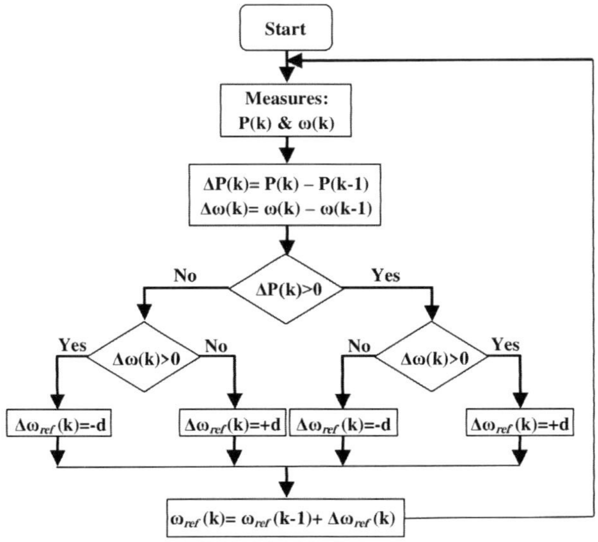

Figure 8. Hill Climbing Search method flowchart.

3.4. Fuzzy logic (FLC) MPPT

The MPPT technique based on fuzzy logic recently introduced one of the intelligent methods used in wind systems to determine and track the operating point corresponding to the maximum power whatever of the wind speed. Compared with conventional control techniques, this control is robust and does not need exact knowledge of the system mathematical model. A FLC controller includes three functional blocks: Fuzzification, Inference engine and Defuzzification. Thus, it is characterized by input variables, output variables, membership functions and fuzzy rules. The fig. 9 shows the block diagram of FLC MPPT controller for wind system:

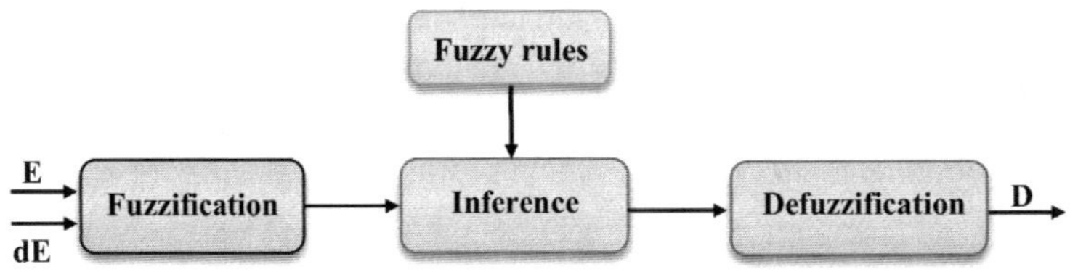

Figure 9. FLC MPPT controller.

The input variables of the fuzzy controller are the error E(k) and the change of error dE(k) which are defined by (11) and (12), while the output is the duty cycle D.

$$E(k) = \frac{P(k)-P(k-1)}{\omega(k)-\omega(k-1)} \tag{11}$$

$$dE(k) = E(k) - E(k-1) \tag{12}$$

Where P(k) and ω(k) are the output power and the rotor speed of the wind generator, respectively.

The principle followed by the fuzzy controller to reach the maximum power point is as follows: In the first step, the controller converts the input and output variables into linguistic variables in the form of membership functions (fuzzification step), which are defined in five fuzzy subsets: NB (negative big), NS (negative small), ZE (zero), PS (positive small) and PB (positive big). Fig. 10 shows the memberships functions of E(k), dE(k) and D(k). In the second step, the controller establishes logical relations between the input variables and the output variable while defining membership rules (inference step), using the MIN-MAX inference method, the defined inference rules are shown in table 3. And in the final step, the controller converts the fuzzy subsets of the output variable into a numerical value, which is the duty cycle applied to the dc/dc boost converter, using the defuzzification method of the center of gravity which is the simplest and fastest.

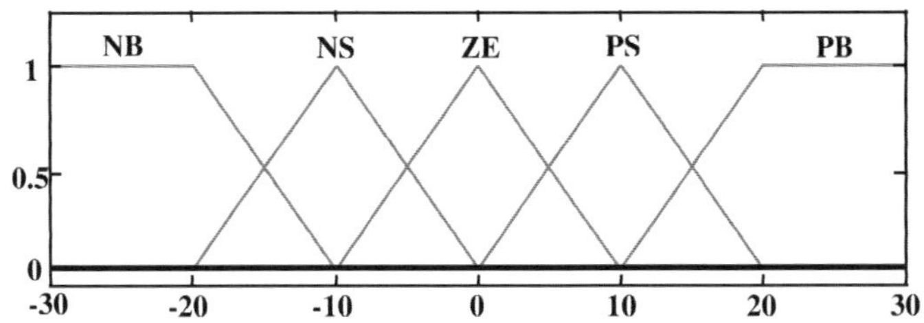

(a)

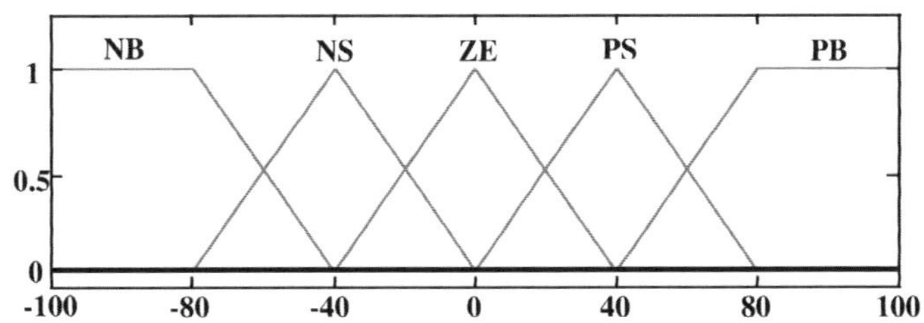

(b)

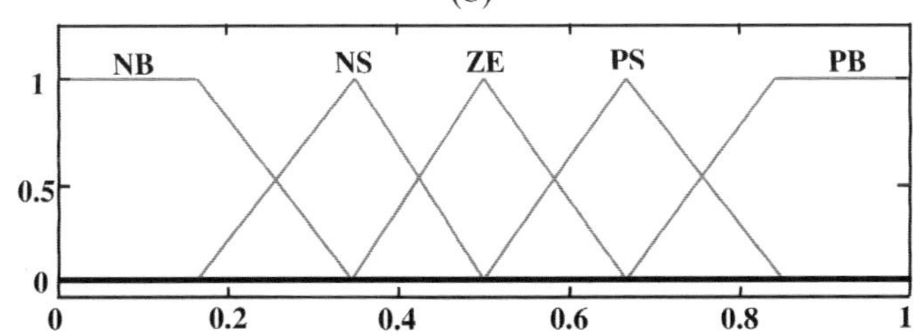

(c)

Figure 10. Membership functions for: (a) Input E, (b) Input dE, (c) Output D.

Table 3. Fuzzy logic rules.

	dE				
E	NB	NS	ZE	PS	PB
NB	NB	NB	NS	PS	PB
NS	NB	NB	NS	PB	PS
ZE	ZE	ZE	ZE	ZE	ZE
PS	NS	NS	NS	ZE	ZE
PB	NS	NB	NB	ZE	ZE

4. Simulation Results and Discussion

In this work, a detailed implementation and simulation of a wind system was carried out using the Matlab/Simulink software, based on the equations and technical parameters of wind turbine and PMSG generator described previously. Also an MPPT controller has been developed to optimize the performance of the proposed system under varying wind speed conditions, with different MPPT approaches that have been simulated, evaluated and compared.

Figs. 11 and 12 show the simulation results corresponding respectively to the power coefficient and output mechanical power for the wind turbine with different wind speed values (12m/s and 10m/s), using the Optimal Torque Control (OTC) method, the Power Signal Feedback (PSF) method, the Hill Climbing Search (HCS) Method and the Fuzzy Logic Control (FLC).

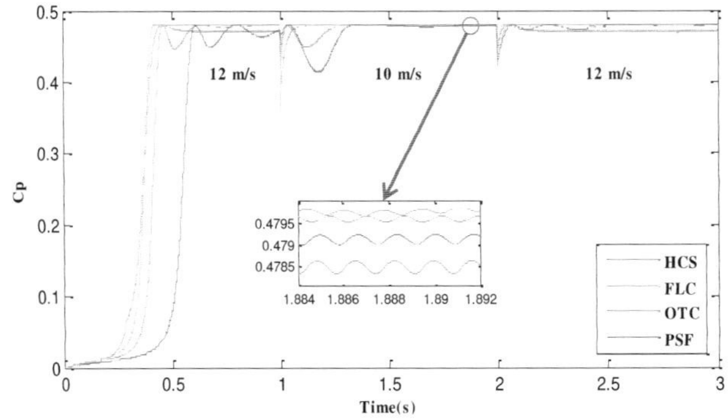

Figure 11. Power coefficient (Cp) for different MPPT methods.

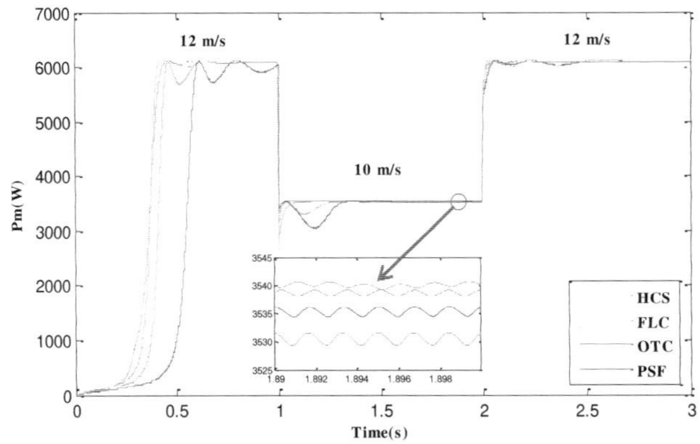

Figure 12. Wind turbine mechanical power for different MPPT methods.

From these results, we noticed that the control based on fuzzy logic is most efficient technique compared to other methods. It achieves maximum power with more stability, precision and better response time. However, the PSO and OTC methods require slow response time to determine the MPP, which remains less than that of the FLC, with more oscillations resulting in a loss of energy. For that, it can be stated that the FLC intelligent control is a robust and efficient approach, it has better performance and a good dynamic response under variable wind speed conditions.

5. Conclusions

In this paper, four MPPT techniques like OTC, PSF, HCS and FLC are developed and compared to optimize the power produced by a wind energy system under varying conditions. According to the performance analysis of each method, we noted that the MPPT controller based on the fuzzy logic allows determining and tracking the maximum power point with more efficiency, fast response and high reliability compared to other controllers based on OTC, PSF and HCS. Therefore, it is concluded that the FLC based approach offers better advantages and good performance for a wind energy conversion system.

References

[1] Bhandari B, Poudel S.R, Lee K.T and Ahn S.H 2014 Mathematical modeling of hybrid renewable energy system: A review on small hydro-solar-wind power generation. *International Journal of Precision Engineering and Manufacturing-Green Technology*. Vol. **1(2)** pp. 157-173

[2] Abdel-Raheem Y, Sayed M.A, Abdelwhab M.N and Shabib G 2016 Generator side converter for five-phase PMSG based wind turbine generators using perturbation and observation technique. *International Journal on Power Engineering and Energy*. Vol. **7(1)** pp. 610-618

[3] Rekioua D 2014 *Wind Power Electric Systems: Modeling, Simulation and Control* (London-England: Springer-Verlag)

[4] Minh H.Q, Frederic N, Essounbouli N and Abdelaziz H 2012 A new MPPT method for stand-alone wind energy conversion system. *IEEE 2nd International Symposium on Environment-Friendly Energies and Applications (EFEA)(Northumbria University, UK)* pp.335-340

[5] Li X-P, Fu W-L, Shi Q-J, Xu J-B and Jiang Q-Y 2013 A fuzzy logical MPPT control strategy for PMSG wind generation systems. *Journal of Electronic Science and Technology*. Vol. **11(1)** pp. 72-77

[6] Javadi S and Hazin M.H 2014 An improved MPPT method of wind turbine based on HCS method by using fuzzy logic system. *International Journal of Smart Electrical Engineering*. Vol. **3(1)** pp. 29-35

[7] Sefidgar H and Gholamian S.A 2014 Fuzzy logic control of wind turbine system connection to PM synchronous generator for maximum power point tracking. *International Journal of Intelligent Systems and Applications*. Vol. **6(7)** pp. 29-35

[8] Rathi R and Sandhu K.S 2016 Comparative analysis of MPPT algorithms using wind turbines with different dimensions & ratings. *IEEE 1st International Conference on Power Electronics. Intelligent Control and Energy Systems (ICPEICES)(Delhi, India)*

[9] El-Sebaii A.M, Hamad M.S and Helal A.A 2013 A sensorless MPPT technique for a grid-connected PMSG wind turbine system. *IEEE 2nd Renewable Power Generation Conference (RPG)(Beijing, China)*

[10] Tiwari R and Babu N.R 2016 Fuzzy logic based MPPT for permanent magnet synchronous generator in wind energy conversion system. *IFAC(International Federation of Automatic Control) – Papers Onligne-Elsevier*. **49(1)**, pp.462-467

[11] Lalouni S, Rekioua D, Idjdarene K and Tounzi A.M 2013 Comparative study between MPPT algorithms applied to wind energy conversion system. *International Conference on Renewable Energies (CIER)(Sousse, Tunisia)*

IOP Conf. Series: Earth and Environmental Science **161** (2018) 012024 doi:10.1088/1755-1315/161/1/012024

Investigation of microinverter based on the two-switch DC-DC flyback converter topology

L EL IYSAOUY[1,4 *], E BIELSKIS[2], M LAHBABI[1], A BASKYS[3] and A OUMNAD[4]

[1]The Signals, Systems and Components Laboratory, Sidi Mohamed Ben Abdellah University FST Fez, Morocco

[2]Center for Physical Sciences and Technology, Sauletekio al.3, LT-10257 Vilnius, Lithuania

[3]Department of Computer Engineering, Vilnius Gediminas Technical University, Naugarduko St. 41, LT-03227 Vilnius, Lithuania

[4]Department of Electrical Engineering, Mohamed V University EMI Rabat, Morocco

[*]E-mail: lahcen.eliysaouy@usmba.ac.ma

Abstract. This paper presents an investigation of a novel scheme of the single stage flyback microinverter based on two-switch **[1]**. The purpose of this topology of microinverter is to transform the DC power generated by the DC voltage input to a standard electric grid AC voltage. The single stage two-switch DC-DC flyback microinverter for the case CL output filter ensures the isolation between the DC power sources and the AC sources by a flyback transformer; has high efficiency (more than 96%) if the parameters of dual two-transistor stage: M1 to M4 are chosen correctly and Total Harmonic Distortion THD less than 5 %. However, the design of this microinverter is characterized by a low number of components.

1. Introduction

Due to their non-sustainability and harmful effects on the environment, fossil fuels no longer stand as attractive sources of energy. Hence, photovoltaic (PV) power management concepts are necessary to derive as much energy as possible from the solar power. PV energy systems are being widely studied because of its profits of environmentally friendly and renewable features. Moreover, solar cells are unique in that they directly convert the incident solar irradiation into electricity. Thereby, photovoltaic (PV) power management insights are necessary to draw as much power as possible from the solar energy. PV energy systems are being widely advised because of its advantage of environmentally friendly and renewable features.

Several PV systems are used in different configurations to achieve high DC voltage [1], [2]. These configurations include following connection methods: parallel series (SP), total-cross-tied (TCT), Honey Comb (HC), bridge linked (BL) and Su-do-ku [1]– [3]. Nevertheless, the PV panels frequently operate in mismatching conditions due to various panel adjustments and shadowing effects. This mismatching problem can cause serious problems for the power-conditioning system. To overcome this drawback, several publications has proposed a module-integrated converter concept [4]. Such a PV systems composed of a PV panels with an individual Photovoltaic DC-AC inverters are called PV flyback microinverters [5]–[8]. For the voltage source inverter (VSI), the mildest output line filter has just one inductor. Toward the better performance of a filter, the CL and LCL filters can be constructed with mixtures of capacitors and inductors. The output filter is a critical component of the microinverter to improve the output power quality and reduce total harmonic distortion (THD). This is because the output current of the system contains high-frequency harmonic caused by the pulse width modulation (PWM) switching. For inverter type voltage source inverter (VSI), the mere output filter has just one inductor. Therefore, in order to get the best performance of the filter, the CL and LCL filter take place and can be configured with mixtures of capacitors and inductors [9].

2. The microinverter with the Two-Switch DC-DC flyback converter topology

2.1. System operation

Figure 1 shows the topology of investigated single stage microinverter based on two-switch DC-DC flyback microinverter [10]. The DC side circuit is made up around the dual two-transistor stage: M1 to M4 and the transformer magnetizing inductance. The Mosfets of single stage microinverter topology are operated at high frequency 25 kHz. The Mosfet transistor M5 side output is on while the entire positive grid cycle, and vice versa for Mosfet transistor M5. The diodes in series with M5 and M6 are inserted in order to block reverse power flow from the grid. Supposing a negative grid voltage, and that all currents are zero (Discontinuous Current Mode DCM): A new cycle is started by switching Mosfet transistors M1 and M4 on (respectively Mosfet transistors M2 and M3 for a negative grid voltage). Besides, this lets the magnetizing current to rise up from zero. The Mosfet transistors are retained on in a pre-specified amount of time (a function of requisite power transfer, and DC voltage). They are switched off then, and the stored power in the magnetizing inductance is carried in the secondary-side of the transformer. Further, the stored power is injected into the grid, within Mosfet transistor M5, where the body-diode of the Mosfet transistor M6, and the filter (vice versa for a positive grid voltage). Beside the high frequency harmonics of the formed AC voltage caused by the switching of converter switches are filtered by the low pass filter. The microinverter output voltage waveforms, Total Harmonic Distortion (THD) spectrum and microinverter yield were reviewed. The investigation was performed for two low pass filter types: LCL and CL. However, in this paper, we will focus just on CL filter.

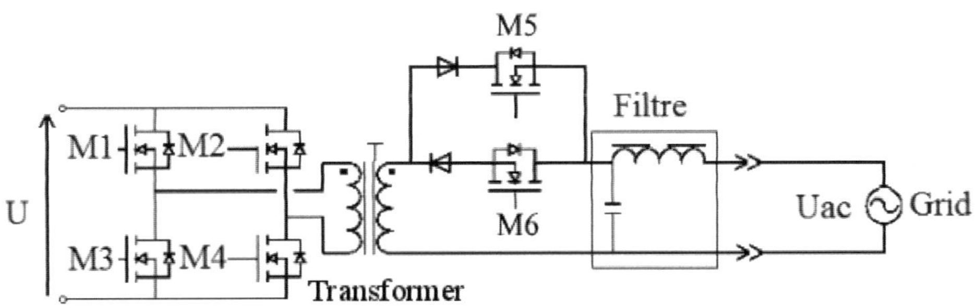

Fig. 1: Two-switch DC-DC flyback microinverter topology

2.2. Operation Principle

The high switching frequency provides high-order harmonics to output [11]. Therefore, an output filter is required for the rejection of switching frequency components. CL filter is extensively used for current source inverters (CSI) and LCL filter which has an extra inductor element to the typical CL filter, is designed in [10]. The interests of LCL filter appear on the system size and the component's capacity. Although, we will focus on the characteristics of CL filter. The investigation was performed for 50 Hz microinverter output voltage frequency 25 kHz switching frequency. Usually, a CL filter is used for both current and voltage inverters grid connected. For the mitigation of switching frequency elements, the CL and LCL filters are designed as an output filter. Figure 1 shows the circuit of the microinverter topology with the CL filter [11]. As the grid impedance remain unknown, to fix up the transfer function of the circuit and parameters of design, the additional inductive reactance (Lg) expression is defined and used with the CL filter parameters (L and C) as shown in Figure 2. The performance of CL filter corresponds to resonant frequency location. Therefore, the key of designing parameters is that the frequency of harmonics that have to be filtered would be behind the resonant frequency of the filter [11]. The transfer function of CL filter as follow:

$$A_{LC} = \frac{U_g}{U_c} = \frac{1}{S^2CL+1} \tag{1}$$

where Uc is the voltage side of flyback converter and Ug is the output voltage side of grid connected of the filter CL. According to the control theory of the second-order system, the frequency is equal to:

$$fc = \frac{1}{2\pi\sqrt{LC}} \tag{2}$$

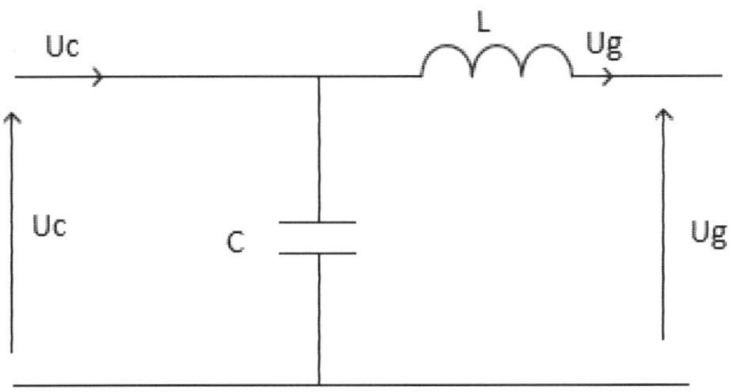

Fig. 2: Circuit diagram of CL Filter

3. Simulation results

The investigation was performed using simulation based on the software Matlab/Simulink. The microinverter output voltage waveforms and spectrum and converter efficiency were investigated. The waveforms of output voltage are reached and their spectrum is explained in Figure 3(a, b, c).

The results show that the Total Harmonic Distortion (THD) of the output voltage for different voltage amplitude (amp) at 25000 Hz switching frequency (fs) is not higher than 5% and respect the requirements of IEEE standard 1574. The results show that the low pass (CL) filter plays an important role in grid-connected converter when trying to reduce switching-frequency voltage ripples injected into the grid.

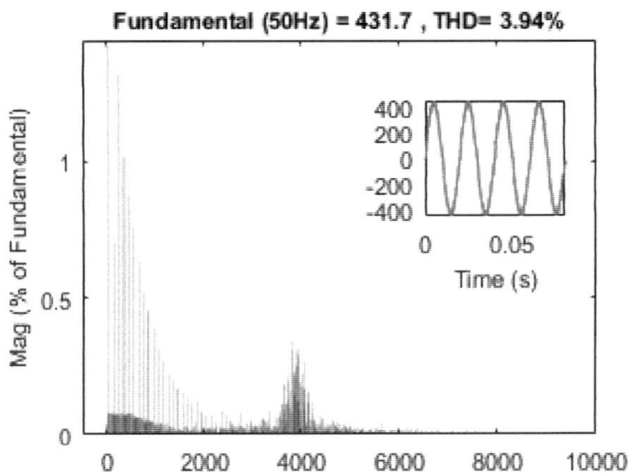

a) THD of the output voltage for amp (400 RMS).

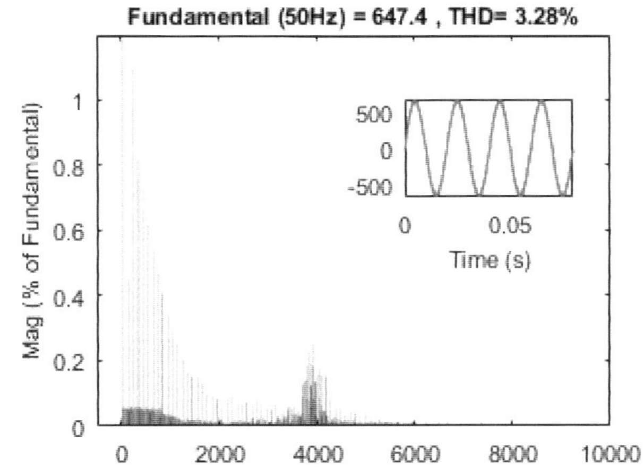

b) THD of the output voltage for amp (600 RMS).

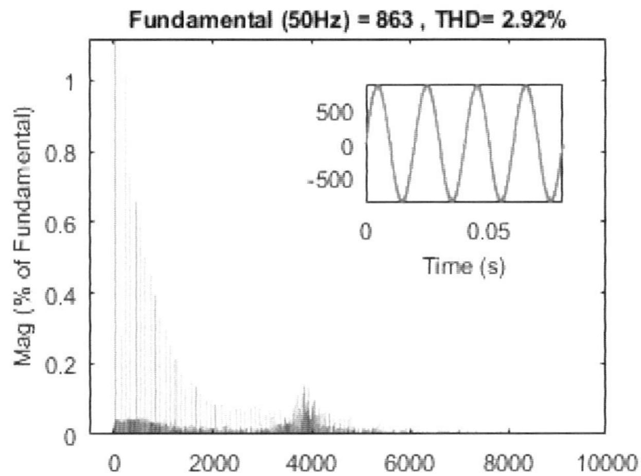

c) THD of the output voltage for amp (800 RMS).

Fig. 3. The spectrum and Total Harmonic distortion THD of the output voltage at switching frequency 25000 Hz

The obtained output voltage waveforms and spectrum are illustrated in Figure 3. The results showed that the THD of the output voltage at 25000 Hz switching frequency is not higher than 4% and respect the requirements of IEEE. The dependences of investigated single stage microinverter efficiency on output power at various internal resistances of flyback converter switches M1-M4 in state ON are shown on Figure 4. It is seen that the efficiency of converter decreases when internal resistance of switch transistors in state ON increases.

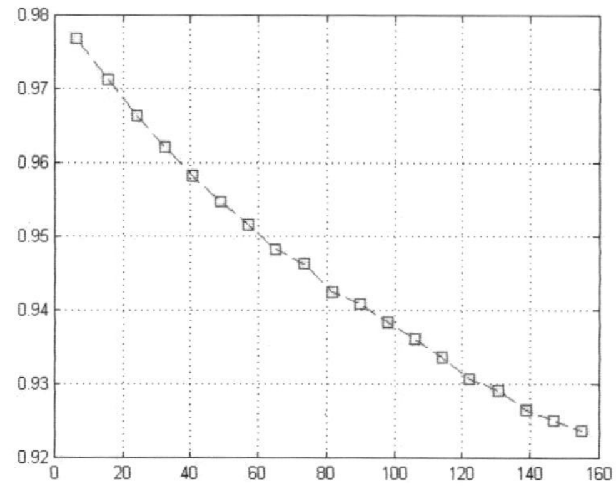

a) Ron = 0.008 Ω

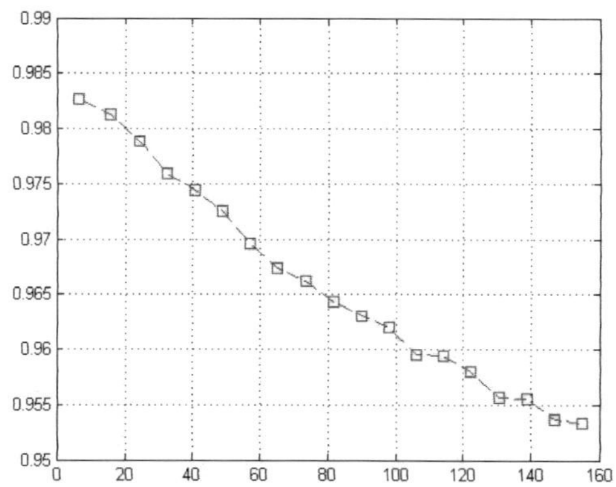

b) Ron = 0.04 Ω

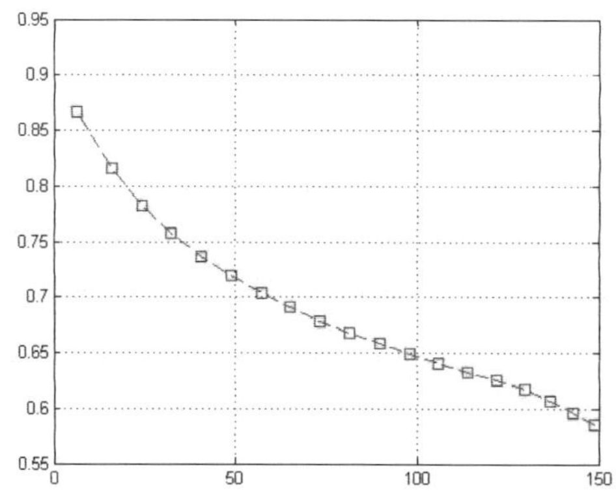

c) Ron = 0.5 Ω

Fig. 4: Dependences of microinverter efficiency on output power at various internal resistances of switch transistors M1-M4 in state ON

4. Conclusion

In this The results of investigation of microinverter based on the couple of two-switch DC-DC flyback converters show that the THD of the output voltage at 25000 Hz is not higher than 4%

and respect the requirements of IEEE. Furthermore, the efficiency of microinverter reaches 0.96 at 100W output power at 25 kHz switching frequency when the internal resistance of MOSFET switches M1-M4 are around 0.008 Ω. Finally, it can be concluding that the investigated grid connected microinverter Is characterized by the high efficiency and provides low THD of output voltage.

Acknowledgment
This work was supported by Vilnius Gediminas Technical University Lithuania.

References

[1] Bielskis E, Baskys A and Sapurov M 2017 Single Stage Microinverter Based on Two-Switch DC-DC Flyback Converter *Elektron. Ir Elektrotechnika* **23** 29–32

[2] El-Dein M Z S, Kazerani M and Salama M M A 2013 Optimal Photovoltaic Array Reconfiguration to Reduce Partial Shading Losses *IEEE Trans. Sustain. Energy* **4** 145–53

[3] Lahcen E I, Lahbabi M and Oumnad A 2016 Effet de l'ombrage partiel sur les différentes configurations de Panneaux Photovoltaïques *ResearchGate* 3ème édition du Colloque international "Modélisation et Analyse en Mécanique Energétique" CIMAME'2016

[4] Srinivasa Rao P, Saravana Ilango G and Nagamani C 2014 Maximum Power from PV Arrays Using a Fixed Configuration Under Different Shading Conditions *IEEE J. Photovolt.* **4** 679–86

[5] Chiu H-J, Lo Y-K, Yang C-Y, Cheng S-J, Huang C-M, Chuang C-C, Kuo M-C, Huang Y-M, Jean Y-B and Huang Y-C 2013 A module-integrated isolated solar microinverter *IEEE Trans. Ind. Electron.* **60** 781–8

[6] Singh B, Singh S, Chandra A and Al-Haddad K 2011 Comprehensive Study of Single-Phase AC-DC Power Factor Corrected Converters With High-Frequency Isolation *IEEE Trans. Ind. Inform.* **7** 540–56

[7] Lo Y K and Lin J Y 2007 Active-Clamping ZVS Flyback Converter Employing Two Transformers *IEEE Trans. Power Electron.* **22** 2416–23

[8] Tacca H E 1998 Single-switch two-output flyback-forward converter operation *IEEE Trans. Power Electron.* **13** 903–11

[9] Bielskis E, Baskys A and Sapurov M 2016 Impact of Transformer Design on Flyback Converter Voltage Spikes *Elektron. Ir Elektrotechnika* **22** 58–61

[10] Tang Y, Loh P C, Wang P, Choo F H and Gao F 2012 Exploring Inherent Damping Characteristic of LCL-Filters for Three-Phase Grid-Connected Voltage Source Inverters *IEEE Trans. Power Electron.* **27** 1433–43

[11] Li F, Zhang X, Zhu H, Li H and Yu C 2015 An LCL-LC Filter for Grid-Connected Converter: Topology, Parameter, and Analysis *IEEE Trans. Power Electron.* **30** 5067–77

[12] Liu J, Zhou L, Yu X, Li B and Zheng C 2017 Design and analysis of an LCL circuit-based three-phase grid-connected inverter *IET Power Electron.* **10** 232–9

IOP Conf. Series: Earth and Environmental Science **161** (2018) 012025 doi:10.1088/1755-1315/161/1/012025

Treatment of landfill leachate from Fez city (Morocco) using Fenton and photo-Fenton processes

I El Mrabet[1], M Kachabi[1], M Nawdali[1], A Harrach[1], F Khalil[2], M Ijjaali[1], M Benzina[3] and H Zaitan[1,*]

[1]Laboratoire de Chimie de la Matière Condensée, Faculté des Sciences et Techniques, Université Sidi Mohamed BenAbdellah. B.P. 2202, Fès, Maroc
[2]Laboratoire de Chimie Appliquée, Faculté des Sciences et Techniques, Université Sidi Mohamed BenAbdellah. B.P. 2202, Fès, Maroc
[3]Laboratoire Eau Environnement Energie, Ecole Nationale d'Ingénieurs de Sfax, B.P. 1173, Route de Soukra, 3038 Sfax, Tunisie

[*]Email: hicham.zaitan@usmba.ac.ma

Abstract. In the present work, the treatment of a leachate from Fez city urban sanitary landfill (Morocco) was evaluated using different advanced oxidation processes (AOPs), such as, Fenton (Fe^{2+}/H_2O_2), photo-Fenton (Fe^{2+}/H_2O_2/UV-A), UV-A and UV-A/H_2O_2. The leachate was characterized by high chemical oxygen demand (COD), low biodegradability and intense dark color. The treatment efficiency was evaluated as a function of different operation variables ([H_2O_2], [Fe^{2+}], UV-A irradiance), resulting in COD removals between 30% and 77%. Removal efficiencies decreased in the following order: photo-Fenton> Fenton> UV-A/H_2O_2> UV-A.Thus, a detailed experimental analysis was carried out to analyze the effect of hydrogen peroxide and iron concentration in the photo-Fenton process. It was observed that: (a) COD removal ranged from 4% to 84% depending on the H_2O_2 dose, (b) COD removal increased by adding the H_2O_2 dose in multiple steps (84%) and (c) iron concentration corresponding to a Fe^{2+}/COD mass ratio = 0.35 was found to be the most favorable. At optimum conditions, COD removal was 70% and 84% for Fenton and photo–Fenton processes, respectively. In addition, the optimum contact time for both processes was 1 hr. Finally, the results of this study showed that the Fenton and photo-Fenton process is capable of achieving high levels of COD removal.

1. Introduction

Landfill of Fez (Morocco) receives a large urban and industrial waste amount (about 800 tons per day), generating high volume of leachates (around 400-450 m^3 per day). Landfill leachate is considered a potential contamination source of water resources, as they may percolate through soils and subsoils, causing serious environmental damage if they are discharged without any kind of treatment [1]. The leachate contains large quantities of organic compounds, heavy metals, a complex variety of materials, and many other hazardous chemicals. The composition and concentration of such effluents is variable,

depending on the type of deposited waste, the origin and composition of the wastes, and the structure, management and age of the landfill.
Additionally, leachate may cause malodors and aerosols, though these effects tend to be temporary and local. Therefore, it becomes necessary to treat landfill leachate and remove pollutants before it is discharged into water resources and land.
Various treatment technologies based on biological treatment, physical-chemical process such as chemical precipitation, coagulation–precipitation, ultrafiltration and reverse osmosis have been applied successfully for the treatment of landfill leachates [2-6]. On the other hand, combined processes including two or three physical–chemical methods have been widely used including ozonation-activated carbon adsorption [7-8], coagulation–ozonation [9], and precipitation–reverse osmosis [10].
Recently, conventional, Fenton and photo-Fenton processes have been investigated for landfill leachate treatment. In the Fenton process, hydrogen peroxide is added to wastewater in presence of ferrous salt, generating a large amount of free radicals such as "hydroxyl" (•OH) which can remove both organic and oxidizable inorganic components due to their ability to change in the chemical structure of the pollutants.

This work presents a study of the remediation of a mature landfill leachate applying several AOPs which include H_2O_2, Fe^{2+}, UV-A radiation and their combination.

This study aims (a) to characterise the leachates of Fez landfill and (b) to examine the efficiency of different advanced oxidation processes (AOPs), such as, Fenton (Fe^{2+}/H_2O_2), photo-Fenton (Fe^{2+}/H_2O_2/UV-A), UV-A and UV-A/H_2O_2 in terms of the removal of COD and colour.

2. Materials and methods

2.1. *Materials*

Leachate samples were collected from a municipal landfill in Sidi Hrazem, located at 11 km from the city of Fez, Morocco (Figure. 1). This site (about 100 ha) has been in activity since 2004, receiving approximately 365 000 ton household wastes per year from different areas of Fez and generates about 350–450 m^3 leachate per day. Raw leachate samples were collected in 25 L plastic containers directly from the landfill site. Then, samples were transported to the laboratory and preserved in a refrigerator at 4 °C in the dark in order to minimise leachate decomposition due to microbial activity.

Figure 1. Leachate from the urban sanitary landfill of Fez city (Morocco).

The materials used in the experiments were $FeSO_4 \cdot 7H_2O$ as source of catalyst, H_2SO_4 96% and NaOH 5 M to adjust pH value and (H_2O_2, 30% (w/w)) as oxidant reagent. All reagents used were analytical grade a, except as noted.

2.2. *Fenton's and Photo-fenton Reaction*

All tests were carried out at ambient temperature in batch mode according to the following sequential steps. Initially, the leachate to be treated with a volume of 200 mL was introduced into a pyrex reactor and magnetically stirred; pH was adjusted to fixed value (3) by adding H_2SO_4 (0.1N). Then, mass of ferrous sulfate was added and the solution was stirred for 5 min using a magnetic plate to obtained a homogeneous solution. This was followed by the addition of oxidant solution (H_2O_2) under vigorous stirring to start Fenton reaction. This moment was considered to be the starting point (t_0) of the dark (no radiation) reaction. The reaction mixture was kept away from any light source for the entire duration of each experimental run. Afterwards, the samples (5 mL) were withdrawn every 5 min during the initial 15 min of the reaction and every 15 min during the following 120 min. The sample was centrifuged for 5 min at 2000 rpm to collect the supernatant for the analysis of COD and the UV_{254nm}. Once each sample was obtained, the reaction was followed until COD analysis showed no significant further variation in concentration or until the 60 min reaction time was completed. These experiments were repeated using different ratios of H_2O_2/COD and H_2O_2/Fe^{2+} values.

When applying the photo-Fenton process, all experimental procedures were the same except that a UV-A lamp of three (20 W, 365 nm) was used. Effluents collected were irradiated with the tubular-shaped photoreactor. Photoreactor (See Figure 2) consists of a three UV lamp located vertically at a distance of about 10 cm from the preserved leachate sample in a Pyrex glass reactor. The working volume of the sample was 0.1 L exposed to UV lamps. The entire assembly was placed in an aluminum cylinder to reflect any UV-A light dispersed in the photoreactor) on a magnetic stirrer to maintain a uniform concentration. The system was continuously cooled with a permanent fan maintained at a very proximal distance from the reactor.

Hydrogen peroxide was added as the lamp was turned on, and this moment was considered to be the start of the degradation process. The effect of different ratios of H_2O_2/COD and H_2O_2/Fe^{2+} values were also tested for the Photo-Fenton reaction. Samples were taken at regular time intervals, centrifuged at 2000 rpm for 5 min to remove suspended solids and analyzed by UV–visible spectrophotometer and analysis and COD measured.

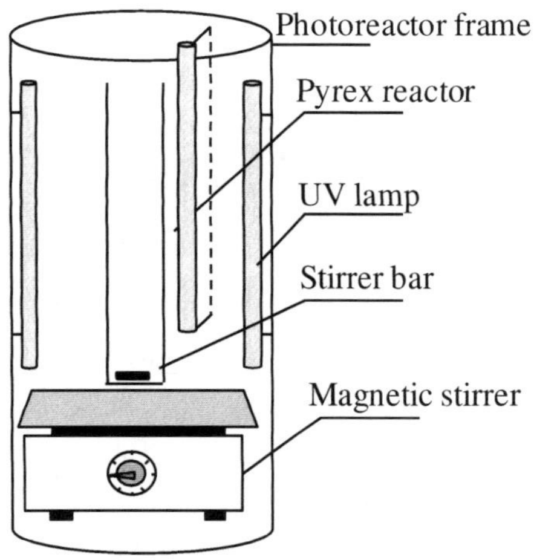

Figure 2. Schematics of photoreactor with suspended UV lamp

2.3. *Analytical determinations*

Leachate samples were characterised before and after the Fenton and Photofenton treatment. Physical-chemical parameters such as pH, conductivity, turbidity, apparent color at 465nm were measured using respectively a pHmeter (JENCO – Electrocinics LTD), a conductivity meter (WTW

inoLab, Cond Level 1), a turbidimeter HI 88713 - ISO HANNA-Instruments and a spectrophotometer (VWR UV-6300PC), after calibration with standard solutions. Chemical oxygen demand (COD), biological oxygen demand (BOD_5), total suspended solids (TSS), nitrate, nitrite, sulfate were quantified according to the Standard Methods for Water and Wastewater Examination [11]. Colour number (CN) was determined using the method described by Tizaoui et al. [12].
UV–visible absorption measurements were performed using a spectrophotometer VWR UV-6300PC. Inductively coupled plasma atomic emission spectroscopy analysis ICP-AES (Activa, Jobinyvon) was used to quantify the concentration of metals after samples were filtered through 0.45 µmfilter. Chloride was quantified by the Mohr method. The COD removal efficiencies were obtained using the following equation:

$$\text{Removal}\,(\%) = \frac{(C_i - C_f)}{C_i} \times 100 \qquad (1)$$

Where C_i and C_f refer to the initial and final COD concentrations in leachate.

3. Results and discussion

3.1. *Landfill leachate*

The main chemical-physical characteristics of the sanitary landfill leachate are summarized in Table 1.

Table 1: Characterisation of Fez landfill leachate

Parameters	Value
Conductivity (mS cm^{-1})	28
T (°C)	24
pH	8
TSS (mg. L^{-1})	15775
Turbidity (NTU)	248
COD (mg. L^{-1})	5653
BOD_5 (mg. L^{-1})	600
BOD_5/COD	0.1
Colour Number	2.22
TA (mg. L^{-1})	0
TAC (mg. L^{-1})	406.7
SO_4^{2-} (mg. L^{-1})	140
NO_3^- (mg. L^{-1})	6.58
NO_2^- (mg. L^{-1})	6.56
Cl^- (mg. L^{-1})	7100
SO_4^{2-} (mg. L^{-1})	140
Cr^{3+} (mg. L^{-1})	1.74
Cu^{2+} (mg. L^{-1})	0.05
Fe^{2+} (mg. L^{-1})	1.83
Mn^{2+} (mg. L^{-1})	0.043
Zn^{2+} (mg. L^{-1})	0.2
Pb^{2+} (mg. L^{-1})	<0.01

The dark colour of the liquid is related to the high values of COD (5653 $mgO_2.\ l^{-1}$), BOD_5 (600 $mgO_2.l^{-1}$) and to the high turbidity content (248 NTU), being all these parameters higher than the values of the Morocco Reject Requirements. The high concentration of COD is due to the presence of different kinds of dissolved organic matter. Another relevant point is the high conductivity (28 mS. cm^{-1}) that could be related to the high salinity value, principally due to the amount of chlorides (7100 mg L^{-1}), nitrates (6.58 mg. L^{-1}), nitrites (6.56 mg. l^{-1}) and sulfates (140 mg. l^{-1}). The BOD_5/COD ratio is close to 0.1 with a pH of 8, indicating that this leachate may be defined as stabilised leachate and difficult to further biodegradation. In addition, the landfill leachate contains Cr, Cu, Fe, Mn and Zn with concentration of 1.74, 0.05, 1.83, 0.043 and 0.2 mg. L^{-1} respectively. The total suspended solids are of the order of 15.77 $g.L^{-1}$.

3.2. *COD removal*

UV photolysis without addition of any catalyst or oxidant reagent showed very low degradation efficiency (<4%) so it is not a suitable treatment. This could be attributed to the low transmittance in the landfill leachate as indicated by the presence of turbidity and deep black colour. When the oxidation was carried out using only H_2O_2 oxidant reagent, the reduction of organic compounds concentration expressed as COD was less than 14% after 1 h of treatment (Table 2, Figure. 3).

Table 2: COD reductions achieved in the leachate treatment

AOP	H_2O_2 (mg. l^{-1})	Catalyst (mg. l^{-1})	COD Removal (%)
H_2O_2	2500	0	14
Photolysis UV	0	0	2
H_2O_2	10000	0	4
UV/H_2O_2	2500	0	32
UV/H_2O_2	10000	0	11
Fenton Fe^{2+}/H_2O_2	2500	2000	58
Fenton Fe^{2+}/H_2O_2	10000	2000	70
Photo-Fenton (UV/Fe^{2+}/H_2O_2)	2500	2000	75
Photo-Fenton UV/Fe^{2+}/H_2O_2	10000	2000	84

The combination of UV radiation and hydrogen peroxide (UV/H_2O_2) enhanced the oxidation of organic compounds compared to direct UV photolysis through the formation of hydroxyl radicals.
Two different concentrations of H_2O_2 were added in the presence of UV-A. Experiments using 2500 mg. L^{-1} and 10000 mg. L^{-1} H_2O_2 (H_2O_2/COD mass ratios = 0.37– 1.48) were carried out achieving COD reductions of 32% and 11% respectively (Table 2). These results showed that the photochemical oxidation enhanced the organic matter removal, although the results should be improved to obtain a cleaner effluent before discharge. Possibly the intensive colour of the solution inhibited light transmittance and H_2O_2 photolysis was not effective enough.

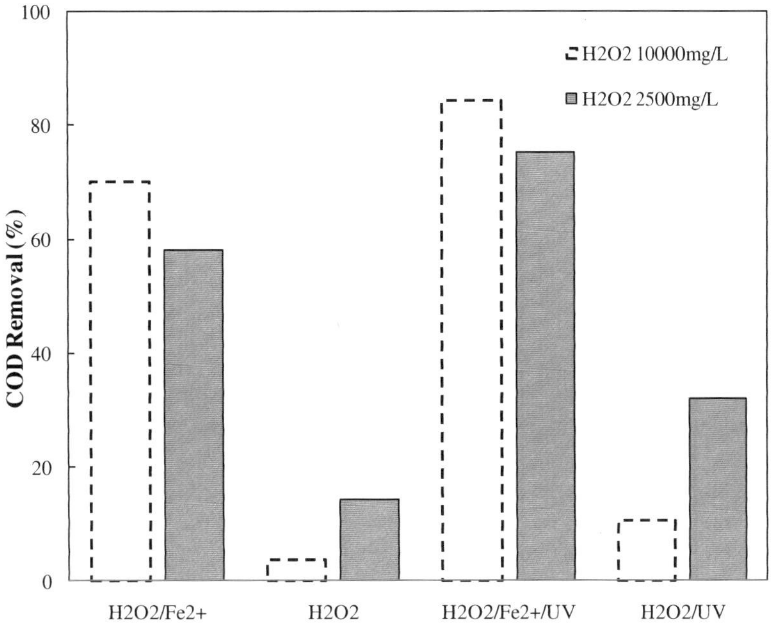

Figure 3. COD removal in the different AOPs (H_2O_2 = 2500 mg. l^{-1}, H_2O_2 = 10000 mg. l^{-1}, Fe^{2+} = 2000 mg. l^{-1}, oxidation time = 60 min).

Fenton oxidation experiments (Fe^{2+}/H_2O_2), with a residence time of 60 min and pH=3 using a fixed concentration of Fe^{2+} of 2000 mg. L^{-1} resulted in 58% of organic matter decrease, expressed as COD, using 2500 mg. L^{-1} H_2O_2. Increasing the hydrogen peroxide concentration up to 10000 mg. l^{-1}, lead to better results (70%).

The degradation efficiency of the oxidation process could be accelerated including UV radiation in the Fenton process. This alternative ($UV/Fe^{2+}/H_2O_2$) is known as photo-Fenton. UV-A radiation leads to faster regeneration of Fe^{2+} ions in the solution.

Fe^{3+} complexes are photoreactive and its photolysis generates Fe^{2+} ions in acidic conditions [13]. Photo-Fenton experiments led to the maximum reduction of COD. A 75% of COD removal was obtained using 2500 mg. l^{-1} of H_2O_2. This percentage increased to 84% when the higher hydrogen peroxide concentration (10000 mg. L^{-1}) was used.

The results showed a high percentage of colour removal in all conditions (>90%). So, in the final solutions, colour was almost negligible. This change in colouration was probably due to changes in the nature of the compounds present in the leachate solution. AOPs were able to degrade aromatic or polymeric organic compounds, which were the main compounds responsible for colour, in much simple compounds. It is important to remark that the UV/H_2O_2 process achieved 60 % colour reduction just in one step as in this AOP sludge was not generated.

These results attest the efficiency of the photo-Fenton oxidation process due to the higher production of hydroxyl radicals, comparing to the Fenton process or UV/H_2O_2, as result of the combination of H_2O_2 and a metallic catalyst in presence of UV radiation.

The efficiency order of the applied advanced oxidation processes in this work regarding COD reduction was: photo-Fenton > Fenton > UV-A/H_2O_2 >UV-A.

3.2.1. Effect of hydrogen peroxide concentration

Hydrogen peroxide dose is the key factor in Fenton and Photo-Fenton process. The theoretical mass ratio of removable COD to that of H_2O_2 is 470.6/1000; that is 1000 mg. l^{-1} H_2O_2 theoretically removes 470.6 mg. l^{-1} COD by oxidation [14]. In this study different initial concentrations ranging from 2500 mg.L^{-1} to 19000 mg L^{-1} (H_2O_2/COD mass ratios= 0.44 to 3.22), were investigated in order to determine the optimum ratio for the treatment of initial COD leachate. pH was kept at 3 and constant Fe^{2+} dosage of 2000 mg.L^{-1} (H_2O_2/Fe^{2+}mass ratio = 0.35),. Aliquot of the sample was taken regularly from the reaction in order to determine the residual COD.

Fig. 3 shows the effect of initial H_2O_2 concentration on the final COD removal efficiency. It was observed that COD removal increased from 55% to 82% when H_2O_2 dosage increased up to 18214 mg.L^{-1}. This may be due to the fact that increased H_2O_2 dosage produces more $^{\cdot}OH$ radicals leading to higher substrate degradation [14]. H_2O_2 dosage from 7285 mg. L^{-1} to 18214 mg.L^{-1} did not enhance significantly the yield of the process. This may be due to scavenging of •OH radicals by H_2O_2 as in the following reaction [15].

$$^{\cdot}OH + H_2O_2 \rightarrow HO^{\cdot}_2 + H_2O$$

This reaction leads to the production of hydroperoxyl radical ($HO^{\cdot}_2$), a species with much weaker oxidizing power compared to $^{\cdot}OH$ radical [16]. Besides, an excess amount of H_2O_2 can cause the auto decomposition of H_2O_2 to water and oxygen, and the recombination of $^{\cdot}OH$ radicals as in the following reactions [17], thereby decreasing the concentration of $^{\cdot}OH$ radicals and reducing degradation efficiency.

$$2\,H_2O_2 \rightarrow 2\,H_2O + O_2$$

$$^{\cdot}OH + {}^{\cdot}OH \rightarrow H_2O_2$$

The optimal concentrationof H_2O_2 was obtained as 7285 mg. l^{-1} for the Fenton process, and photoFenton processes. Thus a concentration of H_2O_2 of 7285 mg. l^{-1} was selected in further experiments.

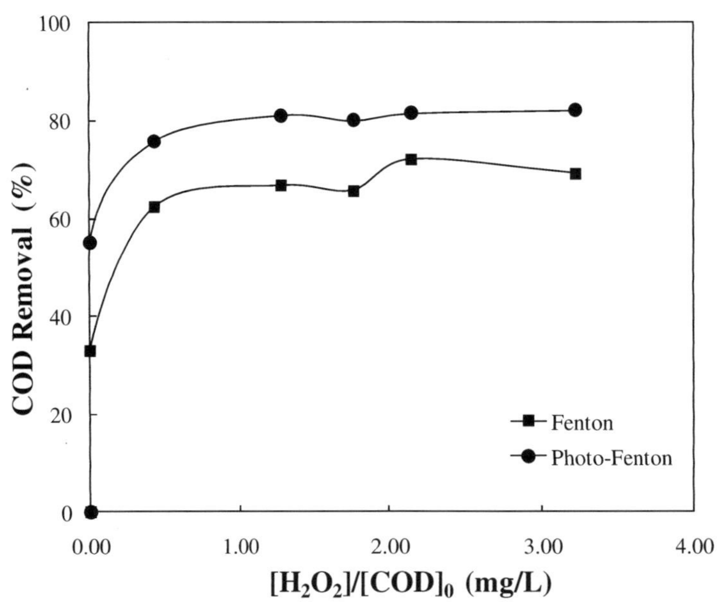

Figure 4. Effect of H_2O_2 concentration on %COD removal by Fenton and Photo-Fenton reaction ($[COD]_0$= 5653 $mgO_2.L^{-1}$, $[Fe^{2+}]$= 2000 mg.L^{-1}, pH= 3 and H_2O_2/COD mass ratios= 0.44 to 3.22).

IOP Conf. Series: Earth and Environmental Science **161** (2018) 012025 doi:10.1088/1755-1315/161/1/012025

3.2.2. Effect of Fe^{2+} concentration

In order to investigate the effect of ferrous ion concentration on Fenton and Photo-Fenton treatment of leachate of COD_0= 5653 mgO_2/L, H_2O_2 concentration was kept constant at 7285 mg. l^{-1} (H_2O_2/COD weight ratio = 1.29) and pH= 3, while ferrous ion concentration has been changed from 200 to 3000 mgO_2/L which corresponds to Fe^{2+}/ COD mass rations 0.04 and 0.53, respectively.
Fig. 4 shows the effect of initial Fe^{2+} concentration on the final COD removal efficiency.

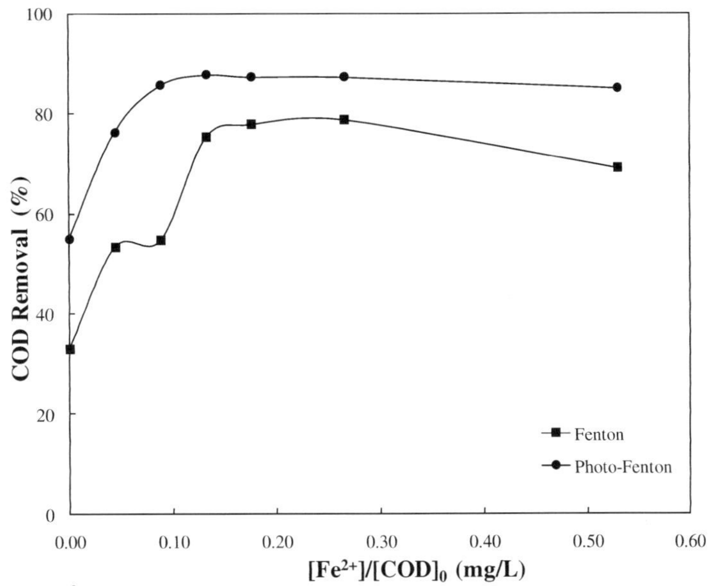

Figure 5. Effect of Fe^{2+} concentration on %COD removal by Fenton and Photo-Fenton reaction ($[COD]_0$= 5653 mgO_2/L, $[H_2O_2]$= 7285 mg.L^{-1}, pH= 3 and Fe^{2+}]/COD mass ratios= 0.04 to 0.53).

Fig. 4 shows that an increase in Fe^{2+}/COD ratio over 0.13 did not lead to a significant improvement of final COD removals. Increasing Fe^{2+}/COD ratio above 0.33 did not enhance the efficiency or rate of the oxidation process. Moreover, when higher Fe^{2+}/COD mass ratios were used, several side reactions would occur decreasing the yield of the process [18-19]. The maximum COD reduction achieved in the Fenton and photo-Fenton processes were 78 and 87 respectively.

4. Conclusion

This study reports the comparison of advanced oxidation processes, photo-Fenton, Fenton and UV-A/ H_2O_2, for removal of landfill leachates. UV photolysis required high time of irradiation and it did not result in a satisfactory COD removal (<13%). UV/H_2O_2 treatment increased considerably the reduction achieved by UV light alone (32%) and 60 % of colour removal was obtained. With Fenton (Fe^{2+}/H_2O_2) process, result is were obtained with final COD reduction of 58%.
The photo-Fenton process (UV/Fe^{2+}/H_2O_2) was found to be the most efficient treatment, combining the use of iron as catalyst and hydrogen peroxide in presence of UV radiation, for reduction of COD (84%) in landfill leachates compared to the other processes investigated (UV/Fe^{2+}/H_2O_2 > Fe^{2+}/H_2O_2 > UV-A/ H_2O_2 > UV-A).
An additional experimental analysis was carried out to find the most favorable operation conditions in the photo-Fenton process. Dosing 7285 mg/LH_2O_2 and using 2000 mg/L Fe^{2+}leading to a 81% COD reduction at the end of the treatment process.

References

[1] Li W, Hua T, Zhou Q X, Zhang S G and Li F X 2010 *Desalination* **264** 56-62
[2] Kurniawan T A, Lo WH and Chan G Y S 2006 *J. Hazard. Mater.* **129** 80–100
[3] Chan G Y S, Chang J, Kurniawan T A, Fu C X, Jiang H and Je Y 2006 *Desalination* **202** 310–317
[4] Zin N S M, Abdul Aziz H, Adlan M N, Ariffin A, Yusoff MS and Dahlan I 2015 *Desal. Water Treat.* **54(11)** 2951–2958
[5] Tatsi A A, Zouboulis A I, Matis K A and Samaras P 2003 *Chemosphere* **53** (2003) 737–744
[6] Chaouki Z, El Mrabet I, Khalil F, Ijjaali M, Rafqah S, Anouar S, Nawdali M, Valdés H and Zaitan H 2017 *J. Mat. Environ. Sci.* **8(8)** 2781-2791
[7] Rivas F J, Beltran F, Gimeno O, Acedo B and Carvalho F 2003 *Water Res.,* **37** 4823–4834
[8] Chaouki Z, Khalil F, Ijjaali M, Valdés H, Rafqah S, Sarakha M and Zaitan H 2017 *Desal. Water Treat* **83** 262–271
[9] Monje-Ramirez I and Orta de Velasquez M T 2004 *Water Res.* **38** 2359–2367
[10] Renou S, Poulain S, Givaudan J G and Moulin P 2008 *J. Hazard. Mater.* **313** 9–22
[11] Standard methods for the examination of water and wastewater. Washington, DC: American Public Health Association (APHA), 2012.
[12] Tizaoui C, Bouselmi L, Mansouri L, Ghrabi A 2006 *J. Hazard. Mat.* **140** 316–324.
[13] Kavitha V and Palanivelu K 2004 *Chemosphere* **55** 1235–1243
[14] Deng Y and Englehardt J D 2006 *Wat. Res.* **40** 3683-3694
[15] Andreozzi R, Canterino M, Marotta R and Paxeus N 2005 *J. Hazard. Mater.* **122** 243-250.
[16] Ting W P, Lu MC and Huang Y H 2008 *J. Hazard. Mater.* **161** 1484-1490.
[17] Mandal T, Maity S, Dasgupta D and Datta S 2010 *Desalination* **250** 87-94
[18] Zhang H, Choi H J and Huang C P 2005 *J. Hazard. Mater.* **B125** 166–174.
[19] Lopez A, Pagano M, Volpe A and Di Pinto C 2004 *Chemosphere* **54** 1005–1010.

Acknowledgment

Authors are gratefully for the financial support provided from the Moroccan and Tunisian higher education ministries (Project 17/TM 12).

Modelling and simulation of a wind system using variable wind regimes with Backstepping control of DFIG

M El Ghamrasni[1*], H Mahmoudi[1], B Bossoufi[2]

[1] Electronic of Power and Control Team (EPCT), Department of Electrical Engineering Mohammadia Engineering School Mohammed V University, Rabat –Morocco

[2] Laboratory of Electrical Engineering and Maintenance, Higher School of Technology, EST-Oujda University of First Mohammed, Morocco.

*madihaelghamrasni@research.emi.ac.ma

Abstract. Nowadays wind turbine power is considered as the most reliable solution for power generation to complement the other power sources. While the majority of the established wind turbines have a constant speed, the use of wind turbines equipped with a variable speed is increasing. The Asynchronous Generator is a machine that has shown a great performance and it is currently used in the wind turbine industry. In the present study we propose a model of wind system with doubly Fed Induction Generator using variable wind regimes modelling. We use Matlab / Simulink software to analyse the performance of this system.

Keywords: Wind-Turbine, wind modeling, spectral decomposition, asynchronous generator, Backstepping control, Matlab / Simulink.

1. Introduction

The intense industrialization in recent decades and the increase in the usage of household electrical devices lead to a considerable global electricity need. To cope with these problems, countries are turning progressively towards the use of clean energy such as renewable energy. Among these renewable energy sources, wind power shows the greatest energy potential. Currently, the wind power used in onshore wind farms is a system based on asynchronous generator.

The variation of the operating regime of an autonomous wind system is caused by two factors:

- The variation of the system load
- The variation of the wind speed

That is why it is important to analyze wind speed properties.

There are two major families of wind turbines: the vertical and the horizontal axis wind turbines. Due to economical reasons that are related to their manufacture and installation, the horizontal axis wind turbines have been widely used compared to vertical axis wind turbines. It has been shown that the doubly Fed Induction Generator (DFIG) is a machine that has excellent performance and is commonly used in the wind turbine industry [1].

There are several advantages to using a DFIG for a variable speed wind turbine, including minimal exhaustion on mechanical parts, noise reduction and the possibility to control active and reactive power [2].

The performance and energy production depend not only on the GADA generator, but also on way of which the both parts of "back-to-back" converter are controlled.

In order to control the speed of DFIG with a specific detection model, we have designed a nonlinear Backstepping control schematic. The asymptotic stability of the resulting closed loop system is guaranteed upon the use of the Lyapunov stability theorem. Due to the complexity and the diversity of the electric control devices, it is difficult to define a general structure for such systems. However, if we consider the elements most commonly encountered in these systems, it is possible to define a general structure of an electric control device.

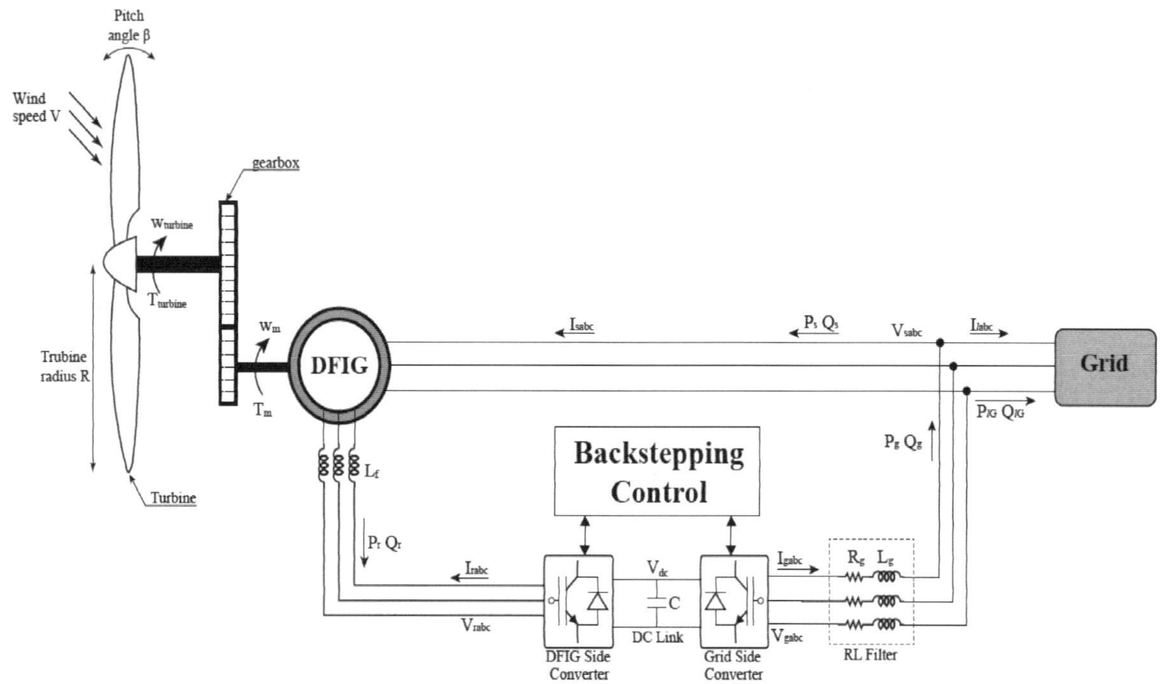

Figure 1. Architecture of the control

2. Wind model

One of the necessary conditions to ensure the feasibility of a wind project is that the average power developed by the wind turbines is at least equal to the power demanded by the local network, including the sum of the powers of the receivers supplied with priority.

Therefore, the energy demands imposed on the site, to ensure the feasibility of a project, refer to:

- The wind power density;

- The behavior of the system during random high frequency variations, given by the turbulence component of the wind speed.

The evolution of the wind speed can be modeled from the model established by Van Der Hoven. It contains the spectral components of the long-term and medium-term variations, as well as those due to the turbulence component [3].

We obtain the model of wind speed:

We reduce the temporal scale of wind variation for the smooth running of the simulation. We get the wind speed shown in the figure 2:

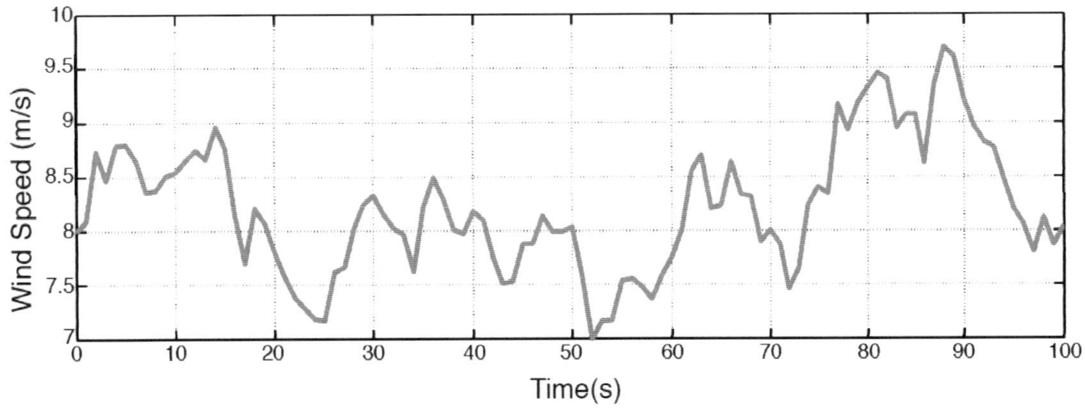

Figure 2. wind speed time scale reduced

3. Wind Turbine Model

We build the model of turbine from the following system of equations [2]:

$$P_{incident} = \frac{1}{2}.\rho.S.v^3 \tag{1}$$

$$P_{extracted} = \frac{1}{2}.\rho.S.C_p(\lambda,\beta).v^3 \tag{2}$$

$$\lambda = \frac{\Omega_t.R}{v} \tag{3}$$

$$C_p^{\max}(\lambda,\beta) = \frac{16}{27} \approx 0.593 \tag{4}$$

$$C_p(\lambda,\beta) = c_1.\left(c_2.\frac{1}{A} - c_3.\beta - c_4\right).e^{-c_5\frac{1}{A}} + c_6.\lambda \tag{5}$$

$$\frac{1}{A} = \frac{1}{\lambda + 0.08.\beta} - \frac{0.035}{1+\beta^3} \tag{6}$$

$$C_{al} = \frac{P_{eol}}{\Omega_t} = \frac{1}{2}.\rho.S.C_p(\lambda,\beta).v^3.\frac{1}{\Omega_t} \tag{7}$$

$$J = \frac{J_{tur}}{G^2} + J_g \tag{8}$$

$$J\frac{d\Omega_{mec}}{dt} = C_{mec} = C_{ar} - C_{em} - f.\Omega_{mec} \tag{9}$$

S: the area swept by the pales of the turbine [m^2]
ρ : the density of the air ($\rho = 1.225 kg / m^3$ at atmospheric pressure).
v :wind speed [m/s].
$C_p(\lambda, \beta)$ is the power coefficient.
λ : the specific speed

β : the angle of orientation of the blades
$P_{extracted}$: the power extracted of the turbine.
Ω_t : Rotational speed of the turbine
C_{al} : Torque on the slow axis (turbine side)
J_{tur} : turbine inertia
J_g : inertia of the generator.
Ω_m : Mechanical speed of DFIG
C_{ar} : Aerodynamic torque on the fast axis of the turbine
$c_1 = 0.5872$, $c_2 = 116$, $c_3 = 0.4$, $c_4 = 5$, $c_5 = 12.5$, $c_6 = 0.0085$

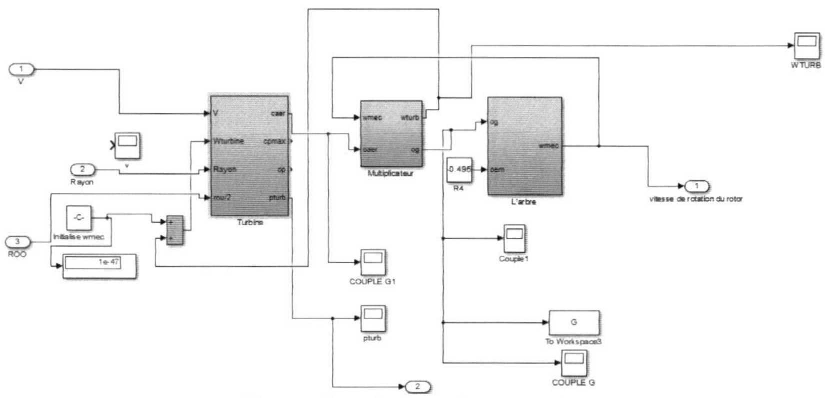

Figure 3. Wind turbine model

4. DFIG Model

We choose a dual feed asynchronous machine in order to allow an optimal production of electricity whatever the conditions of wind and speed of the wind turbine [4].

Using the d-q reference The Power equations DFIG can be written as follows [2]:

$$\begin{cases} v_{ds} = R_s.i_{ds} + \dfrac{d}{dt}\phi_{ds} - \omega_s.\phi_{qs} \\ v_{qs} = R_s.i_{qs} + \dfrac{d}{dt}\phi_{qs} + \omega_s.\phi_{ds} \\ v_{dr} = R_r.i_{dr} + \dfrac{d}{dt}\phi_{dr} - \omega_s.\phi_{qr} \\ v_{qr} = R_r.i_{qr} + \dfrac{d}{dt}\phi_{qr} + \omega_r.\phi_{dr} \end{cases} \tag{10}$$

$$\omega_r = \omega_s - P.\Omega$$

$$\text{With} \begin{cases} \phi_{ds} = L_s.i_{ds} + M.i_{dr} \\ \phi_{qs} = L_s.i_{qs} + M.i_{qr} \\ \phi_{dr} = L_r.i_{dr} + M.i_{ds} \\ \phi_{qr} = L_r.i_{qr} + M.i_{qs} \end{cases} \tag{11}$$

$$L_s = l_s - M_s, \quad L_r = l_r - M_r$$

L_s, L_r: Cyclic inductances of stator and rotor phase;

IOP Conf. Series: Earth and Environmental Science **161** (2018) 012026 doi:10.1088/1755-1315/161/1/012026

l_s, l_r : Inductances of stator and rotor phase;
M_s, M_r: Mutual inductances between stator and rotor phases respectively;
M: Maximum mutual inductance between stator and rotor stage.
The electromagnetic torque of the DFIG is written as:

$$C_{em} = P(\phi_{ds}.i_{qs} - \phi_{qs}.i_{ds}) \tag{12}$$

P is the number of pole pairs of the DFIG.
The active and reactive powers stator and rotor of the DFIG are written as follows:

$$\begin{cases} P_s = v_{ds}.i_{ds} + v_{qs}.i_{qs} \\ Q_s = v_{qs}.i_{ds} - v_{ds}.i_{qs} \\ P_r = v_{dr}.i_{dr} + v_{qr}.i_{qr} \\ Q_r = v_{qr}.i_{dr} - v_{dr}.i_{qr} \end{cases} \tag{13}$$

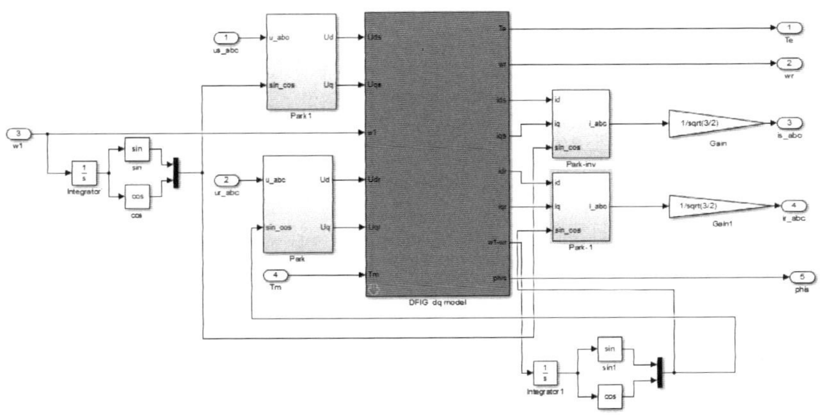

Figure 4. DFIG model

5. Backstepping control

The main objective of controlling the rotor side converter is to operate the wind turbine almost at the nominal power. For this, it is necessary to control the active and reactive powers of stator by calculating and controlling the rotor's voltages. According to the electrical equation of the DFIG, we can calculate the following expressions of the instantaneous variations of the reactive and active powers of the stator [5]:

$$\begin{cases} \dfrac{dP_s}{dt} = \dfrac{3}{2}\left[\dfrac{dV_{sd}}{dt}.I_{sd} + \dfrac{dI_{sd}}{dt}V_{sd} + \dfrac{dV_{sq}}{dt}.I_{sq} + \dfrac{dI_{sq}}{dt}V_{sq}\right] \\ \dfrac{dQ_s}{dt} = \dfrac{3}{2}\left[\dfrac{dV_{sq}}{dt}.I_{sd} + \dfrac{dI_{sd}}{dt}V_{sq} - \dfrac{dV_{sd}}{dt}.I_{sq} - \dfrac{dI_{sq}}{dt}V_{sd}\right] \end{cases} \tag{13}$$

The instantaneous stator current variations can be expressed with the following equations:

$$\begin{cases} \dfrac{dV_{sd}}{dt} = w_s.V_s.Cos(w_s.t) = w_s.V_{sq} \\ \dfrac{dV_{sq}}{dt} = -w_s.V_s.Sin(w_s.t) = -w_s.V_{sd} \end{cases} \tag{14}$$

With: $$\begin{cases} V_{sd} = V_s.Sin\,(w_s.t) \\ V_{sq} = V_s.Sin\left(w_s.t + \pi/2\right) = V_s.Cos\,(w_s.t) \end{cases} \tag{15}$$

So, the study of stability of the system is characterized by the state vector [X] and the control variable [U]:

$$[X] = [P_s \quad Q_s \quad \Omega]^T \text{ : The state vector.} \tag{16}$$
$$[U] = [V_{rd} \quad V_{rq}]^T \text{ : The control variable.}$$

The Lyapunov function is dividing in two steps, one for the control of the speed and the other for the control of the powers.

- Backstepping controller speed

The first step of the Backstepping control defines the error of the state variable by the following calculation:

$$e_\Omega = \Omega_{ref} - \Omega \tag{17}$$

Its derivative gives:

$$\dot{e}_\Omega = \frac{de_\Omega}{dt} = \dot{\Omega}_{ref} - \dot{\Omega} \tag{18}$$

It results:

$$\dot{e}_\Omega = \dot{\Omega}_{ref} - \frac{p}{J.V_s}\varphi_{sq}.P_s + \frac{p}{J.V_s}\varphi_{sd}.Q_s - \frac{1}{J}T_m \tag{19}$$

Subsequently, we define the Lyapunov function by the form:

$$V_1 = \frac{1}{2}\left(e_\Omega^2\right) \tag{20}$$

Its derivative gives:

$$\dot{V}_1 = e_\Omega \dot{e}_\Omega = e_\Omega\left(\dot{\Omega}_{ref} - \frac{p}{J.V_s}\varphi_{sq}.P_s + \frac{p}{J.V_s}\varphi_{sd}.Q_s - \frac{1}{J}T_m\right) = -K_\Omega e_\Omega^2 + e_\Omega\left(K_\Omega e_\Omega - \frac{p}{J.V_s}\varphi_{sq}.P_s + \frac{p}{J.V_s}\varphi_{sd}.Q_s - \frac{1}{J}T_m\right) \tag{21}$$

We use the Backstepping design method to ensure the stability of the sub-system. For this, we need to make the equation (13) negative, we consider the active and reactive powers as virtual inputs of our system and we define the following equations as:

$$\begin{cases} Q_{s\_ref} = Q_s \\ P_{s\_ref} = \dfrac{1}{\left(\dfrac{p}{J.V_s}\right).\varphi_{sq}}\left(K_\Omega e_\Omega + \dfrac{p}{J.V_s}\varphi_{sd}.Q_{s\_ref} - \dfrac{1}{J}T_m\right) \end{cases} \tag{22}$$

With: $K_\Omega > 0$.

We substitute equation (15) in the derivative of the Lyapunov function equation V_1 and we assume that Ω_{ref} is constant. As a result, we have the negativity of the function:

$$\dot{V}_1 = -K_\Omega.e_\Omega^2 \leq 0 \tag{23}$$

- Backstepping Controller power

The objective of the section is to control the active and reactive powers of stator by calculating and controlling the rotor's voltages. For this, we define the following errors:

$$\begin{cases} e_{P_s} = P_{s\_ref} - P_s \\ e_{Q_s} = Q_{s\_ref} - Q_s \end{cases} \tag{24}$$

their derivatives are:

$$\begin{cases} \dot{e}_{P_s} = \dot{P}_{s\_ref} - \dot{P}_s \\ \dot{e}_{Q_s} = \dot{Q}_{s\_ref} - \dot{Q}_s \end{cases} \tag{25}$$

The control's laws of real machine are V_{rd} and V_{rq}. To analyze the stability of this system, we define a new Lyapunov function final V_2 given by the following form:

$$V_2 = \frac{1}{2}\left(e_\Omega^2 + e_{P_s}^2 + e_{Q_s}^2\right) \tag{26}$$

The result of the derivative of the Lyapunov function is:

$$\begin{aligned}\dot{V}_2 =& -K_\Omega e_\Omega^2 - K_{P_s} e_{P_s}^2 - K_{Q_s} e_{Q_s}^2 + e_\Omega\left(K_\Omega e_\Omega \frac{p}{J.V_s}\varphi_{sq}.P_s + \frac{p}{J.V_s}\varphi_{sd}.Q_s - \frac{1}{J}T_m\right) + \\ &+ e_{P_s}\left(K_{P_s} e_{P_s} - \left(\frac{3}{2\sigma L_s.L_r}\right)\left(\frac{2}{3}(R_s.L_r + R_r.L_s)P_s + \right) - \omega_s.Q_s\right) + e_{Q_s}\left(K_{Q_s} e_{Q_s} - \left(\frac{3}{2\sigma L_s.L_r}\right)\left(\frac{2}{3}(R_s.L_r + R_r.L_s)Q_s + \right) + \omega_s.P_s\right)\end{aligned} \tag{27}$$

With: $K_\Omega>0$, $K_{Ps}>0$ and $K_{Qs}>0$.

The expressions of the control voltages V_{rq} and V_{rd} are extracted from equation (21):

$$\begin{cases} V_{rd} = -\dfrac{1}{V_{sd}}\left(\dfrac{1}{M}\left(-\left(\dfrac{2.\sigma.L_s.L_r}{3}\right).(K_{P_s}.e_{P_s} - \omega_s.Q_s)\right)\right) \\ V_{rq} = \dfrac{1}{V_{sd}M}\left(\left(-\left(\dfrac{2.\sigma.L_s.L_r}{3}\right)(K_{Q_s}.e_{Q_s} + \omega_s.P_s)\right)\right) \end{cases} \tag{28}$$

Equation (29) implies the negativity of the Lyapunov function V_2 as following:

$$V_2 = -K_\Omega e_\Omega - K_{P_s} e_{P_s} - K_{Q_s} e_{Q_s} \leq 0 \tag{29}$$

6. Simulation and Results

Applying the backstepping control, we obtain the following results:

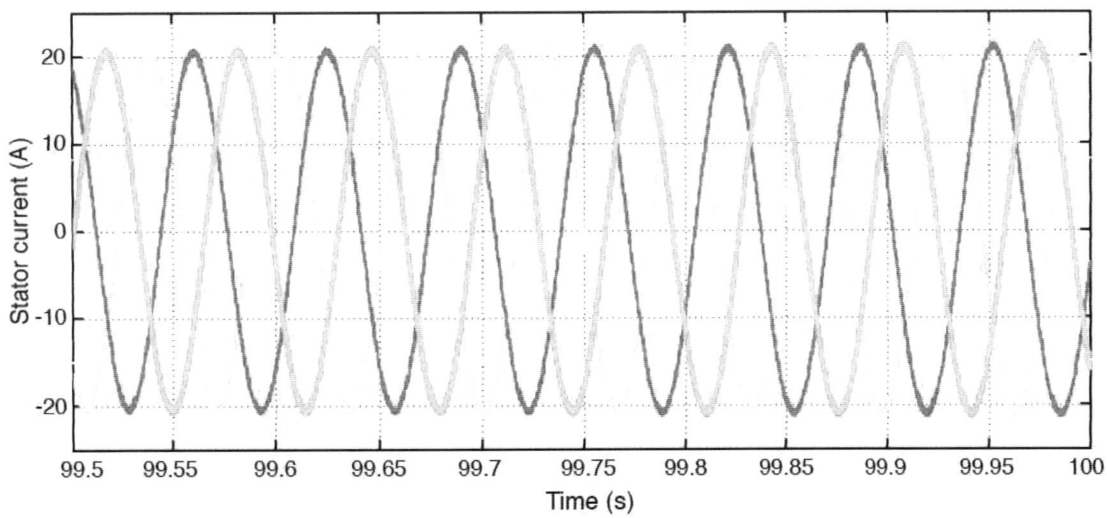

Figure 5. Statoric current of DFIG

IOP Conf. Series: Earth and Environmental Science **161** (2018) 012026 doi:10.1088/1755-1315/161/1/012026

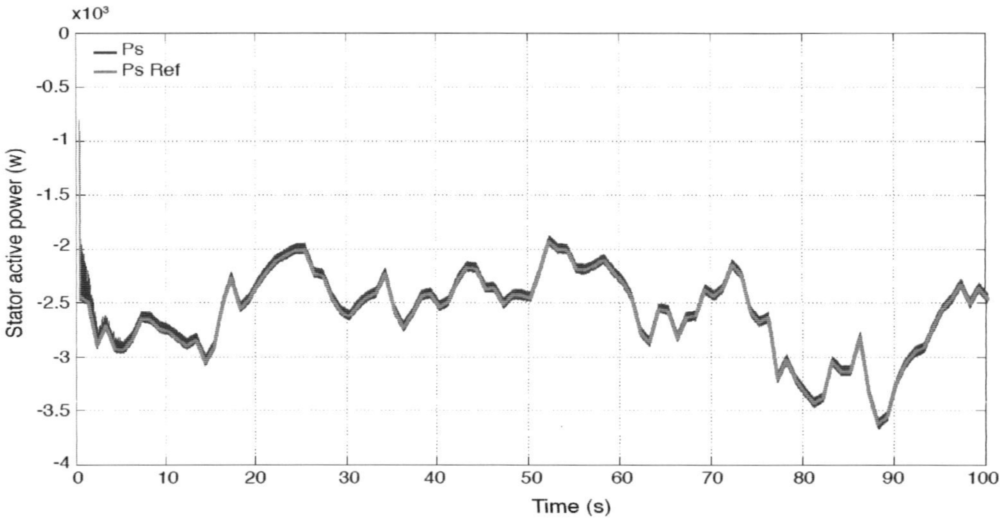

Figure 6. Ps and Ps Ref

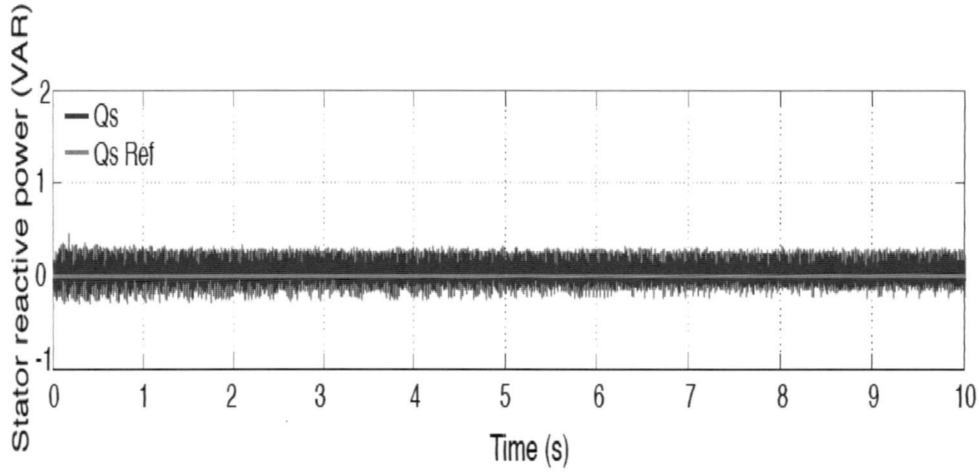

Figure 7. Qs and Qs Ref

The results presented here were obtained from a modeling in MATLAB SIMULINK

In figure 5 we observe that the currents at the outputs of the system are purely sinusoidal with a constant frequency.

We also observe in figure 6 that the active power is negative because it sent to the electrical grid, we notice that the reactive power sent to the grid is near zero in figure 7 , the system is more robust.

7. Conclusion

In this work, we present a model and simulation of a wind turbine system using a variable wind regime. Simulations results are shown interesting obtained control performances of the DFIG.

The simulation of the wind speed was developed with a modeling algorithm realized from measurement files. By breaking it down into a slow and turbulence component.

References:

[1] Mahmoudi H, El Ghamrasni M, Lagroui A and Bossoufi B 2015 *Journal of theorical and appliend information technology* **82** 321.

[2] Bossoufi B, Karim M, Lagrioui A, Taoussi M and Derouich A 2015 *Renewable Energy Journal* **81** 903.

[3] Ackermann T and Soder L 2002 *Renewable and Sustainable Energy Reviews* **6** 67.

[4] Ihedrane Y, El Bekkali C and Bossoufi B 2017 *International Journal of Power Electronics and Drive System* **8** 444.

[5] Bossoufi B, Ionita S, Alami Arroussi H, EL Ghamrasni M and Ihedrane Y 2017 *International Journal of Automation and Control* **11** 451.

[6] Lamnadi M, Trihi M, Bossoufi B and Boulezhar A 2016 *International Journal of Power Electronics and Drive System* **7** 241.

[7] Alami Aroussi H, Ziani El.M and Bossoufi B 2016 *Journal of Theoretical and Applied Information Technology* **83** 426.

[8] Yung-Tsai W 2016 *Journal of Elsevier of Renewable Energy* **94** 383.

[9] Bossoufi B, Karim M, Ionita S and Lagrioui A 2012 *Journal of Journal of Electrical Systems* **8** 236.

[10] Sarwar Kaloi G, Wanga J and Hussain Baloch M 2016 *Journal of Elsevier of Energy Reports* **2** 194.

Development of nickel-cobalt bimetallic/conducting polymer composite used as a catalyst in the oxygen evolution reaction (OER)

Amal ANTAR, Youssef NAIMI and Driss TAKKY

Electrochemistry Team, Laboratory of Physical Chemistry of Materials, Faculty of Sciences Ben M'sik, University of Hassan II Casablanca, Avenue Cdt Driss Elharti, BP 7955, Casablanca, Morocco.

Corresponding authors: antar_amal@yahoo.com ou youssefnaimi@outlook.com

Abstract. This paper discusses results on the development of nickel-Cobalt Bimetallic/conducting polymer composite layer used as a catalyst in the Oxygen evolution reaction (OER). Based on cost, corrosion stability and over-potential, nickel is an advantageous material for the anode in alkaline water electrolysis. However, the use of this metal is associated with two main difficulties: the progressive deactivation of the anode occurs at higher potentials, presumably due to the modification of the active nickel species into inactive ones, and the potential becomes unavoidably increasingly positive over time, thus decreasing the efficiency of the electrode. Nickel-based bimetallic catalysts, such as Ni-Ir, Ni-Co, Ni-Mn, Ni-Fe, usually obtained by electrodeposition, have been widely studied in recent years in the search for catalysts that improve the activity of nickel towards the OER. However, polymeric films were used by several researchers to incorporate various metal complexes, with an aim of improving the activity and stability of electrodes surfaces. Modified electrodes obtained by these polymeric films are characterized by good ionic conductivity and permeability. These as well, we conducted a study of the oxygen evolution reaction in the basic medium on electrodes modified by polymerization of aniline and electrodeposition of (Nickel-Cobalt) bimetallic. In the present work, the electrodeposition technique has been adopted to introduce different amounts of cobalt into the nickel. Surface morphologies of the electrodeposits and their electrocatalytic properties have been studied for the oxygen evolution reaction in NaOH solutions. The Ni-Co bimetallic film is much more reactive than Ni for the OER. Co has a significant effect on the OER. The effect of Co deposited with NiO film depends on its content.

1. Introduction

The world of the scientific research was marked by a renewed interest carried on two types of materials: conducting organic polymers (COP) and inorganic semiconductors [1-2]. The rise in prices of the energy resources, the search for an urban and rural environment durably healthy especially with the pressure of the ecology movements corroborate this orientation. The studies on polymers led to the development of new materials able to replace in certain cases metals, because of their lightness, of their corrosion resistance and their relative facility of manufacturing [3]. Research related to organic polymers became logical and convenient in electrochemistry for manufacturing electrodes [4]. Hydrogen is regarded as one of the most promising energy carriers in the future, among different possible methods; the electrochemical gas evolution during water electrolysis has attracted most of the attention as one of the sustainable and renewable chemical technologies for producing hydrogen. Generally, Water electrolysis is carried out in alkaline medium because alkaline medium creates the possibility for cheap materials to be used as efficient electrodes [5-6]. In alkaline medium. Electrochemical water electrolysis consists of two parts:

The cathodic hydrogen evolution reaction (HER): $2H_2O + 2e^- \rightarrow 2OH^- + H_2$ (1)

The anodic oxygen evolution reaction (OER): $4OH^- \rightarrow 2H_2O + 2e^- + O_2$ (2)

However, the efficiency of water electrolysis in alkaline medium is severely limited by the large anodic overpotential of oxygen evolution reaction (OER). To address this problem, considerable efforts have been devoted to exploring the electrocatalyst with low OER over-potential, and great advances have been archived [7]. Over the past decade, noble metals have been investigated for the OER; Ruthenium (Ru) and iridium (Ir) are the best OER catalysts. However, the price of Ru and Ir is expensive. Therefore, the other noble metals such as Pt, Pd, and Au have been paid attention. It has been commonly accepted that Pt is the best catalyst for the Oxygen reduction reaction ORR, but Pt has an only moderate activity for the OER. Therefore, many researchers have devoted to looking for Pt-based catalysts based on the bifunctional mechanism for water electrolysis [8-9].

Oxidized species of cobalt (Co-O), nickel (Ni-O) and cobalt-nickel (CoNi-O) have been extensively investigated during many years because their interest as electrode materials to favor electrochemical reactions in alkaline media [10–11].
The oxidized species of Co, Ni, and CoNi Bimetallic have been proposed as sensors of biomolecules as glucose [12] or uric acid [13], catalysts for methanol electro-oxidation [14] and electro-synthesis of hydrogen or oxygen [15-16].

The electrodes are the seat of the electrochemical reactions. The choice of catalysts requires the taking into account of several factors:
1) Excellent intrinsic electrocatalytic properties; 2) high activate surface; 3) high electric conductivity; 4) mechanical, chemical and electrochemical stability in the long run; 5) Increased selectivity; 6) less problems caused by the gaseous emission of oxygen evolution reaction (OER); 7) availability and low costs of manufacturing; 8) deprived of harmful effect on health and the environment. All these requirements are met with difficulty in the same material.

2. Water electrolysis technologies

Water electrolyzers' are classified, depending on the kind of electrolyte and thus the type of ionic agent (OH^-, H^+, O^{2-}), and the operation temperature, into three main categories: alkaline, solid oxide and polymer–electrolyte membrane (PEM). The operating principles of the three main types of electrolysis technologies are presented in table 1.

Table 1. Typical characteristics of the main electrolysis technologies [17-20].

	Alkaline electrolysis		Proton Exchange electrolysis	Oxygen ion electrolysis		
	Liquid	Polymer Electrolyte Membrane		Solid Oxide Electrolysis		
	Conventional	Solid alkaline	H^+ -PEM	H^+ -SOE	O^{2-} - SOE	Co-electrolysis
	Low temperature electrolysis			High temperature electrolysis		
▪ charge carrier	OH^-	OH^-	H^+	H^+	O^{2-}	O^{2-}
▪ temperature	20-80°C	20-200°C	20-200°C	500-1000°C	500-1000°C	750-900°C
▪ Electrolyte	Liquid	Solid(polymeric)	Solid(polymeric)	Solid(ceramic)	Solid(ceramic	Solid(ceramic
▪ anodic raction (OER)	$4OH^- \rightarrow 2H_2O+O_2+4e^-$	$4OH^- \rightarrow 2H_2O+O_2+4e^-$	$2H_2O \rightarrow 4H^+ + O_2 + 4e^-$	$2H_2O \rightarrow 4H^+ + O_2 + 4e^-$	$O^{2-} \rightarrow \frac{1}{2}O_2 + 2e^-$	$O^{2-} \rightarrow \frac{1}{2}O_2 + 2e^-$
▪ Anodes	$Ni\rangle Co\rangle Fe\rangle(Oxides)$	Ni-based	$IrO_2 \cdot RUO_2 \cdot Ir_xRU_{1-x}O_2$ $(Supports(TiO_2 \bullet TiC \bullet ITO))$	$La_xSr_{1-x}MnO_3 + ZrO_2$		$La_xSr_{1-x}MnO_3 + ZrO_2$ $CO_2 + 2e^- \rightarrow CO + O^{2-}$
▪ Cathodic reaction (HER	$2H_2O + 4e^- \rightarrow 4OH^- + 2H_2$	$2H_2O + 4e^- \rightarrow 4OH^- + 2H_2$	$4H^+ + 4e^- \rightarrow 2H_2$	$4H^+ + 4e^- \rightarrow 2H_2$	$H_2O + 2e^- \rightarrow H_2 + O^{2-}$	$H_2O + 2e^- \rightarrow H_2 + O^{2-}$
▪ Cathodes	Ni- alloys	$Ni \cdot Ni - Fe \cdot NiFe_2O_4$	Pt/C	Ni	Ni	Ni
▪ Efficien	59-70%		65-82%	Up to100%	Up to 100%	
▪ cyapplicability	Commercia	Laboratory Scale	near-term commercialization	Laboratory Scale	demonstration	Laboratory Scale
▪ Advantages	Low capital cost, relatively stable, mature technology	combination of alcaline and H+ PEM electrolysis	conpact design, fast response/ strat-up ligh-purity H_2	enhanced kinetics thermodynamics lower energy demands, low capital cost	enhanced kinetics thermodynamics lower energy demands, low capital cos	direct production of syngas
▪ Disadvantages	corrosive electrolyte, gas permeation, slow dynamics	Low OH- conductivity in polymeric membranes	High cost polymeric membranes; acidic: noble metals	mechanically unstable electrodes (cracking), safety issues : improper sealing		
▪ Challenges	Improve durability /reliability; and Oxygen Evolution	Improve electrolyte	Reduce noble-metal utilization			

The three types of electrolysis have its own advantages and disadvantages, but all three technologies are appealing and promising for sustainable energy application.

Main advantages of alkaline electrolysis are low cost owing to the use of no-noble electrodes and long-term stability. On the contrary, the acidic environment in PEM electrolyzers prevents the kinetics of the redox reactions and necessitates the use of expensive noble metal catalysts and materials for the bipolar plates. This, together with the high cost of polymeric membranes, is the main limitation for the commercialization of PEM electrolysis in the near term.

Alkaline electrolysis is well established as the most applied commercial technology.
Water electrolysis in liquid alkaline media takes place at temperatures below 100 °C and is the simplest and most mature among the three electrolysis technologies [21].

Electrocatalysts exhibit deactivation after the long-term operation, improved electrocatalytic activity, mainly for the oxygen evolution reaction of which the kinetics are not favored in alkaline media. To meet these needs, research lines follow two main routes the Conductivity and catalysis [22]. The first is related to the conductive polymer (surface, porosity, conductivity) which also exhibit high efficiency and long-term stability. The second route one aims at the development of novel efficient and stable materials, either by doping Ni, Co, or Fe based electrocatalysts with other elements or by investigating alternative materials [23].

3. Experimental

3.1. Electrodes preparation

Before electrodeposition of a Ni-Co/PANI layer, the Glassy Carbon (GC) substrate was carefully prepared by mechanical wet-polishing using grit sandpaper (600 and 1500), followed by thorough rinsing with distilled water and cleaning in an ultrasonic bath for 6–7 min in order to remove polishing residues. Subsequently, the substrate was degreased with ethanol, rinsed with ultrapure water. The Nickel–Cobalt bimetallic hydroxide films were electrodeposited on the pretreated GC substrates in

IOP Conf. Series: Earth and Environmental Science **161** (2018) 012027 doi:10.1088/1755-1315/161/1/012027

nickel (II) nitrate hexahydrate and cobalt (II) nitrate hexahydrate aqueous solutions with composition ratios shown in Table 2. Potassium nitrate (0.1 M) was used as a support electrolyte and as a hydroxyl precursor. All prepared samples with increasing cobalt molar concentration deposition solutions. A three-electrode electrochemical cell, consisting of a pretreated GC, as a working electrode (diameter = 2 mm), a platinum foil counter electrode, and an Ag/AgCl (saturated KCl solution) as a reference electrode. The reference electrode was positioned as close to the surface of the working electrode as possible by means of a Luggin capillary. The electrodeposition experiments were carried out for 90 min. The deposition was carried out at a constant potential of -0.85 V vs. Ag/AgCl reference electrode. Table 1 present the experimental matrix for the synthesis of the Nickel–Cobalt and Nickel/Cobalt bimetallic.

Table 2. Experimental matrix for the synthesis of the nickel-cobalt and nickel/cobalt bimetallic.

Layer	Bath composition		Conditions of deposition
Polyaniline	Aniline, H_2SO_4	0.01 M 0.5 M	Cycles of potential
Ni	$Ni(NO)_3$•$6H_2O$, KNO_3, H_2SO_4	0.01 M 0.1 M 0.5 M	Deposition potential: −0.85VAg/AgCl Deposition time: 90 min
Ni-Co	$Ni(NO)_3$•$6H_2O$ $Co(NO)_3$•$6H_2O$ KNO_3 H_2SO_4	0.009 M 0.001 M 0.1 M 0.5 M	Deposition potential: −0.85VAg/AgCl Deposition time: 90 min
Ni-Co	$Ni(NO)_3$•$6H_2O$ $Co(NO)_3$•$6H_2O$ KNO_3 H_2SO_4	0.005 M 0.005 M 0.1 M 0.5 M	Deposition potential: −0.85VAg/AgCl Deposition time: 90 min
Ni-Co	$Ni(NO)_3$•$6H_2O$ $Co(NO)_3$•$6H_2O$ KNO_3 H_2SO_4	0.002 M 0.008 M 0.1 M 0.5 M	Deposition potential: −0.85VAg/AgCl Deposition time: 90 min

3.2. Morphology of PANI, Ni/PANI, and Ni-Co/PAN
The morphological characteristics of the synthesized Ni–Co bimetallic films and the polyaniline were characterized by scanning electron microscopy (SEM) at the materials Platform of UATRS, CNRST of Rabat, Morocco.

Electrolytically deposited Ni-Co on polyaniline film covered glassy carbon electrode (Ni/PANI/CV)/ (Co/PANI/CV)/ (Ni-Co/PANI/CV) was used as the anode for the evolution oxygen reaction in alkaline medium. The electrochemical behavior and electrocatalytic activity of the electrode studied using cyclic voltammetry and chronoamperometry. The morphology and composition of the modified film were obtained using scanning electron microscope.

Morphology of conducting polymer layers deposited on a glassy carbon substrate was first characterized using a scanning electron microscopy technique.

Figure a-1 shows SEM morphology of PANI film obtained in a solution containing 0.1M aniline and 0.5M H_2SO_4 deposited onto GC. From this figure, the branched PANI fibrils as observed and the fibers are very regular and uniform. Besides, these fibers tend to agglomerate into interconnected networks, which exhibits many multiple pores.

The figure b-1 indicate the SEM morphology of the Ni-polyaniline. As seen there is a quite difference in structural surface morphology between PANI film and PANI film containing Nickel (PANI-NI). Ni particles are clearly seen in PANI. It is to be noted that the incorporation of Ni into PANI does not significantly change the morphology of fiber. Indeed, this expresses the ability of particles to aggregate into larger structures.

Figure c-1 shows an SEM morphology of Nickel-Cobalt Bimetallic/polyaniline. Micrograph of the film containing 50% of Co is very similar to the pure NiOOH films or containing various Co contents Ni film. The Co has a negligible effect on the surface morphology. It can be deposited very homogeneously with Ni in the form of hydroxide.

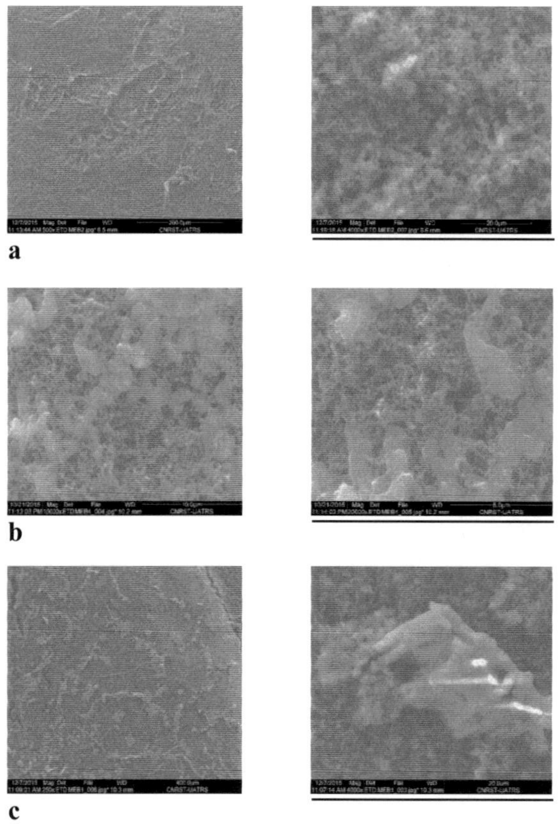

Figure 1. SME images showing a surface morphology of (a) polyaniline (PANI) electrodeposited on a GC substrate, (b) electrodes made by electrodeposition of Ni on a PANI layer on a GC substrate, and (c) electrodes made by electrodeposition of Ni-Co Bimetallic on a PANI layer on a GC substrate.

4. Results and discussion

The process of electrode preparation and modification consisted of two steps:

- Electrodes modified by polymerization of aniline;
- Electrodeposition of (Nickel-Cobalt) bimetallic.

4.1. Electropolymerization

Electropolymerization method such as potential cycling technique offers the possibility of controlling the thickness and homogeneity of the film on the electrode surface. The electrochemical synthesis of polyaniline in the form of emeraldine has been previously described by Park [2]. Electropolymerization method such as potential cycling technique offers the possibility of controlling the thickness and homogeneity of the film on the electrode surface. Figure 2 shows the cyclic voltammograms during the electropolymerization of aniline on the platinum electrode in a solution containing 0.1 M aniline and 0.5 M sulfuric acid for 15 cycles in the potential limits from -0.2 to 1, 2 V at a sweep rate of 50 mV/s. the

peak A obtained at 12 V / (Ag/AgCl) during the first cycle corresponds to the oxidation of aniline. The height of this peak decreases with the number of cycles until its disappearance.
Subsequent cycles (B-C and D) indicate a successive growth of the polyaniline film as evidenced by the increase in the redox current. Two redox couples are readily apparent with E values of 0.25 and 0.74 V in the CVs. Orata and Buttry have proposed that the wave at 0.13 V represents an initial oxidative process in which the polymer is converted from an insulator to a conductor [14]. Upon repeated potential cycling, the peak currents associated with the process around 0.13 and 0.72 V increase in magnitude, indicating the growth of polyaniline film.

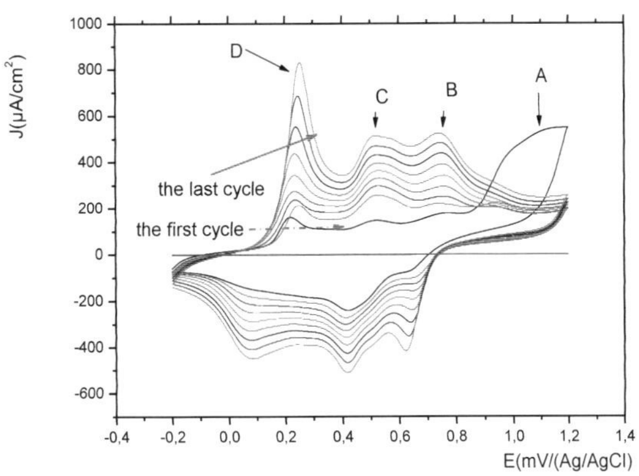

Figure 2. Cyclic voltammograms (9 cycles) showing the electropolymerization of aniline prepared from a 0.5 M sulphuric acid and 0.1 M aniline solution.

4.2. Electrodeposition
The generally accepted mechanism for the OER on nickel anode is written as:

$$OH^- \rightarrow (OH)_{ads} + e^- \quad (3)$$

$$OH^- + (OH)_{ads} \rightarrow (O^-)_{ads} + e^- + H_2O \quad (4)$$

$$2\beta - NiOOH + (O^-)_{ads} \rightarrow 2NiO_2 + H_2O + e^- \quad (5)$$

$$(OH)_{ads} + NiO_2 \rightarrow \beta - NiOOH + (O)_{ads} \quad (6)$$

$$2(O)_{ads} \rightarrow O_2 \quad (7)$$

The electrodes were cycled at 50 mV/s between -3.5 and 6 V/(Ag/AgCl). Figure 3, 4, and 5 show respectively, the voltammogram of Ni-Co/PANI/CV, Co/PANI/CV, and Ni/PANI/CV. The oxidation peak around 380 mV and the reduction peak around 200 mV are attributed to the OER.
The Ni-Co bimetallic film is much more reactive than Ni for the OER. Co has a significant effect on the OER. The effect of Co deposited with NiO film depends on its content. In addition, Co can reduce the peak potential of Ni-Co/PANI and the higher the Co content the more the potential is reduced.

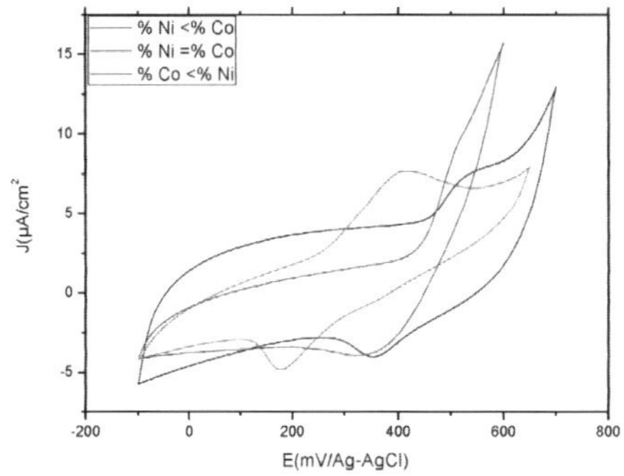

Figure 3. Voltammograms for Ni-Co deposited on the rotating disk electrode glassy carbon coated with a layer of polyaniline (PANI/CV).

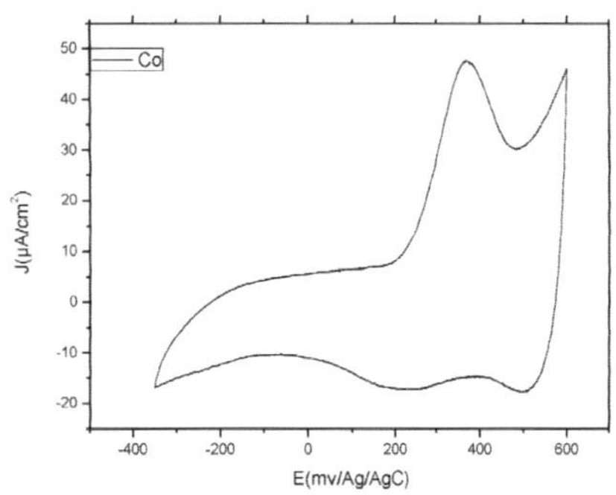

Figure 4. Voltammogram for Cobalt deposited on the rotating disk electrode glassy carbon coated with a layer of polyaniline (PANI/CV)

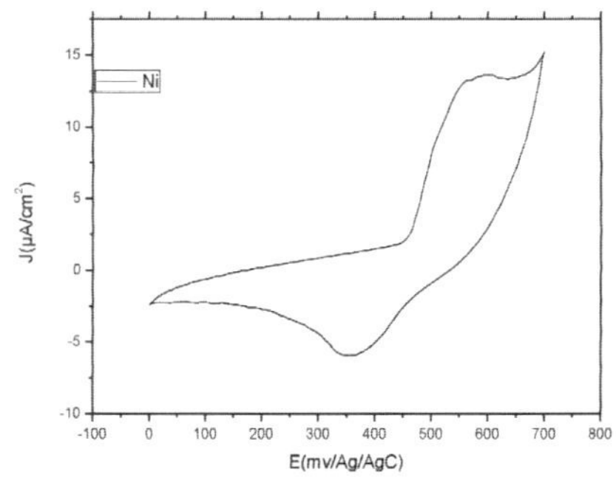

Figure 5. Voltammogram for Nickel deposited on the rotating disk electrode glassy carbon coated with a layer of polyaniline (PANI/CV)

4.3. Curve Koutecky-Levich

For a research on the electrode reaction mechanism and kinetics, particularly those of oxygen reduction reaction (OER), it is necessary to design some tools that could control and determine the reactant transportation near the electrode surface and its effect on the electron-transfer kinetics. The rotating disk electrode (RDE) technique has been widely used for this purpose, particularly for the OER. Limiting currents obey the Levich equation:

$$I_L = 0.62 \times nFSC \cdot D^{2/3} \cdot v^{-1/6} \cdot w^{1/2} \qquad \text{Eq. 1}$$

Or in short form:

$$I_L = B \cdot w^{1/2} \qquad \text{Eq. 2}$$

Where n is the number of electrons transferred, F is Faraday's constant, S the electrode surface, C the concentration of the electroactive substance in the solution, D the diffusion coefficient, ν the kinematic viscosity of the solution, w the rotation rate of RDE and B the Levich slope.

When cathodic limiting currents were plotted vs. w a slightly curved line was obtained, but a plot of I_L^{-1} vs. $w^{-1/2}$ gave a straight line with nonzero intercept (Fig. 4). This plot is known as a Levich-Koutecky plot, based on the equation [24]:

$$\frac{1}{I_L} = \frac{1}{I_{Kin}} + \frac{1}{B \cdot w^{1/2}} \qquad \text{Eq. 3}$$

The nonzero intercept for limiting currents is explained by the presence of kinetic control of the reaction. The relationship between current and rotation rate for a film-covered electrode is the following:

$$\frac{1}{I_L} = \frac{1}{I_{Kin}} + \frac{1}{I_{dif}^{film}} + \frac{1}{B \cdot w^{1/2}} \quad \text{Eq. 4}$$

In accordance with Eq. 3, the dependence of the current density (I) has been plotted in inverse coordinates, as shown in Fig. 4, resulting in a linear interdependence. The extrapolation of this linear dependence to the zero value of $1/w^{1/2}$, to indefinitely high electrode rotation speed, enable to calculate the maximum kinetic current density I_{max}. The thickness of the polymer film is independent of the rate of rotation. The reaction of the evolution of oxygen in medium NaOH (0,1M) is governed by a mixed mode (diffusion and adsorption).

The OER overall current density (I) can be expressed in terms of the kinetic adsorption current density, I_{ads}, and the boundary-layer diffusion limited current density, I_{dif}^{film} by the equation:

$$\frac{1}{I} = \frac{1}{I_{dif}^{film}} + \frac{1}{I_{ads}} \quad \text{Eq. 5}$$

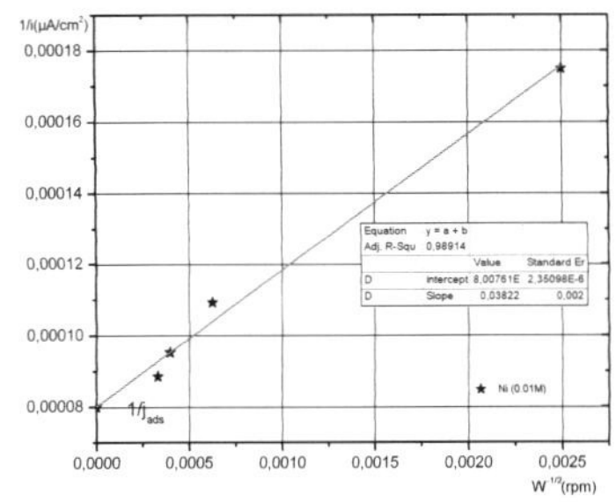

Figure 6. Dependence of the current density (I) with the rotation rate of RDE (w) in inverse coordinates (Levich-Koutecky plot) for the OER on Ni 0.01M electrode.

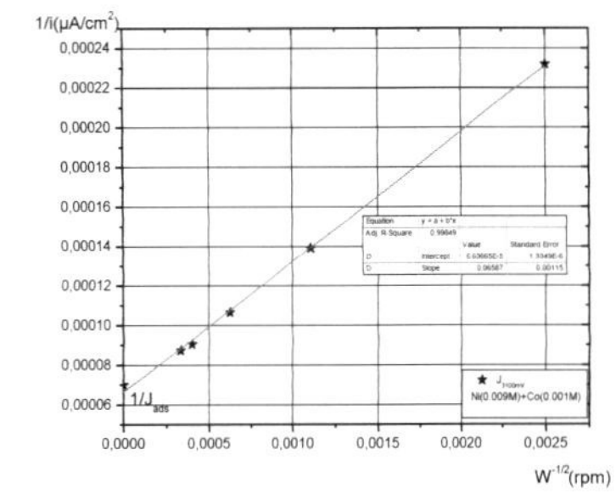

Figure 7. Dependence of the current density (I) with the rotation rate of RDE (w) in inverse coordinates (Levich-Koutecky plot) for the OER on Ni (0.009M) +Co (0.01M) electrode.

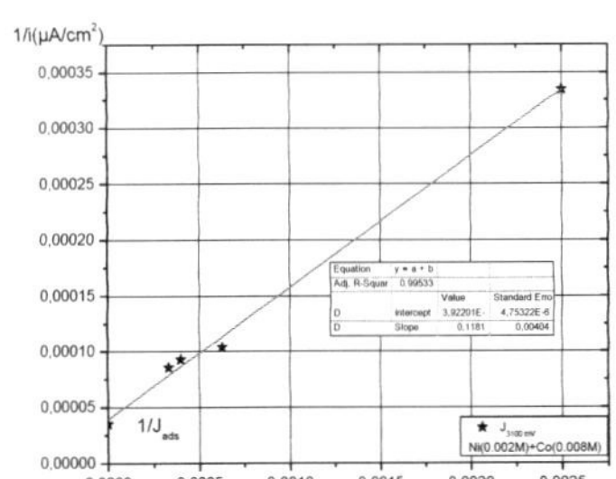

Figure 8. Dependence of the current density (I) with the rotation rate of RDE (w) in inverse coordinates (Levich-Koutecky plot) for the OER on Ni (0.002M) +Co (0.008M) electrode.

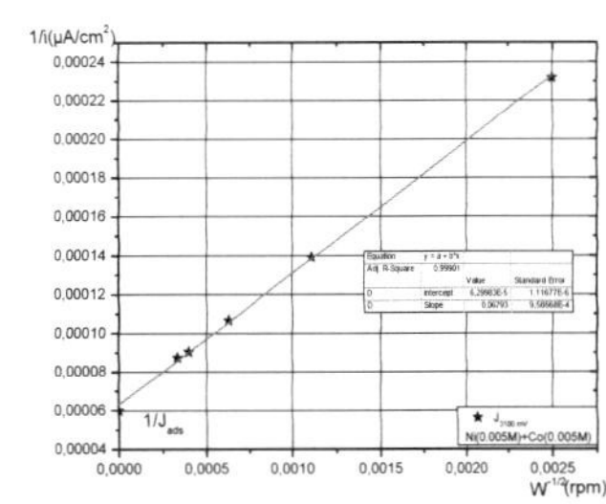

Figure 9. Dependence of the current density (I) with the rotation rate of RDE (w) in inverse coordinates (Levich-Koutecky plot) for the OER on Ni (0.005M) +Co (0.005M) electrode.

To calculate the number of electrons exchanged during the reaction of the evolution of oxygen on the two modified electrodes incorporated into the polyaniline, we used the law Koutecky-Levich equation, which gives the current in the region mixed control $\left(1/_{J_{ads-dif}}\right)$.

5. Conclusions

In this work, we conducted a study of the oxygen evolution reaction in the basic medium on electrodes modified by polymerization of aniline and electrodeposition of (Nickel-Cobalt) bimetallic. The electrodeposition technique has been adopted to introduce different amounts of cobalt into the nickel. Surface morphologies of the electrodeposits and their electrocatalytic properties have been studied for the oxygen evolution reaction in NaOH solutions. Then, the data obtained have been treated according to Koutecky–Levich equation, and the values of a maximum kinetic current density. The present experimental study successfully demonstrates that the results of these tests have led to the conclusion that the catalyst Co in the film has an evident effect on the OER which varies with its content but does not change the mechanism.

Co has a significant effect on the OER. The effect of Co deposited with NiO film depends on its content. The electrocatalytic activity may be related, to the distribution and state of Ni-Co particles incorporated in polyaniline film matrix and the synergistic effects between the dispersed Ni-Co particle and the polyaniline film.

References

[1] Aricò A S, Siracusano S, Briguglio N, Baglio V, Di Blasi A, Antonucci V 2013 *Journal of Applied Electrochemistry* **43** 107–118.
[2] Zhang Q, Wei Z D, Liu C, Liu X, Qi X Q, Chen S G, Ding W, Ma Y, Shi F, Zhou Y M 2012 International Journal of Hydrogen Energy **37** 822–830.
[3] Zeng K, Zhang D 2010 Progress in Energy and Combustion Science **36** 307–326.
[4] Varcoe J R, Atanassov P, Dekel D R, Herring A M, Hickner M A, Kohl P A, Kucernak A R, Mustain W E, Nijmeijer K, Scott K, Xu T, Zhuang L 2014 Energy Environ. Sci. **7** 3135–3191.
[5] Tseung A C C, Jasem S 1977 Electrochimica Acta **22** 31–34.
[6] Tarrús X, Montiel M, Vallés E, Gómez E 2014 International Journal of Hydrogen Energy **39** 6705–6713.
[7] Singh B, Laffir F, Dickinson C, McCormac T, Dempsey E 2011 Electroanalysis **23** 79–89.
[8] Barbir F, 2005 Solar Energy **78** 661–669.
[9] Parrondo J, Arges C G, Niedzwiecki M, Anderson E B, Ayers K E, Ramani V 2014 RSC Advances **4** 9875.
[10] Park S, Shao Y, Liu J, Wang Y 2012 Energy & Environmental Science **5** 9331.
[11] Papadimitriou K D, Geormezi M, Neophytides S G, Kallitsis J K 2013 Journal of Membrane Science **433** 1–9.
[12] Pan J, Chen C, Li Y, Wang L, Tan L, Li G, Tang X, Xiao L, Lu J, Zhuang L 2014 Energy Environ. Sci. **7** 354–360.
[13] Paganin V A , Ticianelli E A, Gonzalez E R, Kumar G S, Raja M, Parthasarathy S 1998 J. Electrochem. Soc. **145** 8.
[14] Orata D, Buttry D A 1987 Journal of the American Chemical Society **109** 3574–3581.
[15] Laguna-Bercero M A 2012 Journal of Power Sources **203** 4–16.
[16] King R L, Botte G G 2011 Journal of Power Sources **196** 2773–2778.
[17] Sapountzi FM, Gracia JM, Weststrate CJ (Kees-J, Fredriksson HOA, Niemantsverdriet JW (Hans) janv 2017. Energy and Combustion Science 58 1 35.
[18] Kanan M W, Nocera D G 2008 Science **321** 1072–1075.
[19] Hauch A, Ebbesen S D, Jensen S H, Mogensen M 2008 Journal of Materials Chemistry **18** 2331.
[20] Grigoriev S A, Kuleshov N V, Grigoriev A S, Millet P, 2015 Journal of Fuel Cell Science and Technology **12** 031004.

[21] Christinelli W A, da Trindade L G, Trench A B, Quintans C S, Paranhos C M, Pereira E C, 2017 Energy **141** 1829–1835.
[22] Carmo M, Fritz D L, Mergel J, Stolten D 2013 International Journal of Hydrogen Energy **38** 4901–4934.
[23] Baranton S, Coutanceau C 2013 Applied Catalysis B : Environmental **136–137** 1–8.
[24] Gojković S Lj, Zecevic S K and Savinell R F 1998 J. Electrochem. Soc. **145** 3713.

IOP Conf. Series: Earth and Environmental Science **161** (2018) 012028 doi:10.1088/1755-1315/161/1/012028

Efficiency of a small wind turbine using BEM and CFD

Y Khalil[1,2], L Tenghiri[1,2], F Abdi[2] and A Bentamy[1]

[1]School of Science and Engineering, Al Akhawayn University, 53000, Ifrane, Morocco.
[2]Laboratory of signals, systems and components, Electrical Engineering Department, Faculty of Science and Technologies P.O. box 2202 Route d'Imouzzer, Fez, Morocco.

E-mail: yassine.khalil@usmba.ac.ma

Abstract. The wind turbine is a mechanism which converts the mechanical energy to an electrical energy. The aerodynamic efficiency of the wind turbine is defined by the kinetic energy captured from the wind. A higher aerodynamic efficiency depends only on the design of the rotor blades. Theoretically, this efficiency (which is known by the power coefficient) is restricted by the Betz-Joukowski limit. Generally, to evaluate the power coefficient of a wind turbine, the Blade Element Momentum theory (BEM) have been used because it is fast and gives accurate results. In this work, a design of a small wind turbine is presented. The evaluation of the power coefficient of this wind turbine is estimated using the BEM theory. In order to minimize time and cost generated by the experimental tests, a computational fluid dynamic (CFD) approach is adopted to estimate the efficiency of the wind turbine. This method simulates the flow around the rotor blades to estimate the pressure and velocity distributions of the airflow, and then the aerodynamic performance. This CFD method can be conducted using many models (inviscid, laminar, k-w model, and Spalart Allmaras). In order to find the best model that should be used, a 2D simulation of the airflow around an airfoil validated experimentally was performed using Fluent. Then, the established methodology would be adopted for a 3D simulation of the airflow around the rotor. The results obtained were satisfactory and accurate compared to the results given by the BEM theory.

1. Introduction

The power output for Wind turbines varies from a few Watts to tens of megawatts. The IEC safety standard for small wind turbines, IEC 61400-2, defines a small turbine as producing a power less than 50 kW [1]. This power corresponds to a rotor swept area less than 200 m^2. Generally, the rotor has three blades in order to perform the trade-off between the cost and the aerodynamic performance. The blade is defined by its radius, the chord length, the twist angle and the aerodynamic airfoil. The design of the wind turbine blade is crucial as it is the key factor of the determination of the aerodynamic efficiency (power coefficient) of the wind turbine [2]. The restrictions on power coefficient is given by the Betz-Joukowski limit (~0.59), but for a realistic wind turbine, its value $C_p < 0.5$ [3]. The evaluation of the power coefficient of a wind turbine have been conducted by different aerodynamic models such the Vortex method, the panel methods [4], the Blade Element Momentum (BEM) [1, 4, 5], and the methods based on the resolution of Navier-Stokes equations [4, 6]. The last two methods are widely used. The first one is known by its accuracy and its fastness, and the second one by its accuracy, but it requires a long computational time, so sophisticated software are required. This paper aims to evaluate the power coefficient of a small wind turbine by these two methods. The BEM theory which divides the blade into a number of sections between 10 and 20 [1], is conducted by a 2D estimation of the airflow over the airfoil. The second method which requires the resolution of the Navier-stokes equations, is performed by a 3D simulation of the airflow around the blade using Ansys-Fluent software. The second method represents a new and interesting approach as it can replace the experimental tests known by their high cost and time. In order to perform a simulation of the computational fluid dynamics (CFD), many parameters should be mastered (geometry, flow domain, mesh quality, and the model adopted to solve the Navier Stokes equations). The later one has a great impact on the accuracy of results. In this paper, four models (inviscid, laminar, k-w model and Spalart Allmaras) are used and compared with experimental values found for a NACA 0012 airfoil. The best model is adopted for a 3D simulation of the airflow around the blades. Finally, the results will be compared to the output of the BEM method.

2. Description of the wind turbine

The wind turbine designed in this study is conducted to cover the need of a house in isolated area. It represents a fixed pitch variable speed (FPVS) with a design tip speed ratio of 6. The diameter of the rotor is 7m, and the chord length and the twist angle distributions (figures (1) and (2)) are approximately linear in order to make the manufacture of the blade easier [7]. The geometry of the blade is designed using the SWRDC (Small Wind turbine Rotor Design Code) tool which is created in the Calgary University. The objective of this tool is the optimization of the design and the efficiency of small wind turbines using Genetic Algorithm and BEM method.

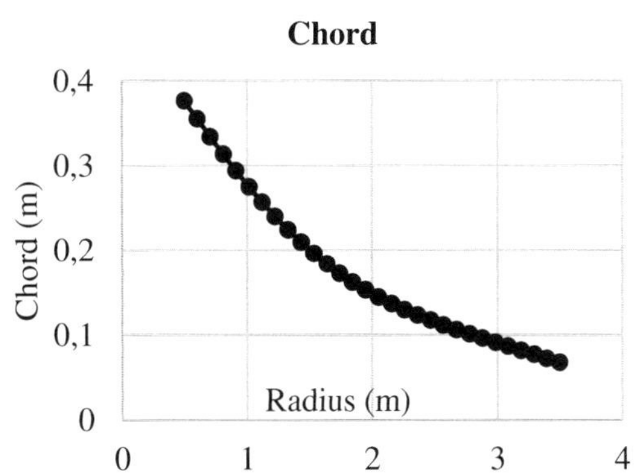

Figure 1. Chord distribution of the blade

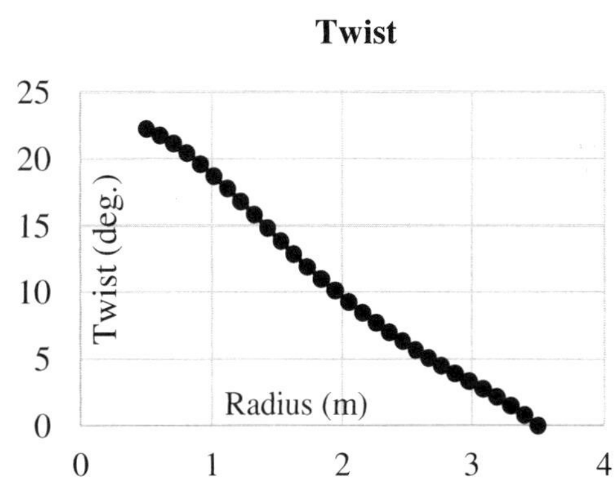

Figure 2. Twist distribution of the blade

3. Methodology and results

The rotor of the wind turbine designed has a diameter of 7 m. The airfoil used is the DU93W210, which is developed in the Delft University in Netherland. The twist and the chord length distributions are given using the SWRDC tool. An estimation of its power coefficient is also given by the SWRDC tool which adopts the BEM theory.

As the BEM theory still a theoretical method and the experimental tests require high cost and time, a computational fluid dynamics (CFD) method was conducted in this work. This method is known by its accuracy, but it requires sophisticated software with specific skills. Besides the software used, other parameters should be chosen correctly. The geometry and meshing should be conducted with respect to some criteria, and the model that simulates the behaviour of the flow should give satisfactory approximation of the reality. In this paper, the simulation of the flow is done using the Ansys-Fluent software which is known by its accuracy to solve Navier-Stokes equations.

3.1. BEM theory

The theory of the blade element momentum combines the method of the blade element and the theorems of conservation of momentum and kinetic energy. The blade element method considers that the blade is divided into N sections. Each section is defined by the radius, the chord length, the twist angle and the airfoil. In the method of blade element, the aerodynamic forces are calculated at each element from the drag and the lift coefficient of the airfoil. The total force is given by the sum of all the elements of a blade multiplied by the number of blades. Figure (3) shows a representation of the aerodynamic forces applied on airfoil. Equations (1) and (2) give the expressions of the thrust and the torque applied to the blade.

$$dT = \frac{1}{2}\rho U_T^2 cN(C_l cos\varphi + C_d sin\varphi)dr \quad (1)$$

$$dQ = \frac{1}{2}\rho U_T^2 cN(C_l sin\varphi - C_d cos\varphi)rdr \quad (2)$$

In addition, the expressions of the thrust and the torque obtained from the conservation of momentum and energy are shown in equations (3) and (4).

$$dT = 4\pi r\rho U_0^2 a(1-a)dr \quad (3)$$

$$dQ = 4\pi r\rho a'(1-a)\Omega U_0 r^3 dr \quad (4)$$

The combination of the blade element method and the conservation of momentum derive to an iterative method from which the power coefficient can be estimated. Figure (4) show the flowchart of this method.

The power coefficient found by applying this iterative method (BEM theory) to the design of the small wind turbine described above is 0.35, and the output power is about 9.5kW. This value should be confirmed by simulating the airflow of the rotor blades.

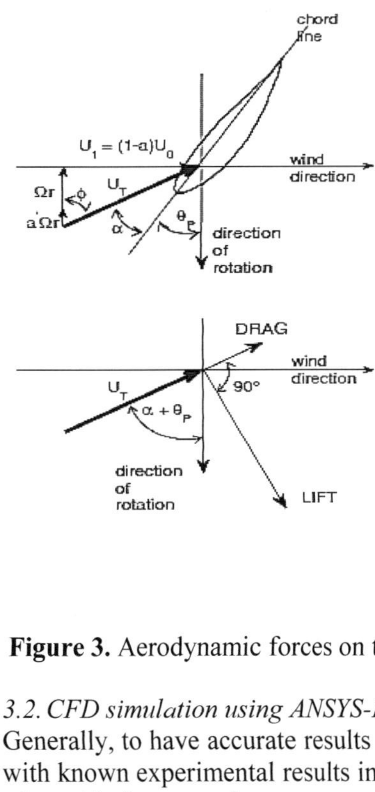

Figure 3. Aerodynamic forces on the airfoil

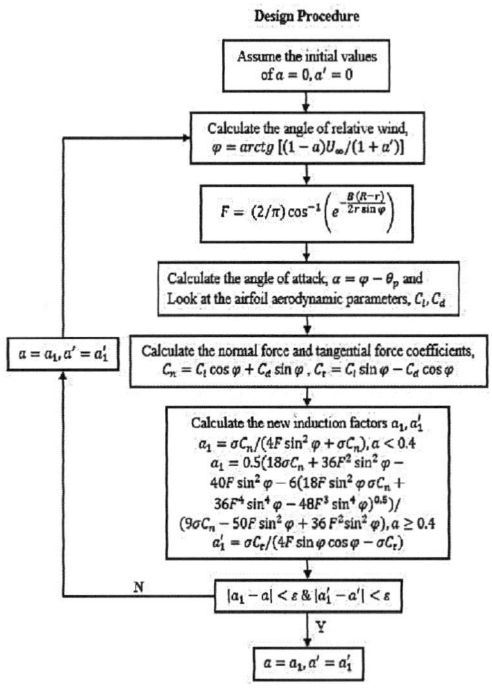

Figure 4. Flowchart of BEM theory

3.2. CFD simulation using ANSYS-Fluent

Generally, to have accurate results using any software, a simulation should be conducted on geometry with known experimental results in order to validate the used approach. Then, the best methodology is adopted in future work.

In this work, the simulation of the airflow is applied in the beginning on an airfoil (NACA 0012) which is widely used, and their aerodynamic performances are known.

First, the geometry of the airfoil is imported to the design modeler in Ansys-Fluent. Second the flow domain (Figure 5) around the airfoil is created respecting the criteria that require high dimensions (the chord length multiplied by 12.5). Then, a Boolean function is applied in order to subtract the airfoil from its domain. In addition, the geometry obtained is meshed (Figure 6), and the parameters of the skewness and the orthogonal quality of the mesh have accepted values.

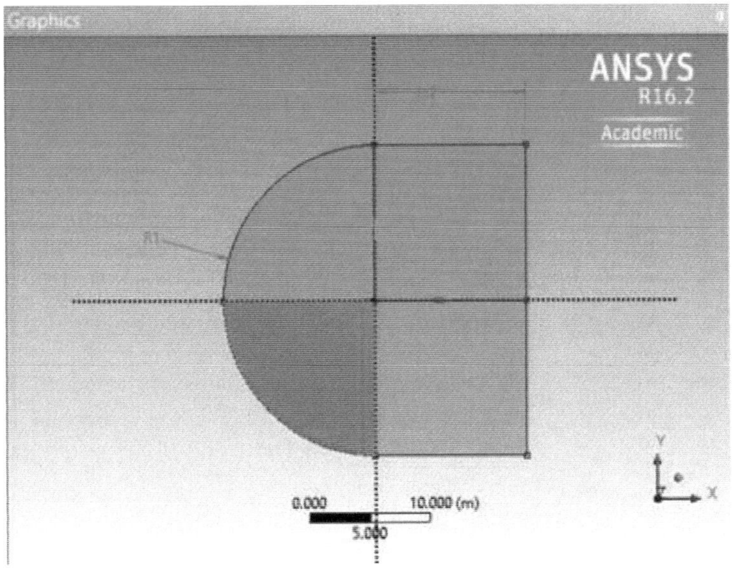

Figure 5. NACA 0012 airfoil and its flow domain

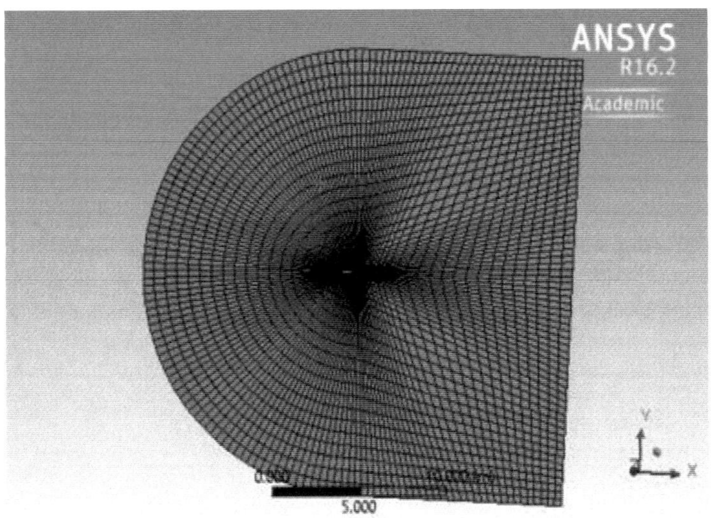

Figure 6. Mesh of the flow domain around the NACA 0012 airfoil

Moreover, many models are compared in this study in order to find the best ones that fit results found experimentally.

First, the inviscid model is adopted. This model adopts the assumptions that the fluid is inviscid and there is no turbulence (perfect fluid). It is clear that this model doesn't simulate the reality. The results here are represented by the lift coefficient which has a value of 0 at an angle of attack of 0 degree (experimentally). Figure 7 shows the results obtained by this model (C_l=7.10-5).

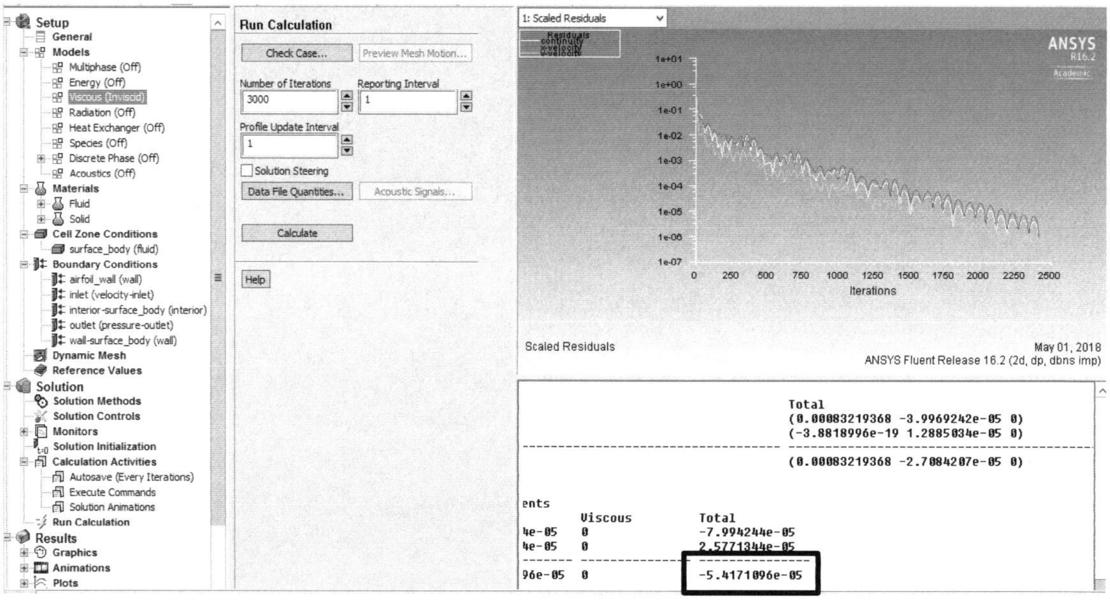

Figure 7. Results obtained by the inviscid model

Second, a laminar model is used. This model takes into account the viscosity of the fluid, but it considers that the flow is laminar (no turbulence). Figure 8 shows the results obtained by this model (C_l= 0.037)

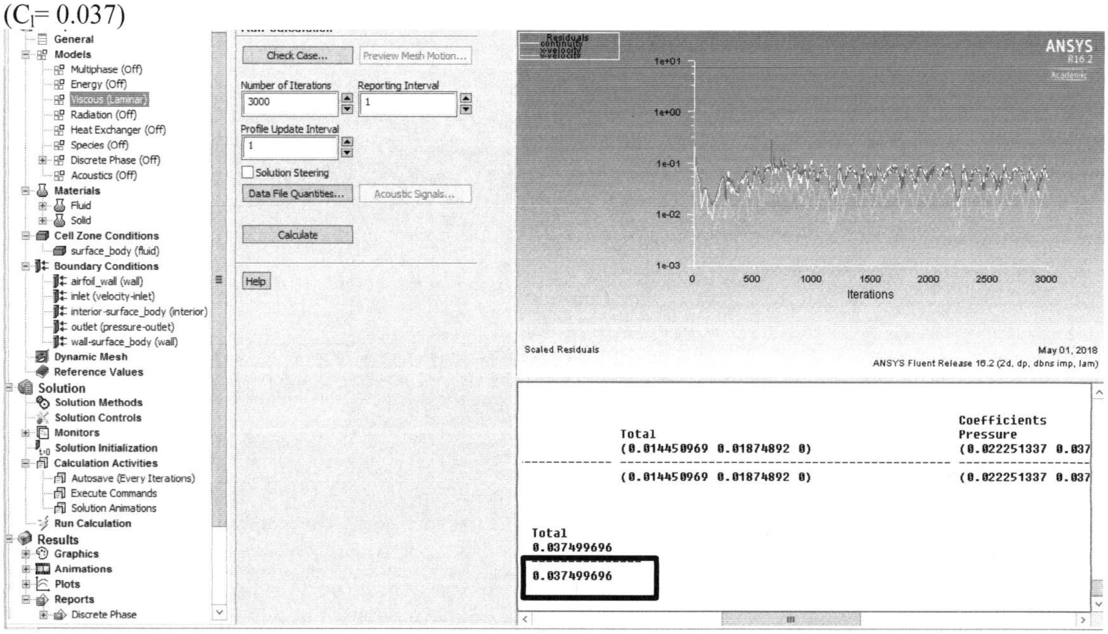

Figure 8. Results obtained by the laminar model

Third, the spalart Allmaras model is also adopted in this study to simulate the airflow around the NACA 0012 airfoil. This model takes into account the viscosity and the turbulence, and it adds one equation to solve the Navier Stokes equations. The model is widely used in the industry, and it gives accurate results. Figure (9) shows the results obtained by this model (C_l= 3.10^{-5}).

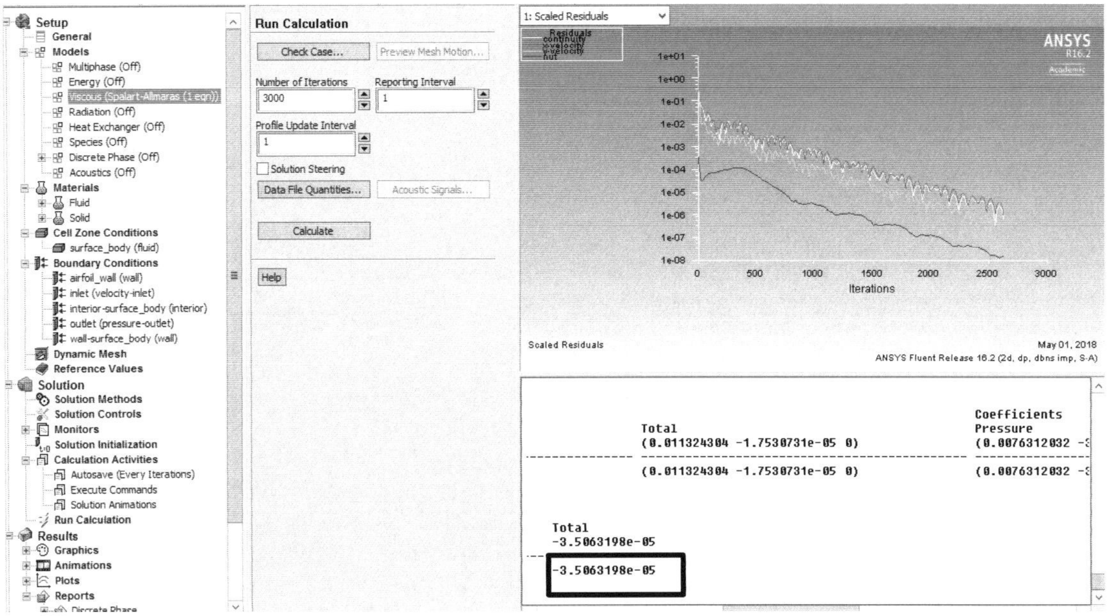

Figure 9. Results obtained by the spalart Allmaras model

The fourth model used here is the k-w model which simulates the turbulence of the flow and the dissipation of the fluid. This model, which adds two equations of the kinetic energy and the dissipation to the Navier Stokes equations, is also very used and accurate [8, 9]. This result is shown in figure 10 ($C_l=10^{-4}$).

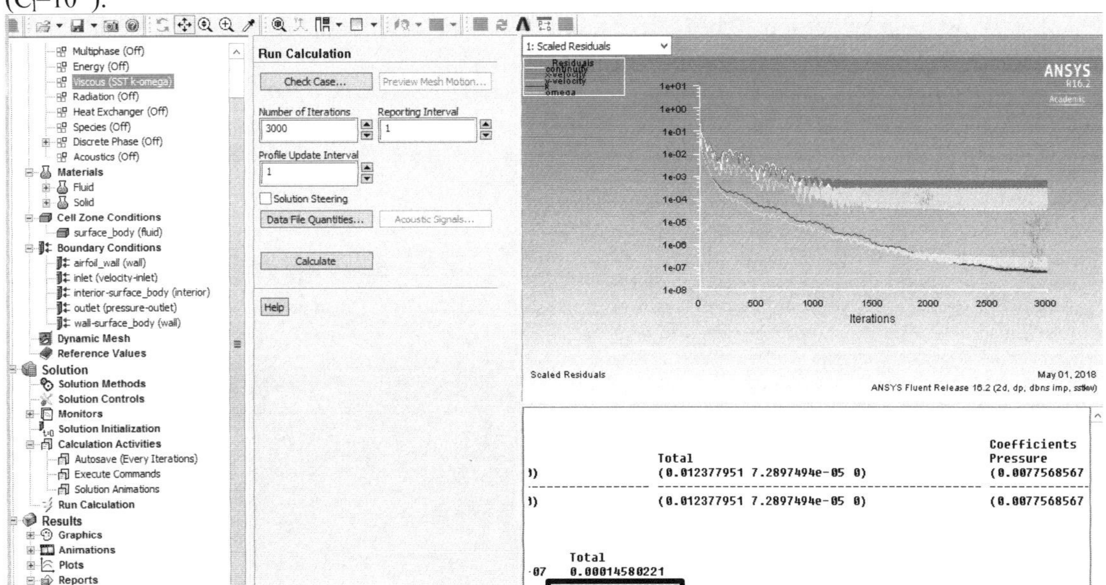

Figure 10. Results obtained by the k-w model

From the results obtained with all models, we can deduce that the two last ones give accurate results because they consider the turbulence of the flow. Therefore, these models would be adopted in the simulation of flow around the rotor especially the flow in this case is more complex.

For doing a CFD simulation using ANSYS software, four steps are conducted. First, a 3-D modeling of the blade is performed using Solidworks software. Second, a flow domain of the blade simulating the airflow is designed in ANSYS as it is shown in figure (11). Third, the geometry of the flow domain is meshed. Figure (12) shows the mesh of this geometry. The flow domain should be large enough, and the mesh should be refined in order to have accurate results [8, 10]. Finally, a model of computational fluid dynamic (CFD) is used to extract the aerodynamic performance of the blade [11].

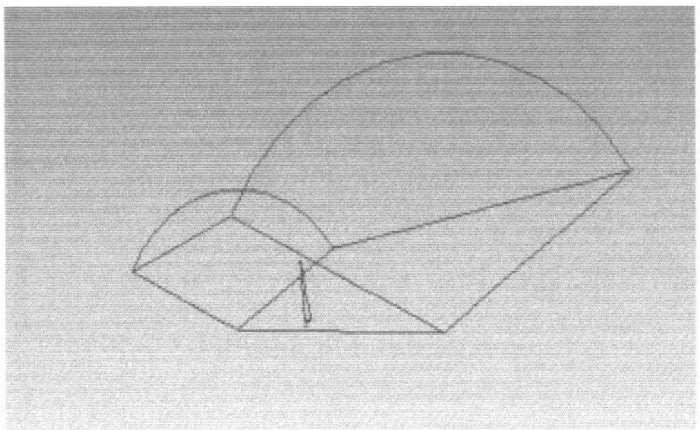

Figure 11. Flow domain of the blade

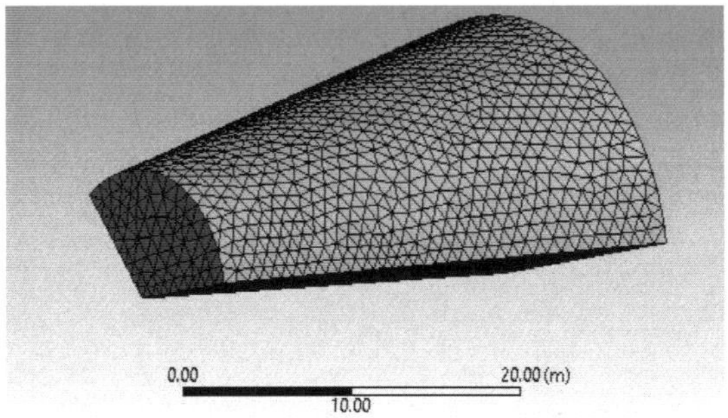

Figure 12. Mesh of the flow domain

After meshing the blade, the geometry is exported to the fluent model in ANSYS. The velocity of the air is taken to be 10.5 m/s, and the rotational velocity of the blade is 18 rad/s (172 RPM).
Although the blades are taken static and not rotating, the flow behind the blades derives vortex behind the rotor as it is shown in figure (7). ANSYS-fluent takes as input the rotational velocity of the rotor which means that the movement of the blades is taken into account in the resolution of the problem. Indeed, the rotational velocity modifies the relative velocity (the velocity of the wind compared to the velocity of the blades), and then the Reynolds Number which modifies the aerodynamic performance of the flow around the blades, Therefore, the static study can give accurate results as those given by a dynamic study which is more complex, and it involves a moving mesh method.
The flow around the rotor of the wind turbines is viscous and turbulent. In order to simulate accurately this flow, a turbulent model should be used. The simulation of the flow in the case of this wind turbine is conducted using the spalart Allmaras and the k-omega models as they are widely used, and they give accurate results for turbulent flows [9]. These models solve the equations of Navier-Stokes equations by an iterative method. When the methods converge, many parameters can be derived such as the velocity, the pressure and the torque applied on the blade. As it is shown in figure (13), the methods converge, and then all parameters can be determined.

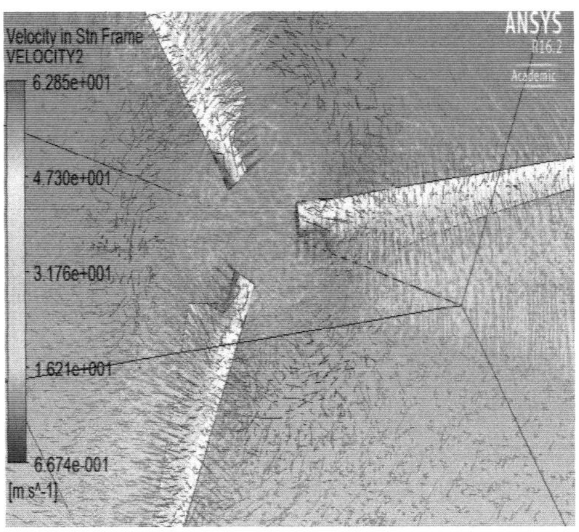

Figure 13. Velocity field on the rotor blades

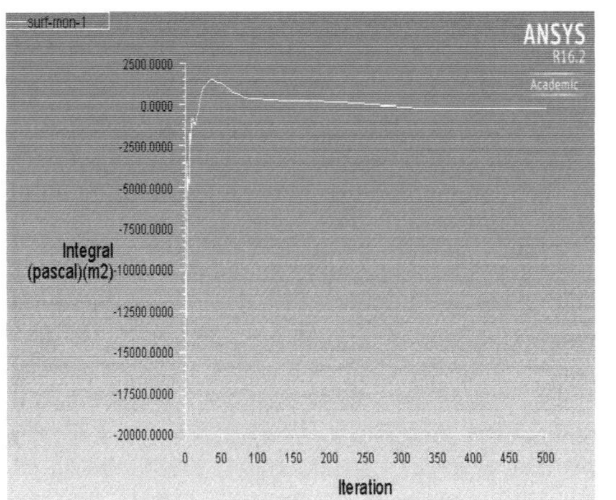

Figure 14. Pressure on the blade along the iterations

The aim of the study is the determination of the power coefficient of the blade. This parameter can be determined using equation (5):

$$C_p = \frac{Q\Omega}{\frac{1}{2}\rho A U_0^3} \tag{5}$$

The torque on the blade around the axis of rotation (OZ) is determined using this CFD simulation on ANSYS-fluent. The value obtained by the k-w model is about 182.2 N.m as it is shown in figure (15). Therefore, the power coefficient of the rotor blades can be calculated according to the equation (5). The value of this coefficient is 0.36, and the output power is 9.8 kW. Whereas, the value obtained by the Spalart Allmaras model is about 179.22 N.m as it is shown in figure (16), and the power coefficient is about 0.35. The power coefficients obtained by both methods are approximately similar, and they give a good estimation compared to the BEM theory.

It should be noticed that, by using software of CFD, the output power depends on many parameters such as the wind speed, the quality of the mesh, and the turbulence model. If one of these parameters is not chosen correctly, the results will be wrong.

The results show that both models used (Spalart Allmaras and k-omega) give approximately the same aerodynamic performance which explains that these methods could be adopted in future works.

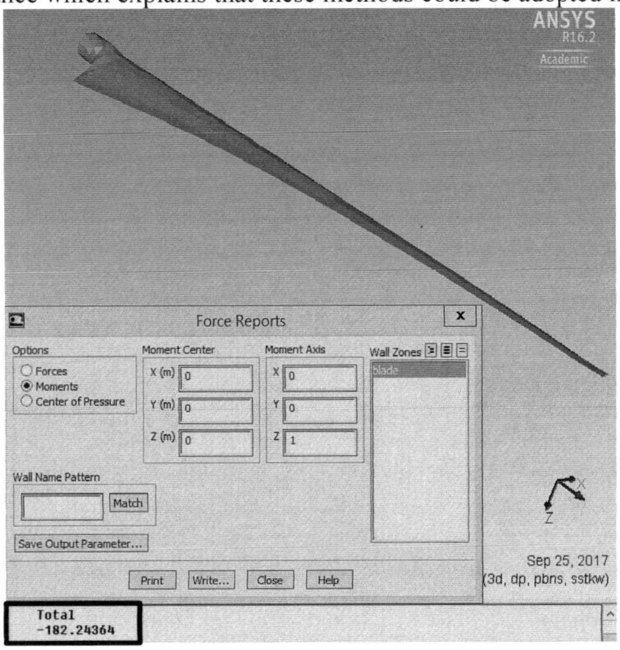

Figure 15. Torque applied on the blade using k-w model

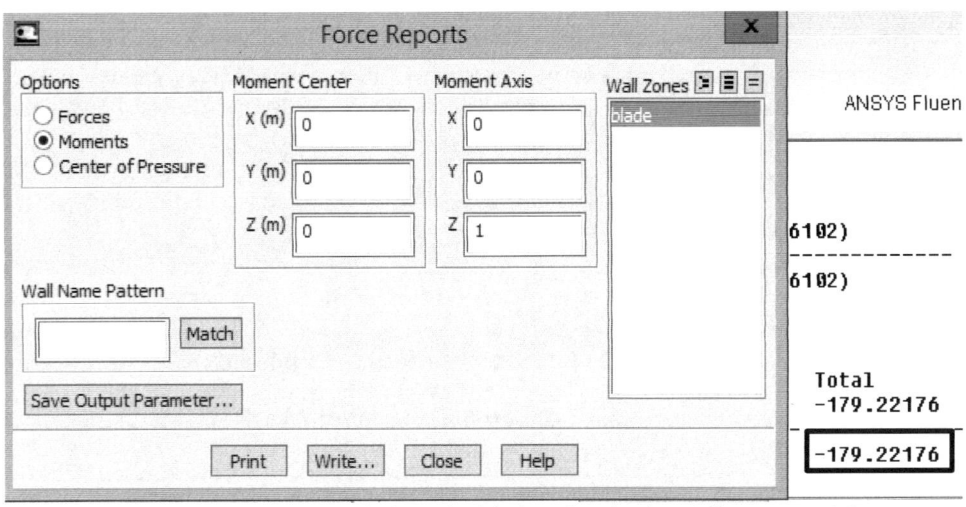

Figure 16. Torque applied on the blade using Spalart Allmaras model

4. Conclusion

This paper targets the estimation of the aerodynamic efficiency of the wind turbines using many models of the computational fluid dynamics (CFD). First, a design of a small wind turbine is conducted using the SWRDC tool which can also determine the power coefficient of the rotor blades by the Blade Element Momentum theory, and the value obtained is 0.35. This method evaluates the power coefficient by estimating the lift and the drag coefficient for the (2D) airfoil. Second, a 2D simulation of the airflow around the NACA 0012 airfoil is performed in order to determine the best model that can be adopted for the 3D simulation of the flow around the blades. Finally, two models that can simulate the turbulence are adopted. Both models give accurate results. For the k-ω model, a power coefficient of 0.35 is obtained, and for the Spalart Allmaras, a value of 0.35 is given. Both results are compared to the BEM method, which shows that the models can be adopted in future works to replace the experimental tests.

Acknowledgments

This work comes within a project of design and manufacture of a small wind turbine (Innowind 2013), which is funded by the Research Institute for Solar Energy and New Energies-IRESEN. We would like to thank them for their support.

References

[1] D. Wood, Small wind turbines. New York: Springer, 2011.
[2] Q. Song and W. David Lubitz, "Design and Testing of a New Small Wind Turbine Blade", Journal of Solar Energy Engineering, vol. 136, no. 3, p. 034502, 2014.
[3] U. Eminoglu and S. Ayasun, "Modeling and Design Optimization of Variable-Speed Wind Turbine Systems", Energies, vol. 7, no. 1, pp. 402-419, 2014.
[4] M. Hansen, J. Sørensen, S. Voutsinas, N. Sørensen and H. Madsen, "State of the art in wind turbine aerodynamics and aeroelasticity", *Progress in Aerospace Sciences*, vol. 42, no. 4, pp. 285-330, 2006.
[5] Manwell, J. & McGowan, J. Wind energy explained: theory, design and application, second eddition. John Wiley & Sons Inc, 2009.
[6] I. Dobrev and F. Massouh, Modèle hybride de surface active pour l'analyse du comportement aérodynamique des rotors éoliens à pales rigides ou déformables, 1st ed. [S.l.]: [s.n.], 2009.
[7] X. Liu, L. Wang and X. Tang, "Optimized linearization of chord and twist angle profiles for fixed-pitch fixed-speed wind turbine blades", Renewable Energy, vol. 57, pp. 111-119, 2013.
[8] S. Corentin, "Optimisation d'une Aile d'Avion à Profil Adaptable : Etude Numérique et Expérimentale, '' M. thesis, département de génie mécanique, Ecole Polytechnique de Montréal, Canada. Dec. 2009.
[9] H. Cao, "Aerodynamics Analysis of Small Horizontal Axis Wind Turbine Blades by Using 2D and 3D CFD Modeling,'' MSc. thesis, School of Computing, Engineering and Physical Sciences, University of Centrale Lancashire, Preston, England, May 2011.
[10] G. Sahu*p*, R.K. Rathore, "Determination of Torque Produced by Horizontal Axis Wind Turbine Blade Using FSI Analysis for Low Wind Speed Regime, '' International Journal of Innovative Science, Engineering & Technology, Vol. 2 Issue 5, May 2015.
[11] "Wind Turbine Blade FSI (Part 1) - Geometry - SimCafe - Dashboard", *Confluence.cornell.edu*, 2017. [Online]. Available: https://confluence.cornell.edu/display/SIMULATION/Wind+Turbine+Blade+FSI+%28Part +1%29+-+ Geometry. [Accessed: 08- Jun- 2017].

IOP Conf. Series: Earth and Environmental Science **161** (2018) 012029 doi:10.1088/1755-1315/161/1/012029

Performance analysis of non-imaging Fresnel lens as a primary stage for CPV units

S. EL HIMER[1,2], A. AHAITOUF[1], S. EL-YAHYAOUI[1], A. MECHAQRANE[1]

[1]Laboratory of Renewable Energies and Intelligent Systems Electrical Engineering department, Faculty of Science and Technology, Sidi Mohamed Ben Abdellah University -Fes PO. Box 2202 Fez Morocco

E-mail: [2]sarah.elhimer@usmba.ac.ma

Abstract. *Fresnel lens is widely used in concentrating photovoltaic (CPV) systems as a primary optical element. It focuses sunlight on small solar cells or on the entrance apertures of secondary optical elements. In this work, we present a performance analysis of a non-imaging Fresnel lens as a primary optical element for different values of focal length. Results show , among others, the necessary use a secondary stage of concentration in order to achieve good performances in terms of the uniformity of the flux and a large acceptance angle compared with those of a Fresnel lens used alone.*

1. Introduction

In recent years, concentration photovoltaic systems have attracted more interest of industrials and researchers working in the field of renewable electricity. These systems needs optical elements to focus direct sunlight onto multi-junction solar cells that have record efficiencies above 44% [1] . There are generally two optical elements in CPV modules: a primary and a secondary elements. The primary is used to collect the sunrays, and the secondary, placed near the receiver, is used to redirect the sunlight onto the cell as well as to improve the energy uniformity.

Two types of optical elements can be used as a primary optical element. Mirrors [2] [3] [4], and Fresnel lenses [5]. The gains that can be achieved with a Fresnel lens or a parabolic mirror were previously compared [6]. Even more, results coming from each way still comparable and the two configurations were developed competitively. In the case of a system using a Fresnel lens, the chromatic aberration phenomenon causes non-uniformity of the concentrated distribution on the solar cell surface [7]. Localized hot spots and non uniformity increases the cell temperature, cell resistance, and lowers the efficiency. Secondary optics helps to solve this problem and can be grouped into reflective and refractive optical elements of various forms. Refractive secondaries are more widespread, they consist of a transparent dielectric (glass or plastic) that redirects the beam at the cell.

In the present work, inspired by the work of Kwangsun Ryu-et-all [8] who analyzed the performances of Fresnel lens only for the low concentrations ratio, we propose to extend the study to the high concentration levels. Deep analysis of the performances of the Fresnel lens highlights the necessity of using a secondary optical element, the pyramid here. The considered

Fresnel lens is assumed to be manufactured from PolyMethyl MethAcrylate (PMMA).
The concerns of discussion are the flux homogeneity on the receiver, the optical efficiency and the acceptance angle. All our results are performed by simulation using the ray tracing TacePro software.

2. Concept and design of Fresnel Lens

For an optical concentration system as shown on fig. 1, the geometrical concentration is given by [9]:

$$C_{geo} = \frac{a^2}{a'^2} = \frac{n_2^2 \sin\theta_f^2}{n_1^2 \sin\theta_s^2} \tag{1}$$

n_1 and n_2 are the refractive index of the entrance and exit medium respectively,a and a' are the entry and the exit area of the concentrator system, θ_s and θ_f are the angles of incoming and outgoing beams respectively. For solar systems, θ_s corresponds to the half angular aperture of the sun.
The Fresnel lens consists of several prism rings [10] arranged as shown in fig.2a.

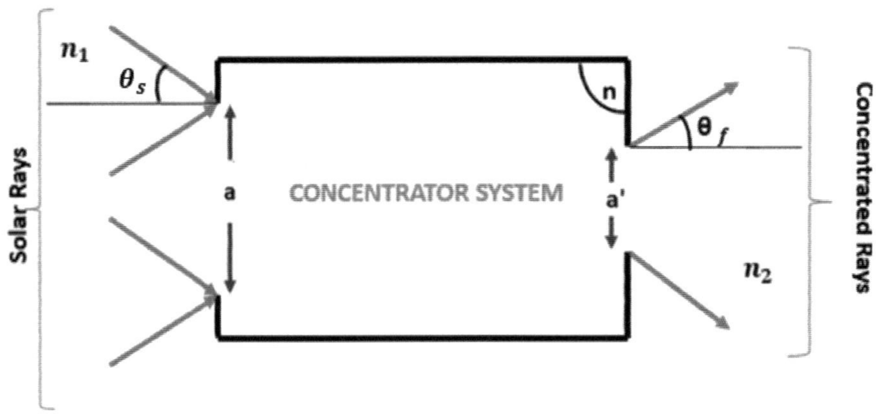

Figure 1: Optical system with an entrance aperture a, a half angle of the incoming radiation θ_s, an output aperture a' and the half exit angle θ_f

Fig.2b shows a single prism a typical the path of a light ray. r_p and r'_p are respectiveley the angles, to the surfaces normal, formed by the refracted ray at the entrance of the prisme and the incident at its exit, β_p is the angle of the refracted ray, θ_s is the incidence solar angle and f is the focal length of the Fresnel lens. Each prism p, of the lens is described by the following equations [11]:

$$\sin\theta_s = n \sin r_p \tag{2}$$

$$n \sin r'_p = \sin\beta_p \tag{3}$$

$$\theta_p = r_p + r'_p \tag{4}$$

$$\beta_p = \theta_a + r'_p \tag{5}$$

$$\tan\theta_a = \frac{d}{2f} \tag{6}$$

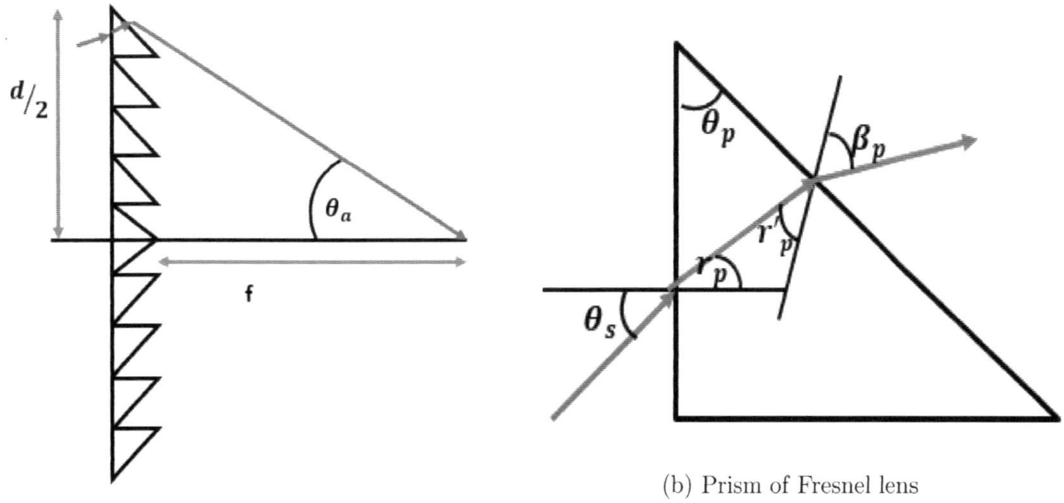

(a) Fresnel lens

(b) Prism of Fresnel lens

Figure 2: Design of Fresnel Lens

From the angle of the prism θ_p,and using the equations 2,3, 4and 5, it is therefore possible to determine the angles of refraction β_p and the opening angle θ_a of the Fresnel lens. Fig.3 shows the variations of both the opening angle of the Fresnel lens and refraction angle of each prism θ_a and β_p as a function of the angle of the prism θ_p for a refractive index $n_f = 1.493$. This figure illustrates that the incidence angles of the Fresnel lens is limited to 45° as a function of the prism angle. This is due to the reflection which undergo the rays once the refraction angle of each prism 90° is reached.

For an accurate simulation, refractive index data, $n(\lambda)$, versus wavelength λ is desirable. in

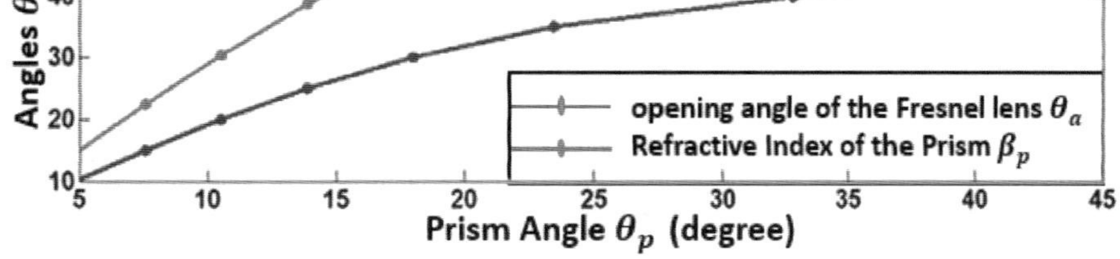

Figure 3: Opening angle θ_a and refraction angle β_p as a function of the angle of the prism θ_p for a refractive index $n_f = 1.493$

the range or interest, in this case, from 0.4 to 0.7µm. Refractive index values are taken from the model reported in [12] for PMMA. In fig.4, we can see that the refractive index decreases with the increment of the wavelength and we get for $n_f = 1.49$, $\lambda = 0.5\mu m$.

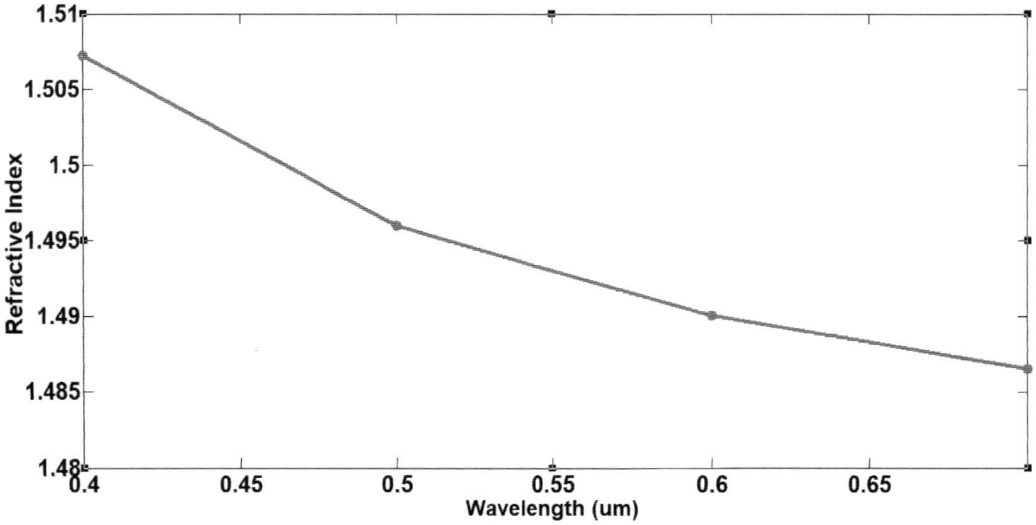

Figure 4: Refractive index characteristic of PMMA used in the simulations

3. Results and Discussions

3.1. Optical Performances of the Fresnel lens

The first test concerns the effect of the focal lenght of the Fresnel lens f, on the flux distribution, the optical efficiency and the acceptance angle. For that, The lens used in this study is a typical a circular Fresnel lens made of PMMA with a diameter of 350mm and the ring facet spacing is 0.38mm. The considered receiver is a squared form of 10mm side, and we varied the focal length of Fresnel lens, made from PMMA, from 175mm to 700mm. Fig. 5 presents the evolution of the flux distribution with the variation of the Fresnel focal length. We notice that the focal length has a big effect on the spot diameter, the spot diameter increases with the increment of the focal length as describe eq.7 with D_s being the spot diameter and λ is the wavelength.
We also notice that the flux distribution can be more uniform if the focal is very long but it will make the CPV system bigger and very wider. This is a new added contrainst to the designed pannel design.

$$D_s = 1.22\frac{\lambda}{\sin\theta_a} \tag{7}$$

As we target a good homogeneity with a higher optical efficiency, we investigate the effect of a secondary stage which is the pyramidal cone in the next part of this work.

3.2. Effect of a secondary stage

The integration of secondary optical elements into a CPV system must allows for higher tolerances in the system and homogenize the irradiance on the solar cell, then for that, we proposed the pyramid as secondary stage made from Fused silica with its aperture exit fixed

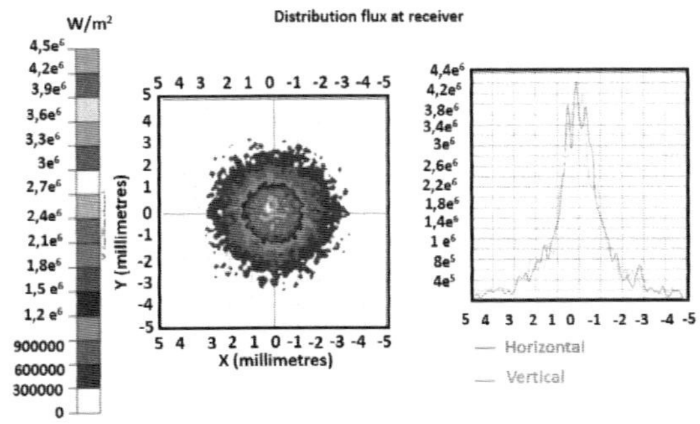

(a) $f = 175$mm

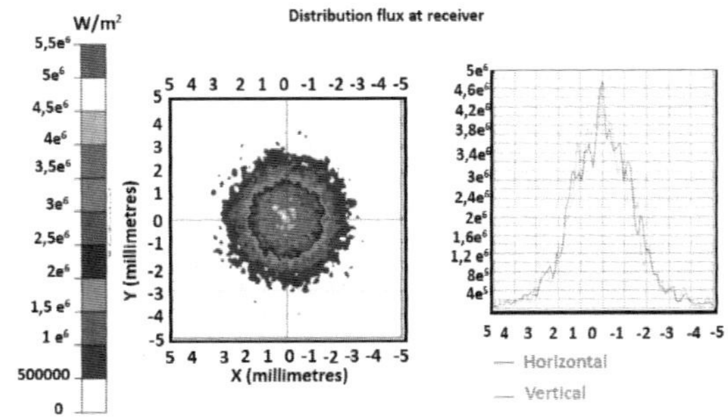

(b) $f = 265$mm

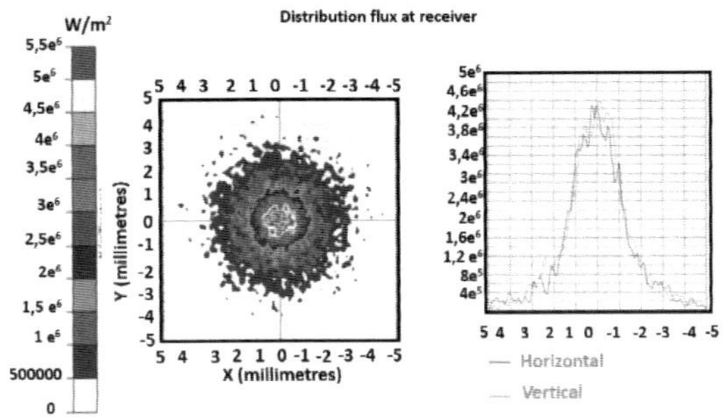

(c) $f = 350$mm

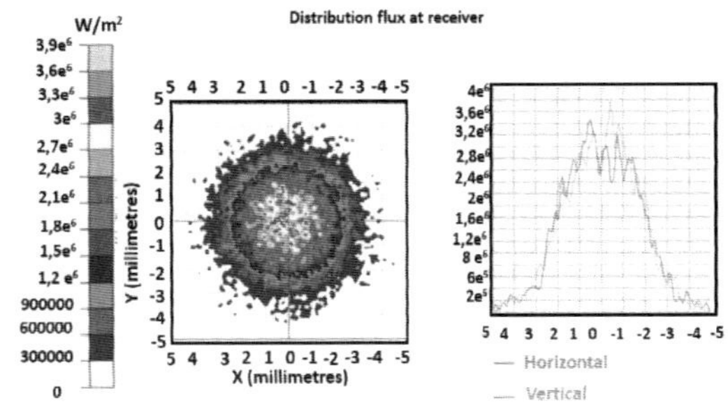

(d) $f = 525$mm

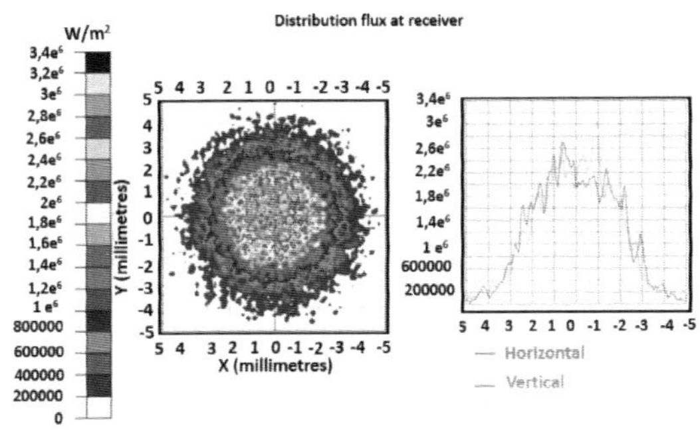

(e) $f = 612$mm

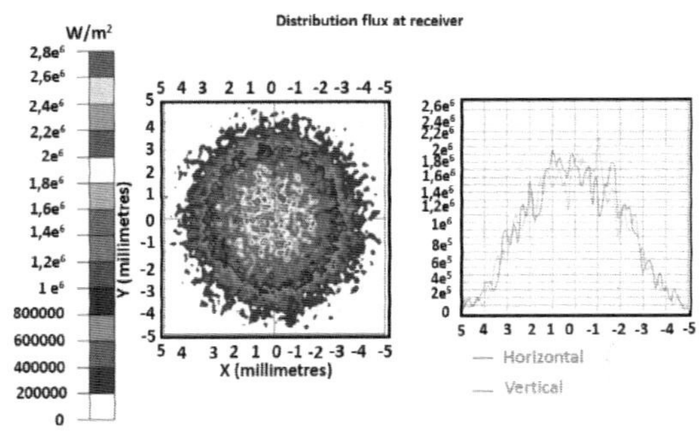

(f) $f = 700$mm

Figure 5: Evolution of flux distribution vs Fresnel lens focal length

to 10mm. This secondary is associated to the same Fresnel lens studied above. The geometry of the pyramid (fig.6) is inspired from its resemblance to the cone; it is defined by the output and input radius r and R, the output and input angle θ_f and θ_a, the angle of pyramid α and its length L [2]. The length of the pyramid depends on the inside number of ray reflections, N. These parameters are dependent and are related by the following equations [2]:

$$\theta_f = \theta_a + 2N\alpha \tag{8}$$

$$R = r + \sum_{1}^{N} a_N \tag{9}$$

$$L = \frac{\sum_1^{N+} a_N}{\tan\alpha} = \frac{R - r}{\tan\alpha} \tag{10}$$

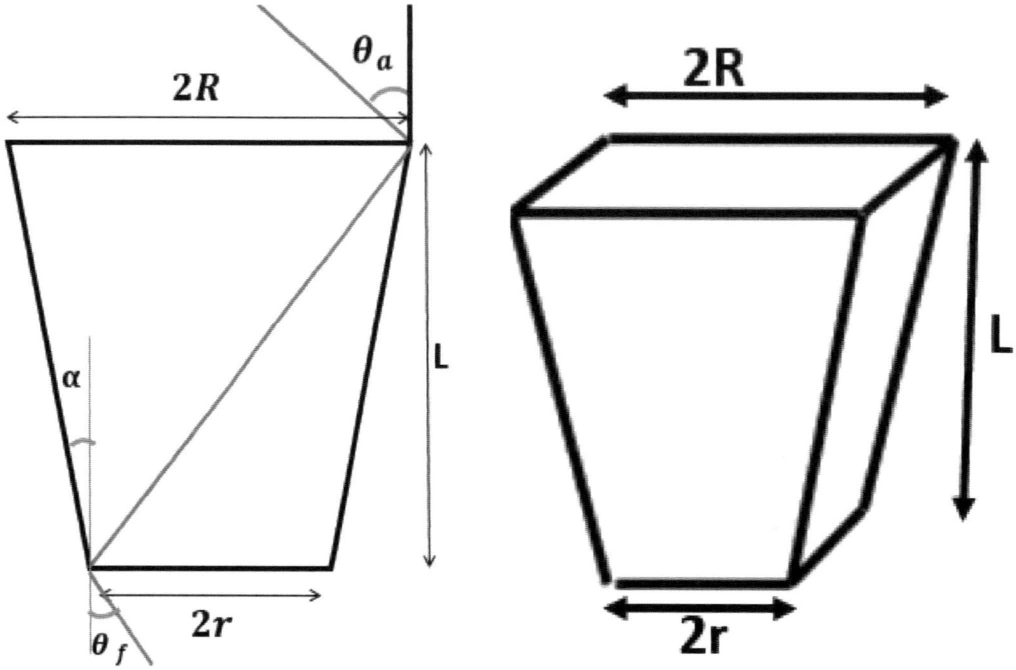

(a) (b)

Figure 6: (a) Section of the Pyramid (b) 3D-view of a pyramid

By varing the focal length of the Fresnel lens and using the equations (7, 8 and 9, we obtain the pyramid input diameter $2R$ evolution as shown on Fig.7. The maximum value of $2R$ tends to 15mm for very large focal distance f, of more than 1m. which is too large for a convenient CPV setup. Fig. 8 shows the net improvement of the flux homogeneity on the exit of the pyramid for different focal length of Fresnel lens compared to the flux distribution observed on fig.5. The flux is distributed on the whole exit of the pyramid, however, it starts by a high peak in the center, and becomes more uniform with the increment of the focal length of the Fresnel lens until $f = 700$mm and starts to be inhomogeneous again .
Fig.9 shows a comparison of the optical efficiency, for the two optical systems: The Fresnel lens and the Fresnel lens associated with the pyramid. We notice that the optical efficiency of the system using only the Fresnel lens, increases with the increment of the focal length until $f = 600$mm and start to decrease whereas for, the second system, with two stage of concentration, it starts by increasing and becomes stable from 350mm.

Table 1 resumes the effect of focal length and the secondary element on the acceptance angle of the optical system, we notice also that we get a large acceptance angle in the case of short focal length, and this acceptance angle becomes largest when the secondary stage is used as shown in the table 1.

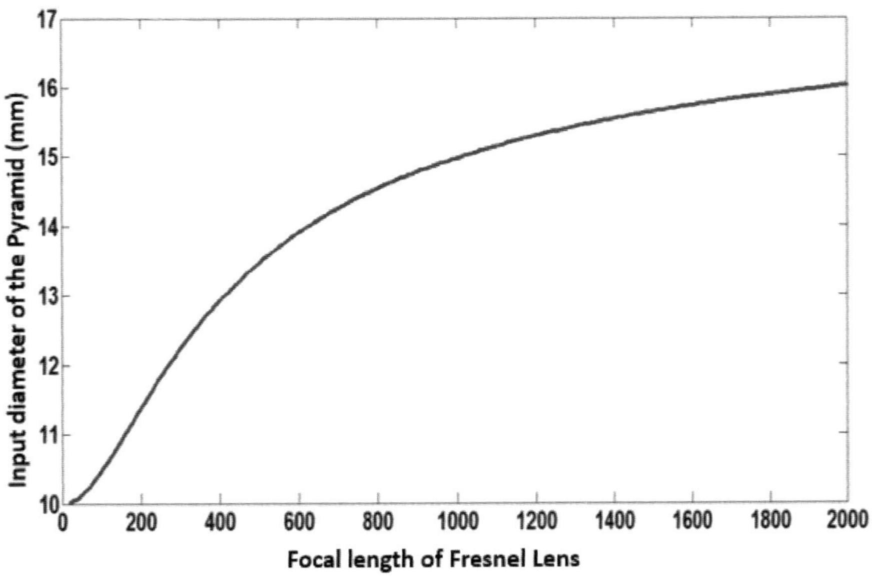

Figure 7: Input diameter 2R of the pyramid versus focal length ,f, for a Fresnel lens diameter of 350mm

Table 1: Acceptance angle variation versus the focal length

	175mm		350mm		700mm	
	Optical Efficiency in (%) at 0°	Acceptance angle (°)	Optical Efficiency in (%)at 0°	Acceptance angle (°)	Optical Efficiency in (%) at 0°	Acceptance angle (°)
Fresnel lens	56.58	0.6	86.566	0.4	88.3	0.2
Fresnel lens + Pyramid	51.4	1	86.27	0.8	88.29	0.4

4. Conclusion

A comparative study of the CPV system performance composed of the Fresnel lens alone and CPV system composed of the same Fresnel lens associated with the pyramid as secondary optical element is performed. We found that the secondary optical element has an important rule to improve the homogeneity on the solar cell, and allow for high tolerance. We found also that the focal length decreases the acceptance angle whatever the configuration of the CPV system.

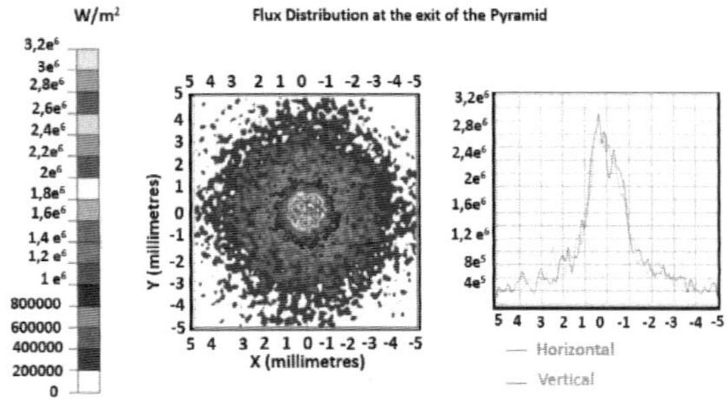

(a) $f = 175\text{mm}$

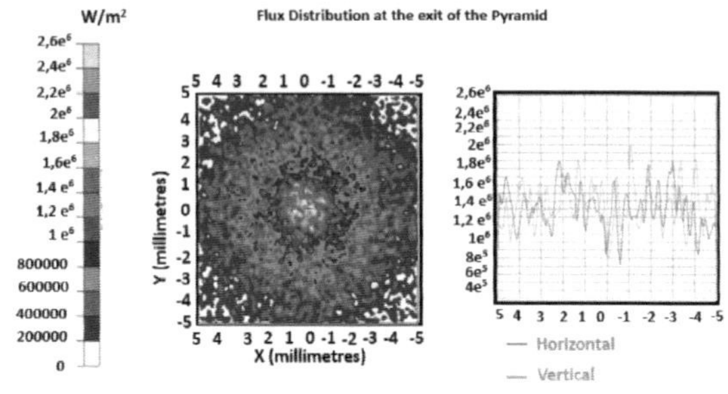

(b) $f = 265\text{mm}$

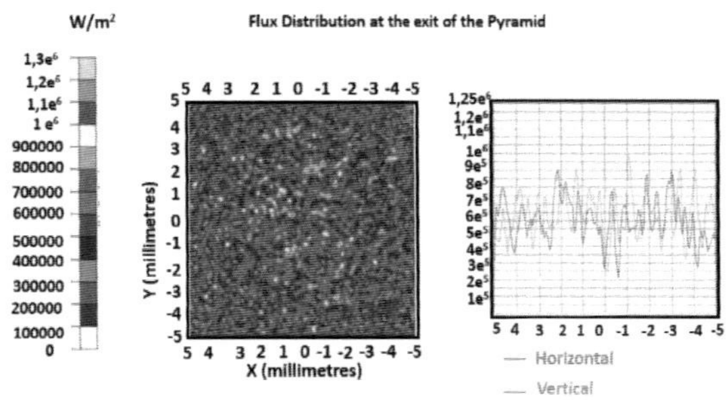

(c) $f = 350\text{mm}$

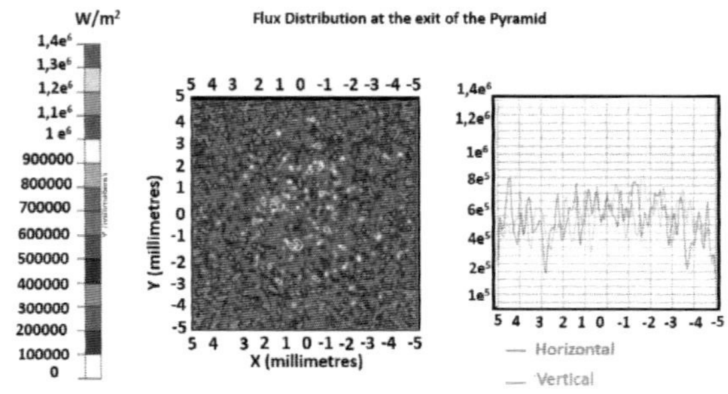

(d) $f = 525\text{mm}$

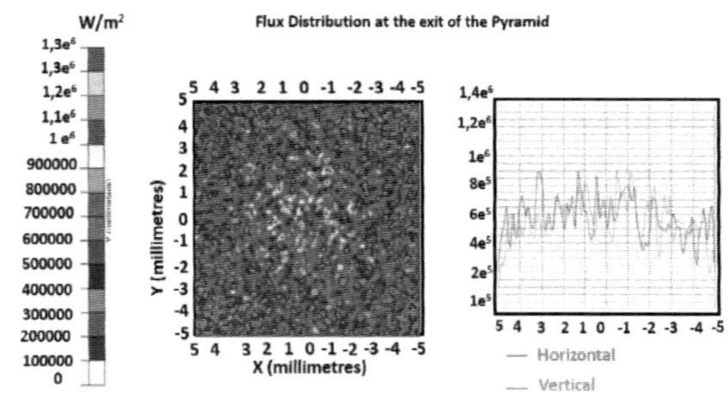

(e) $f = 612\text{mm}$

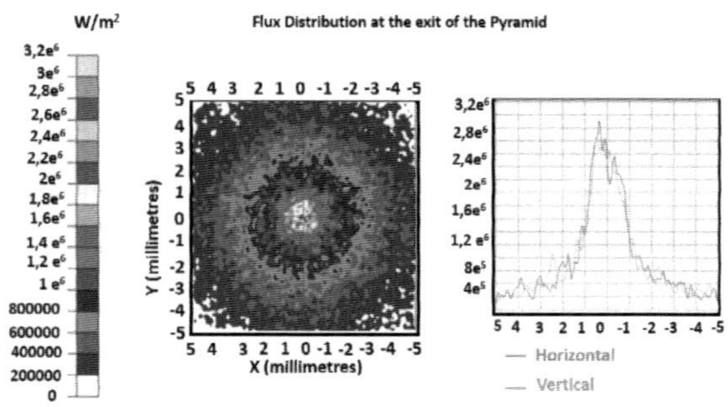

(f) $f = 700\text{mm}$

Figure 8: Evolution of flux distribution using two stages of concentration

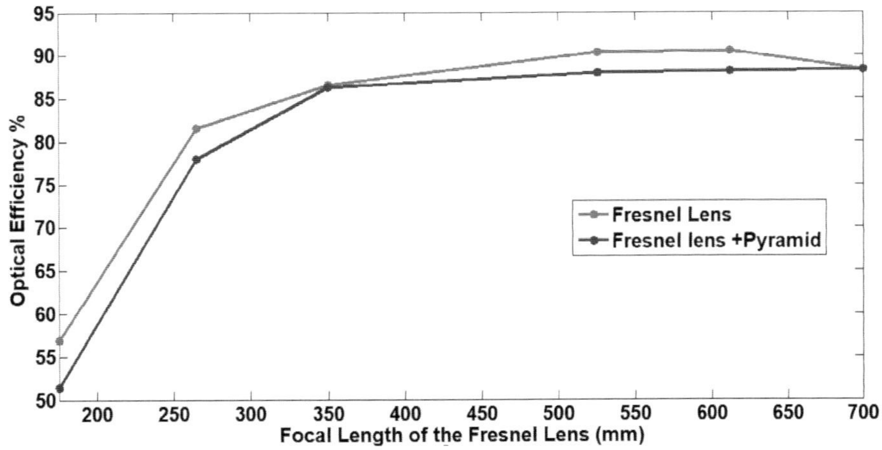

Figure 9: Evolution of the Optical Efficiency of the two CPV system vs Fresnel lens focal length

5. References

[1] Green M.A., Emery K.,Hishikawa Y.,Warta W., and Dunlop E. D., Solar cell efficiency tables (version 42),*Prog. Photovolt. Res. Appl.* **21(1)**, 827–837 (2013).
[2] Feuermann D, Gordon J M, High-concentration photovoltaic designs based on miniature parabolic dishes. *Solar Energy* **70**, 423–430, 2001.
[3] Kritch is very important man E M, Friesem A A, Yekutieli G, Highly concentrating Fresnel lenses. *Applied Optics* **18**, 2688– 2695 2015.
[4] Larbi A B, Godin M, Lucas J, Analysis of two models of (3D) Fresnel collectors operating in the fixed-aperture mode with a tracking absorber. *Solar Energy* **69**, 1–14.2000
[5] Leutz R, Suzuki A, Akisawa A, Kashiwagi, Design of a nonimaging Fresnel lens for solar concentrators. *Solar Energy* **65**, 379–387.1999.
[6] Fu L, Leutz R, Annen H P, Secondary optics for Fresnel lens solar concentrators, *Proc. SPIE* **7785** 7785-09.2010.
[7] Victoria M, Dominguez C, Anton A, and Sala S, Comparative analysis of different secondary optical elements for aspheric primary lenses, *optics Express***17(8)**, 6487-6492, 2009.
[8] Ning X, Winston R, and O'Gallagher J, Dielectric totally internally reflecting concentrators. *Applied Optics*, **26(2)**; 300-305, 1987.
[9] Ryu K , Rhee J G, Park K M, Kim J, Concept and design of modular Fresnel lenses for concentration solar PV system,*Solar Energy***80** 1580–1587, 2006.
[10] Winston R, Principles of Solar Concentrators of a Novel Design, *Solar Energy* **16**, 89-95, 1974.
[11] Zhuang Z, Yu F, Optimization design of hybrid Fresnel-based concentrator for generating uniformity irradiance with the broad solar spectrum, *Optics & Laser Technology***60** 27–33, 2014
[12] https: // refractiveindex.info/shelf=organic & book=poly(methyl-methacrylate)& page=Sultanova.